Forecasting International Migration in Europe: A Bayesian View

THE SPRINGER SERIES ON

DEMOGRAPHIC METHODS AND POPULATION ANALYSIS

Series Editor

KENNETH C. LAND

Duke University

In recent decades, there has been a rapid development of demographic models and methods and an explosive growth in the range of applications of population analysis. This series seeks to provide a publication outlet both for high-quality textual and expository books on modern techniques of demographic analysis and for works that present exemplary applications of such techniques to various aspects of population analysis.

Topics appropriate for the series include:

- General demographic methods
- Techniques of standardization
- Life table models and methods
- Multistate and multiregional life tables, analyses and projections
- Demographic aspects of biostatistics and epidemiology
- Stable population theory and its extensions
- Methods of indirect estimation
- Stochastic population models
- Event history analysis, duration analysis, and hazard regression models
- Demographic projection methods and population forecasts
- Techniques of applied demographic analysis, regional and local population estimates and projections
- Methods of estimation and projection for business and health care applications
- Methods and estimates for unique populations such as schools and students

Volumes in the series are of interest to researchers, professionals, and students in demography, sociology, economics, statistics, geography and regional science, public health and health care management, epidemiology, biostatistics, actuarial science, business, and related fields.

For further volumes:
http://www.springer.com/series/6449

Jakub Bijak

Forecasting International Migration in Europe: A Bayesian View

(with contribution by Arkadiusz Wiśniowski)

Springer

Jakub Bijak
School of Social Sciences
Centre for Population Change and S3RI
University of Southampton
Highfield Campus,
SO17 1BJ Southampton, UK
j.bijak@soton.ac.uk

ISSN 1389-6784
ISBN 978-90-481-8896-3 e-ISBN 978-90-481-8897-0
DOI 10.1007/978-90-481-8897-0
Springer Dordrecht Heidelberg London New York

Library of Congress Control Number: 2010936457

Printed on acid-free paper

Springer is part of Springer Science+Business Media (www.springer.com)

To my Kasia

– JKB

Acknowledgements

The majority of this book was prepared as a part of the work of the Central European Forum for Migration and Population Research, Warsaw (CEFMR), within the research grant number 03–34 of the Foundation for Population, Migration and Environment, Zurich (BMU-PME). CEFMR is a research partnership of the International Organisation for Migration (IOM), Institute of Geography and Spatial Organisation of the Polish Academy of Sciences and the Warsaw School of Economics, and until 2008 the consortium also included the BMU-PME Foundation. I am very much indebted to Prof. Hans-Rudolf Wicker from the University of Berne and the BMU-PME for the continuous support towards this project. In the period 2005–2006 the work was also partially co-financed from the Annual Stipend for Young Scientists received from the Foundation for Polish Science (FNP).

The book was finalised after I have started working in the School of Social Sciences, University of Southampton. In particular, the research on the users' perspectives on migration forecasting was prepared within the framework of the Centre for Population Change (CPC), funded by the ESRC Grant number RES-625-28-0001. The CPC is a joint initiative between the University of Southampton and a consortium of Scottish Universities in partnership with the Office for National Statistics (ONS) and the General Register Office for Scotland (GROS).

Both CEFMR and Southampton proved to be very stimulating environments, for which I would like to cordially thank all my colleagues and supervisors. Needless to say, the views, interpretations and conclusions presented in this book are those of the author and do not necessarily reflect the particular stance of any of the institutions with which he is or was affiliated, or which supported this project at its various stages.

With respect to the subject matter, I am extremely grateful to Prof. Jacek Osiewalski from the Cracow University of Economics and Prof. Marek Kupiszewski from CEFMR. Their time and expertise no doubt significantly contributed to the development of, respectively, Bayesian modelling and migration research aspects presented in the book. The insightful comments of Prof. Marek Męczarski, Prof. Jan Paradysz, Prof. Frans Willekens and two anonymous reviewers helped me a lot to improve the earlier drafts. The editorial process, overseen by Evelien

Bakker and Bernadette Deelen-Mans at Springer, Dordrecht, was a sheer pleasure. I am very grateful for their support and giving me gentle, yet firm motivation for completing the book within a definite timeframe. I am also enormously indebted to Sophie N'Jai and the editors at Springer for detailed proofreading and smoothing the language of subsequent versions of the manuscript. All the remaining errors are exclusively mine.

The study benefited from discussions and exchange of thoughts with several people. In particular, I would like to thank (in alphabetical order): Katarzyna Bijak, Izabela Koryś, Jerzy Marzec, Błażej Mazur, Anna Pajor and Mateusz Pipień, and all participants of the seminars of the Division of Econometrics, Cracow University of Economics. Special thanks for his contribution and fruitful co-operation on other projects go to Arkadiusz Wiśniowski.

Last, but not least, I am indebted to Jolanta Semeniuk-Bijak and Wojciech Bijak for the reasons significantly exceeding the bequest of their Bayesian philosophical viewpoint.[1] The ultimate thanks go to Kasia, for her continuous support towards this work, going far beyond the Kalman filters and countless statistical discussions on various occasions.

[1]See, for example, W. Bijak (1990) – I am especially grateful for mini-lectures on subjective probability around the time of publishing that paper.

Contents

Part IV Perspectives of Forecast Makers and Users

About the Author

Jakub Bijak is a Lecturer in Demography in the Division of Social Statistics, School of Social Sciences, the ESRC Centre for Population Change (CPC) and the Southampton Statistical Sciences Research Institute (S3RI), University of Southampton. He is also an Associate Member of the Central European Forum for Migration and Population Research (CEFMR) in Warsaw.

About the Contributor

Arkadiusz Wiśniowski is a Research Fellow in the S3RI, University of Southampton, and also an Associate Member of CEFMR.

List of Figures

List of Tables

Part I
Introduction

Chapter 1
Introduction and Background

The current book is devoted to known (although seldom used by demographers) Bayesian statistical methods, applied in one of the most under-researched areas of population forecasting; that is, predictions of international migration. The study aims to keep up to date with the methodological state-of-the-art migration forecasting by presenting a thorough review of methods, while at the same time advocating for the use of the stochastic approach within the Bayesian framework, which allows for a formal incorporation of expert judgement. The current, introductory chapter presents the background, aims, scope and structure of the book, as well as terminological and notational conventions.

1.1 Migration Forecasting as a Research Problem

1.1.1 Role of International Migration and Its Forecasts

Migration, and in particular international migration, is one of the key factors that are anticipated to be increasingly influential in shaping population dynamics in the coming decades. This is especially important in the more developed regions of the world, such as Europe, which are already facing zero or negative natural population growth (e.g., van der Gaag & van Wissen, 1999; Kupiszewski, 2002b). Furthermore, the impact of migration is not limited to demography. Migration also affects many other areas of social life, including economy and labour relations, politics and culture; and, *vice versa*, is also influenced by them. All these effects are nowadays gaining in importance, as migratory processes in a globalising world are becoming more and more dynamic and complex (King, 2002).[1]

One of the links between migration, demography and other social processes is the impact of migration on age structures of populations (United Nations, 2000) and labour force resources (Bijak, Kupiszewska, Kupiszewski, Saczuk, & Kicinger, 2007). These relationships indirectly influence labour markets, social

[1] For a summary of issues related to globalisation and mobility, see e.g. the *World Migration Report* (IOM, 2008).

J. Bijak, *Forecasting International Migration in Europe: A Bayesian View*,
The Springer Series on Demographic Methods and Population Analysis 24,
DOI 10.1007/978-90-481-8897-0_1, © Springer Science+Business Media B.V. 2011

security systems, health expenditures, fiscal balance and national savings, as well as other macroeconomic indicators (Roseveare, Leibniz, Fore, & Wurzel, 1996). With respect to the labour market adjustments through migration, it is worth stressing that they affect the economies of both sending and receiving countries. Movements of unskilled workers on one hand and highly-trained professionals on the other are a way to reduce structural labour market imbalances: surpluses at source and deficits at destination. In turn, with respect to the social aspects of migration, an increasingly important issue is the integration of immigrants into the host society. Besides, migration can also influence other areas of social life, like for example education systems.

Given the impact of migration on various areas of life, there is an increasing need for more accurate forecasts of population movements, especially at the international level. This matter is however very delicate, not least because migration is the most complex and most difficult to predict component of population change, bearing high levels of forecast errors (NRC, 2000; Kupiszewski, 2002b). This is due to several factors. Firstly, migration data on which the predictions are based are inconsistent, incomplete and generally low quality. Secondly, migration is extremely sensitive to government actions and other political factors, which themselves are difficult to forecast (NRC, 2000, pp. 177–182). With respect to the latter, of special importance is crisis migration, which generates high waves of refugees, asylum seekers and displaced persons. Some late twentieth century examples of crisis migration flows include conflicts in the former Yugoslavia, Rwanda and Chechnya, to name but a few (UNHCR, 2000). Such tragic events and their demographic consequences are very difficult to predict, both in terms of timing of their occurrence, as well as of their magnitude.

All these factors contribute to the very high ex-post errors of migration forecasts. A comprehensive overview of different projections and forecasts made in various European countries has been provided by Salt and Singleton (1995), who showed that errors in the relative magnitude about 100% were not unusual, with the extremities reaching as much as 1,350%. In some cases even the dominant direction of flows between two countries was erroneously predicted, with a notable example of the recent migration history of Ireland.

Nevertheless, given the increasing impact of international migrations, a need for accurate, high-quality forecasts of population flows is becoming more and more significant despite high error levels. On one hand, migration forecasts themselves are of interest for the policy makers from a purely operational point of view, concentrating on the numbers of migrants and the impact of migration for example on labour markets. On the other hand, such predictions constitute an important and inevitable part of population forecasts in general, which are crucial for many aspects of socio-economic planning. For this reason, many issues related to population forecasting apply directly to international migration predictions. However, as remarked by Duchêne and Wanner (1999, p. 6), it should be noted that demographic forecasts are not exclusively aimed at predicting the future as such, but rather, more importantly, at providing necessary input to the political decision making process. Improving forecasts can contribute to better policy decisions, which in turn may have significant consequences for the societies. Clearly, the consequences do not need to be limited to their economic dimension (Ahlburg, Lutz, & Vaupel, 1998, p. 192).

1.1.2 Problems with Errors in Migration Forecasting

Some level of migration forecasting error is always inevitable, as any inference about the future is made under uncertainty: an issue which is further elaborated on Chapter 2. Nevertheless, as noted by Dawid (1984, p. 278), 'one of the major purposes of statistical analysis is to make forecasts about the future [and] to offer suitable measures of uncertainty associated with unknown events or quantities'. Along similar lines, Robert (2001, p. 2) suggested that adding the uncertainty component allows for more meaningful interpretations of the phenomena under study, not being limited to their explanations *ex post*.

There have been several suggestions about possible ways to improve the accuracy of migration and population forecasts. For example, de Beer (1990b) proposed to examine average migration over several years rather than yearly values. Ahlburg (1995) and Smith (1997), following Armstrong (1985), advocated the benefits from combining (averaging) forecasts in order to reduce their errors. Gjaltema (2001) discussed the advantages of rule-based forecasting, originally developed by Collopy and Armstrong (1992). As an alternative, Sanderson (1998) suggested considering causal relationships between various socio-economic variables and the elements of population change, including international migration. As a prerequisite, the quality of migration statistics, on which the forecasts are based, needs to be enhanced.

In a sample study of some European and North American countries it has been shown that naive forecasts of zero or constant migration flows would usually generate higher prediction errors than forecasts with more sophisticated assumptions (NRC, 2000, pp. 318–325). Moreover, as suggested by Rees, Kupiszewski, Eyre, Wilson, and Durham (1999), methodological advancements in migration forecasting have already led to some declines in the ex-post prediction errors. However, there seem to be inherent limits to the predictability of migration. As argued by several authors from the domain of forecasting, adding more complexity to the models beyond a certain point need not lead to further error reductions (e.g., Makridakis & Hibon, 2000; Orrell, 2007). All these issues together constitute a rationale for the current book, which aims to contribute to the methodology and philosophy of migration forecasting, whilst clearly showing its inherent limitations.

Already over half a century ago, Hajnal (1955) in his seminal paper noted that predictions of future populations are doomed to be uncertain, and the only mitigating factor in that respect is the inertia of demographic processes and the information already embodied in the age structure of the population. He identified the main success of demographic forecasting in the survival analysis of pre-existing cohorts (that is, in mortality predictions), whereas fertility and especially migration have to be seen as much less predictable. As to the development of more accurate methods of forecasting, Hajnal (1955) was utterly skeptical. Nonetheless, he acknowledged that there will most likely be a continuous demand for forecast from their users, for various reasons related to policy making and planning. This perspective is shared throughout the book.

From this point of view, regardless of whether there is still space for further error reductions in population and migration forecasting, there is definitely scope for developing a more robust methodology of making predictions. Given the imperfect data, such a methodology would have to allow for combining statistical information on past trends with expert judgement on migratory processes in order to obtain a plausible model with credible assumptions. Notably, there have already been several attempts to combine both sources of information; for example within a framework of 'expert-based population forecasting,' developed under the auspices of the World Population Program of the International Institute for Applied Systems Analysis (IIASA) in Laxenburg, Austria (Lutz, Sanderson, & Scherbov, 1996, 1998, 2000, 2004). As an alternative, this book proposes to apply the mechanisms of Bayesian statistical inference, for the reasons set forth in Chapter 2.

1.2 Aim and Structure of the Book

1.2.1 Aim and Scope

The aim of the book is to evaluate the existing methods of forecasting international migration and to propose an alternative methodological framework, which would acknowledge the presence of uncertainty and enable the inclusion of judgemental knowledge in the models. Such forecasts, with all caveats concerning their inherent limitations, would ultimately serve as information input for decision making. The solution proposed in this book, following the suggestions of Willekens (1994, pp. 26–30), is to employ the methods and tools of Bayesian statistics, where combining the subjective prior knowledge with statistical information constitutes a natural way of inference. In order to make the book more accessible for broader readership, no prior knowledge of Bayesian statistics is assumed. In that respect, the book offers a brief primer both in the philosophy of Bayesian inference and also in some practical computational issues.

There exists a wide variety of approaches to international migration forecasting. Since migration is a very complex and multi-dimensional phenomenon, its models and predictions stem from various scientific disciplines: demography, geography, economics, statistics, sociology and political science. Some existing attempts to forecasting even involve application of advanced methods of theoretical physics. This book also provides a discussion of various methods, with an overview of migration theories and their possible role in setting the forecast assumptions.

As an empirical illustration, the book presents Bayesian forecasts of migration flows between Germany and three selected European countries: Italy, Poland and Switzerland. In terms of geography, the analysis focuses on countries being important migration sources or destinations of intra-European flows. Also the statistical information for these flows is available and relatively complete. A detailed discussion of data issues, focusing on the selected countries, is presented in Chapter 2. The empirical analysis is based on historical migration data limited to the period 1985–2004, in order to emulate forecasting uncertainty at the onset of a different regime of European migration after the 2004 enlargement of the European Union (EU). The forecast horizon is mid-term, spanning the years 2005–2015. The proposed Bayesian migration forecasting models are compared with their counterparts based on the traditional (frequentist, sampling-theory)[2] statistical approach, with respect to their theoretical characteristics and *ex-post* prediction errors.

Speaking of European migration, the scope of the presented illustrations is obviously incomplete in geographic terms. The main reason for this is the currently limited availability of data conforming to a uniform definition of migration for all flows under study. Hopefully, the recent Regulation of the European Parliament and the Council[3] will contribute to enhancing the statistical knowledge base of migration within the European Union. Promisingly, the Regulation allows sound modelling techniques to be used to estimate the missing data and to ensure the comparability of the data according to one common definition.

As more and better macro-level data on migration flows become available in Europe in the future, attempts can be made to answer many additional research questions, such as those related to population flows following the successive EU enlargements, and to country-specific gross migration inflows and outflows in Europe. Alternatively, with better data, relationships between various origins or destinations of migration can be also examined, which would allow testing the hypothesis about complementary versus substituting character of different directions of flows. However, as good-quality comparable data are a clear prerequisite of all such analyses, the projects aiming at establishing a sound methodology for their estimation should be given priority in the European migration research agenda in the coming years, if various forecasting methods are to be of real use in demographic and policy-oriented research at the EU level.

1.2.2 Structure of the Book

The book is structured in five parts, altogether comprising twelve chapters.

Part I contains the basic introductory information. The current introduction (Chapter 1) starts with background information concerning the study, defining its rationale, aims and scope. Subsequently, terminology and symbols applied in the text are introduced. In Chapter 2, four issues are briefly addressed, setting the scene for the further discussion. These are: problems with migration data, general

[2]Despite the subtle differences, in the current book, the terms 'traditional', 'frequentist' and 'sampling theory' will be used interchangeably to denote the statistical paradigm based on the works of R. A. Fisher, K. Pearson and J. Spława-Neyman (see Chapter 2). 'Sampling theory' in this case refers to the general idea of inference based on the sample data rather than being limited, for example, to representative sampling from finite populations.

[3]*Regulation of the European Parliament and Council No. 862/2007 from 11 July 2007 on the Community Statistics on Migration and International Protection*, Official Journal OJ L 199/23.

issues concerning uncertainty, subjectivity and expert judgement in migration and population forecasting, the Bayesian paradigm in statistical inference, and finally the numerical algorithms used in Bayesian analysis, based on the example of the Markov chain Monte Carlo simulations. More detailed information on Bayesian computations in practice is also presented in Chapter 9, in Part IV.

Part II of the book discusses the existing framework for explaining and forecasting international migration. First of all, in Chapter 3, a brief outline of selected international migration theories is given, with a discussion of their applicability in prediction-making. Then, in Chapter 4, the most important models and methods used in migration forecasting are presented as a practical alternative to theory-based forecasting. The overview follows the distinction between the deterministic and probabilistic paradigms in scientific research. The methods in the former category include migration scenarios, surveys, Delphi methods, and mathematical models of population flows. The probabilistic methods encompass econometric models of migration and stochastic forecasts of migration time series. A few existing Bayesian examples of demographic applications are also listed. The chapter concludes with a typology, comparison and evaluation of the presented methods of migration modelling from the perspective of forecasting applications.

Part III of the book contains several suggestions for developing migration forecasting models within the Bayesian approach. The propositions focus on three groups of methods, described in subsequent chapters and illustrated with forecasts of migration flows among four selected European countries. Thus, Chapter 5 is devoted to the Bayesian model selection and forecast averaging based on the posterior odds criterion, applied here for simple stochastic processes. In Chapter 6, the approach 'from general to specific' is presented in the context of vector autoregression (VAR) models. Such models have the potential to include theoretical information on relations between migration and its likely determinants. In turn, Chapter 7 explores, among others, the Bayesian hierarchical modelling in an attempt to accommodate for the possible discontinuity in migration trends. Of interest are models based on historical analogies and solutions involving non-constant conditional variance. Subsequently, Chapter 8 includes basic remarks on the robustness of forecasts against selected changes in the prior distributions. In addition, a slightly more thorough analysis of both *ex-ante* and *ex-post* empirical performance is offered for selected Bayesian forecasts and their traditional equivalents. Finally, computational issues related to the Bayesian approach are addressed from a practical point of view in Chapter 9, contributed by Arkadiusz Wiśniowski, whereby a short survey of available Bayesian software is presented. This overview is followed by a description of WinBUGS, an environment specifically devoted to Bayesian analysis. Additionally, simple examples of Bayesian computations in the R language are provided. R is very flexible, comprehensive free software designed for statistical analysis, recently gaining in popularity among demographers.

Part IV of the book offers a general discussion of forecasts and the associated uncertainty from the perspective of population forecasters and forecast users, usually, political decision makers. The discussion is largely based on the selected recent literature on forecasting and decision analysis, while providing some basic examples directly pertaining to migration and population studies. Thus, Chapter 10 is predominantly concerned with the theory-based forecasting and the use of migration predictions in the models of population dynamics. As an example of methods, which allow controlling the plausibility of outcomes in such models, the Bayesian melding approach is discussed. Moreover, the issue of microfoundations of forecasts is addressed from the point of view of the notion of 'imperfect knowledge', being a contemporary response in the theory of economics to the empirical failures of the mainstream forecasting. The chapter concludes by sketching a possible research agenda for the future. In a policy-oriented Chapter 11, in turn, a brief introduction to the decision analysis is presented, including a primer in its Bayesian version. A stylised Bayesian decision analysis based on migration forecasts is also presented, aimed to derive optimal decisions under various preferences of the decision makers. At the end, the chapter aims at providing the forecast users with some indications as to which policy questions can be answered by migration forecasts, and how.

Finally, Part V consists of Chapter 12, featuring a summary of the findings and the most important conclusions from the study. Together with a brief synopsis of the main results, some recommendations for future research in international migration forecasting are addressed. The discussion includes selected implications for the forecasters and forecast users (decision makers). The book concludes by putting the obtained results in the context of theoretical debate on population forecasting, with focus on uncertainty issues, the use of expert judgement, and the dilemma of simplicity versus complexity of forecasting models.

The study is accompanied by three Annexes containing graphs and tables, both with the input for the analysis and the empirical illustrations. Annex A contains an inventory of sources, time series of observations and a brief description of methods used to prepare and harmonise the data. Annex B provides the listing of WinBUGS programme code used to calculate Bayesian forecasts presented in Chapters 5–8. The empirical illustrations of the analysis – predictions of population flows between the countries under study – are subsequently dealt with in more detail in Annex C, extending the empirical material offered in the subsequent chapters of Part III.

The book can be read in a variety of ways. Nevertheless, several sequences of chapters are particularly recommended for different readership. A policy-oriented reader, likely to be less interested in technicalities and theorising, might like to gain a general insight into the problems listed in Part IV, preceded by the introduction in Chapters 1 and 2, and with the results of the research summarised in general terms in Chapter 12. A theory-oriented reader can skip the whole of Part III containing modelling and technical details. Finally, the whole book can be of potential interest – at least that is the hope of the author – to demographers, and in particular to statistical demographers (for relevant definitions, see Alho & Spencer, 2005, p. 3), as well as to postgraduate students of demography and migration.

1.3 Terminology and Symbols

1.3.1 Basic Terms Used in the Study

Terminology used in this book generally follows the established conventions in demography. Therefore, *migration* is a form of spatial mobility, understood as

> the movement of people across a specified boundary for the purpose of establishing a new or semi-permanent residence; divided into international migration (migration between countries) and internal migration (migration within a country).[4]

Throughout the book, the term *migration* generally depicts *international migration*, unless explicitly noted otherwise. A *(crude) migration rate* is an intensity measure of migration risk, obtained by dividing the overall number of migrants by the size of *population at risk* – all persons who potentially may experience migration in a given period. Such intensity measures of risk are usually referred to in demography as the *occurrence-exposure rates* (e.g., Rees & Willekens, 1986). A convenient estimate of the population at risk for a given year is the mid-year population, calculated either as the population size on 30 June or as an average of the stocks at the beginning and at the end of the year. Importantly, rates should not be confused with probabilities of experiencing demographic events, such as migration, within a specific timeframe. Probabilities measure, on a scale between zero and one, how likely is it that something can be expected to happen (see Chapter 2 for more details).

An important distinction concerns the possible ways of inference about the future characteristics of demographic phenomena. Thus, a population *projection* usually refers to 'computation of future changes in population numbers, given certain assumptions about future trends in the rates of fertility, mortality, and migration'.[5] The projection is thus directly conditional on the assumptions, which is the reason why in most cases several *scenarios* of a projection are prepared, conventionally baseline, high and low. However, the popularity of population 'projections' among national and international statistical agencies, and individual researchers can raise suspicions that they are in fact perceived as something more than merely the hypothetical paths of development conditional on the assumptions. Indeed, as noted by Booth (2004, p. 10), 'demography appears to be alone among disciplines in its insistence that its projections are somehow not forecasts of the future'. As an alternative, some demographers prefer to use the label 'projection' for computations of future population trajectories assuming constant age-specific fertility, mortality and migration rates (Rogers & Willekens, 1978, p. 62).

---

[4]The Population Reference Bureau (PRB) Glossary of Population Terms (www.prb.org/Content/NavigationMenu /PRB/PRB_Library/Glossary2/Glossary.htm, accessed on 25 May 2005). Despite of the fact of crossing national boundaries, other criteria to distinguish various types of migration flows can be set, for example, by type of change in the place of residence, duration, distance, or socio-demographic characteristics of migrants (Paradysz, 2006, p. 232).

[5]The Population Reference Bureau (PRB) Glossary of Population Terms (*idem*).

Following Keilman (1990, p. 7) a *forecast* is unconditional, as a result of the process, in which 'based on current scientific insights, a forecaster gives his best guess of what the future population will be'.[6] Further, Keilman distinguishes between forecast, which is 'based on relatively poorly developed behavioural theories', and *prediction*, 'based on a calculation with an explicit model or a verified theory' (*idem*, p. 8). However, contemporarily this difference becomes less clear with respect to some modern forecasting techniques, for the reasons that will be further explored in discussing the vector autoregressive model features in Chapter 6. For this reason, throughout the book the terms *forecast* and *prediction* will be used interchangeably. It is also worth noting that although all quoted definitions refer to population projections or forecasts in general, they can be easily adopted for the specific needs of a study of international migration.

Another distinction that needs to be made is the one between *knowledge* and *judgement*. Although, especially with respect to the former one, many competing definitions exist, in the current study *knowledge* is understood as 'the fact or condition of knowing something with familiarity gained through experience or association; [...] of having information or of being learned'. Respectively, *judgement* is defined as 'a proposition stating something believed or asserted'.[7] According to these descriptions, judgement is an external expression of a piece of knowledge possessed by a researcher, and hence can be directly used, for example, to form assumptions about future patterns of the phenomenon under study.

Finally, as far as the properties of (stochastic) processes are concerned, *stationarity* refers to exhibiting the same features at different points in time. In a stronger sense, this is usually defined as generating observations from the same probability distributions, and in a weaker sense, to time-independence of the autocovariance function of the process (cf. Granger & Morris, 1976).

1.3.2 Mathematical Notation

Similarly to the terminological framework, also the mathematical notation generally follows the established conventions. In the formulae, symbols for variables and functions are typed in italics, while for vectors, matrices and sets – in boldface. With respect to the latter, the most common is $\mathbf{R}$ – set of real numbers. Probability notation follows the usual convention: $p(A)$ denotes the probability of an event A, and $p(A|B)$ – conditional probability of A given B; the same rationale applies to

[6]With respect to the dispute on making population 'forecasts' or 'projection', Keyfitz (1972, p. 363) observed that 'a demographer makes a projection, and his reader uses it as a forecast' (after: Ahlburg & Land, 1992, p. 290).

[7]Both citations after the Merriam-Webster Online Dictionary (www.m-w.com, accessed on 25 April 2006). 'Knowledge' can be alternatively defined for example as 'acquaintance with or understanding of a science, art, or technique; the fact or condition of being aware of something; the range of one's information or understanding; the circumstance or condition of apprehending truth or fact through reasoning; the sum of what is known: the body of truth, information, and principles acquired by mankind' (*idem*).

densities of random variables: $p(x)$ and $p(x|y)$. In the description of the statistical theory, a Bayesian notation is applied after Bernardo and Smith (2000). Hence, θ depicts the unknown model parameter subject to estimation, and x – the observations (data). The observations are modelled as random variables, and the actual data are the realisations thereof.

In multidimensional problems, the respective symbols for the vector of parameters and the data vector (or random vector) are θ and x. Vectors are column-wise, and the transposition of a vector $\mathbf{x}$ or a matrix $\mathbf{A}$ is noted $\mathbf{x}'$, respectively $\mathbf{A}'$. The common probability distributions are: the Normal (Gaussian) distribution with mean μ and variance σ^2, denoted[8] as $\mathrm{N}(\mu, \sigma^2)$, its multivariate, k-dimensional, generalisation with mean $\boldsymbol{\mu}$ and covariance matrix $\boldsymbol{\Sigma} - \mathbf{N}_k(\boldsymbol{\mu}, \boldsymbol{\Sigma})$, and the Gamma distribution[9] with parameters μ (scale) and r (shape) $-\Gamma(r, \mu)$. Other distributions, such as Student's t, chi-squared, or Wishart, are introduced where they appear, together with the parameterisation used. The tilde (~) can stand either for 'is distributed' or 'follows a probabilistic model', for example $\alpha_{i,j} \sim \mathrm{N}(0, 5^2)$ or $Y_i(t) \sim \mathrm{AR}(1)$. Further, '$\propto$' denotes proportionality and the abbreviation 'iid' means that random variables are independent and identically distributed.

In the overview of literature on the existing migration models and forecasts, presented in Chapter 4, the notation follows the ones applied in particular original studies. In each case, the formulae are accompanied by appropriate explanatory notes. In all chapters of Part III devoted to the construction and verification of forecasting models, as well as in the data-related Annex A, $M_{i-j}(t)$ stands for the volume of migration (number of migrants) from country i to country j in period t, and $m_{i-j}(t)$ – migration rates per 1,000 of population at risk, the latter denoted as $P_i(t)$. The natural logarithms of $m_{i-j}(t)$ are labelled as $\mu_{i-j}(t)$. Further, except for Chapter 9, $Y_i(t)$ represents the gross domestic product (GDP) per capita in country i in period t, calculated on the basis of the purchase power parity (PPP), and $U_i(t)$ – unemployment rate in country i in period t. The remaining, less common variables are explained as they are introduced.

1.3.3 Bibliographical Notation

In bibliographical notes, references are given to original publications wherever possible, as for example in Bayes (1763). However, if a later edition or a translation of the original text was consulted, both of them are listed in reference notes separated by a slash symbol; for example Rao (1989/1994) denotes the 1994 Polish translation of the 1989 English original. In such cases, the page numbers refer to the edition that

---

[8]The density function is thus $p(x\,|\mu, \sigma^2) = (2\pi\alpha)^{-0.5} \cdot \exp[-(x-\mu)^2/2\sigma^2]$. Alternative parameterisation can be obtained using the precision parameter, a reciprocal of the variance: $t = \sigma^{-2}$. The same remark applies to the k-dimensional Normal distribution with density $p(\mathbf{x}\,|\boldsymbol{\mu}, \boldsymbol{\Sigma}) = [(2\pi)^{k} \cdot \det(\boldsymbol{\Sigma})^{-0.5} \cdot \exp[-0.5 \cdot (\mathbf{x} - \boldsymbol{\mu})' \, \boldsymbol{\Sigma}^{-1}(\mathbf{x} - \boldsymbol{\mu})]$, where precision matrix $\mathbf{T} = \boldsymbol{\Sigma}^{-1}$ can be alternatively used.

[9]For parameterisation of the Gamma distribution, see Chapter 5.

was the actual source of information, which is the one mentioned as last. In the list of references, the additional editions or translations are included in square brackets in the entry for the original source. The multi-author books and articles are listed first in alphabetic order by the surname of the first author, then by surnames of the remaining authors, and only then by the year of publication.

Chapter 2
Preliminaries

In the current chapter, four issues are briefly addressed, which are vital to the further discussion, yet not sufficiently within the scope of the topic of this book to be presented in full. Firstly, the problems with migration data are dealt with, focusing on the diversity of definitions, measurement errors, and possible ways to overcome the inconsistencies within the statistical information. Secondly, the issues concerning uncertainty, subjectivity and expert judgement are discussed, together with their role in migration forecasting. Thirdly, general remarks on the Bayesian statistical inference are presented, with the aim of serving as reference throughout the remaining parts of the book. Finally, numerical algorithms used in Bayesian computations are briefly discussed, based on the example of Markov chain Monte Carlo simulations.

2.1 Definitions and Measurement of International Migration

2.1.1 Data Sources and Definitions

This book focuses on modelling and forecasting *long-term registered migration* flows, disregarding all types of *short-term, unregistered*, or *irregular* population movements. Ideally, a *long-term migrant* should be defined in accordance with the international recommendations of the United Nations, as 'a person who moves to a country other than that of his or her usual residence for a period of at least a year (12 months), so that the country of destination effectively becomes his or her new country of usual residence' (United Nations, 1998, p. 18). However, given the inconsistency of definitions applied in various European countries, as set forth in more detail in the next subsection, this postulate is very difficult to fulfil in practice.

There are two basic concepts of measuring international migration, concerning either assessment of magnitude of population *flows* between particular countries, or estimation of the size of immigrant (usually foreign-born) population *stocks*. The current study is based on the former concept, focusing on forecasting migration flows between the countries of origin and destination. Data of this type can be collected either following the *movement approach*, which involves a continuous registration of migratory events, or based on the discrete-time *transition*

J. Bijak, *Forecasting International Migration in Europe: A Bayesian View*,
The Springer Series on Demographic Methods and Population Analysis 24,
DOI 10.1007/978-90-481-8897-0_2, © Springer Science+Business Media B.V. 2011

approach, which involves recording place of residence on a periodical basis (Rees & Willekens, 1986, p. 22; Willekens, 2008). Whereas the data of the former type are usually available from population registers, the latter ones can be obtained from retrospective censuses or surveys.

Data on migration flows and population stocks used in this book predominantly come from NewCronos, the official database of Eurostat (statistical office of the European Communities). The figures originate from the national statistical authorities of the countries under study that are obliged to periodically submit the information to Eurostat. The time series of potential economic explanatory variables, like GDP or unemployment, have been taken from various publications of international organisations (United Nations, 2000b, 2005b, as well as World Bank, 2003, supplemented by Internet sources). Unless noted otherwise, the time series cover the period 1985–2004 for migration from Italy and Switzerland to Germany, or 1991–2004 for the remaining flows; in the latter cases due to the presence of socio-economic transition in Poland and the unification of Germany. In the case of flows from Germany to Italy and Switzerland, a brief analysis of sensitivity of the results to a longer sample (1985–2004) is also offered, in addition to the main results. A detailed listing of sources and a description of algorithms used in data preparation are provided in Annex A.

2.1.2 Quality and Comparability of Migration Data

Despite the presence of international recommendations, as indicated above, European countries strongly differ with respect to the definitions of migrants that are reflected in the official statistical registers. The period spent abroad, after which a person becomes a long-term migrant, may be self-declared and even undefined in terms of time, as in the case of the 'permanent migration' concept. The latter one, although in practice not well defined and difficult to operationalise, is still in use in several post-socialist countries of Central Europe, including Poland. On the other hand, Germany applies the broadest definitions of long-term migrants in Europe, registering all legally-resident foreigners who dwell in the country, which results in serious discrepancies when the German data are compared with the ones of its migration partner countries (Kędelski, 1990). Differences between European countries with respect to definitions of long-term migrants are discussed in detail in Poulain (1994) and Nowok, Kupiszewska, and Poulain (2006), together with recommendations for the future with respect to the possible harmonisation of migration data in Europe. In general, it has to be noted that various reasons for discrepancies in migration data between the countries include not only different definitions of international migrants, but also the varying extent to which population flows are underestimated, which is determined by legal, technical, organisational and other factors (Bilsborrow, Hugo, Oberai, & Zlotnik, 1997).

One way to assess the magnitude of differences between the numbers of migrants according to different definitions is to study population censuses or additional data sources (cf. Bijak & Koryś, 2009). The example of Poland shows that there are large differences in size between the permanently registered population and the 'usual residents', the latter being recorded in line with the recommendations of the United Nations (1998). For the 1988 census, Sakson (2002), who based her estimations on the border crossing statistics, estimated that this difference (i.e. permanent residents less usual residents) equalled about 590,000 persons. According to the Central Statistical Office of Poland (*Główny Urzad Statystyczny*, GUS), in 2002 this gap totalled about 610,000 persons (GUS, 2003). The register-based population stocks are thus not fully covered by the censuses, which, given a relatively good quality of births and deaths registration, leads to a conclusion that the error is mainly due to under-registration of migration. This is most likely true also, although to a varying extent, for other post-socialist countries, where, despite recent improvements in collection of data on international migration, the statistical information covered by the registers is still far from complete (Eurostat, 1997).

Some problems with the quality and completeness of data on international migration, like under-registration of the population flows, are, to a varying degree, common for most of Europe, perhaps with the exception of the Scandinavian countries. A thorough discussion of the reasons for the inconsistencies in data has been provided by Bilsborrow et al. (1997), as well as more recently, by Poulain et al. (2006) with particular focus on the European Union (EU). The most visible under-registration of international migration flows can be observed for the post-socialist countries of Central and Eastern Europe (Nowok, 2005).

2.1.3 Ways of Dealing with Deficient Statistics

In theoretical demography, there have been many attempts to overcome the problems of incomplete data on population processes, initiated on a large scale by a work by Coale and Demeny (1967) which concentrated on fertility and mortality issues. Since then, numerous studies have been published devoted to estimation of incomplete migration data. Since the 1970s, advanced mathematical methods have been introduced, including the multiregional life tables approach of Rogers (1973) and methods of iterative proportional fitting of Willekens (1982). The 'population accounting' perspective, based on the idea of consistence of migration data in various dimensions (by origin, destination, sex and age), has been advocated for example by Rees (1977) and Rees and Willekens (1986). Methods for estimating parameters of distributions of the number of migrants by age have been developed by Rogers and Castro (1981) following the pioneering study of Rogers, Raquillet, and Castro (1978). The multidimensionality of migration is also present in several other modelling approaches, including the *migrant pool* model, or the bilinear models, discussed in more detail in Alho and Spencer (2005, p. 187).

Poulain (1993) presented an algorithm for estimation of international migration within the European Union, assuming that the data for all the countries are distorted in a similar fashion. However, according to Kupiszewski (2002b, p. 106), this method cannot be used universally and requires careful selection of countries for which estimation is performed. Other developments in dealing with deficient migration data comprise several ideas summarised by Rogers (1999), including the entropy maximisation (EM) algorithms (Willekens, 1977, 1999; Willekens, Pór, & Raquillet, 1981). In addition, Rogers and Raymer (2005) analysed the intensity of subsequent migratory moves, following the ideas of primary and secondary (repeat or return) migration rates introduced by Ledent (1981). A general review of methods and techniques used to combine migration data from various sources using a stochastic approach has been also presented by Willekens (1994).

As the current study focuses on migration forecasting and not on the data issues as such, it is based on applying a very simplistic approach in dealing with data deficiencies. Firstly, it has to be noted that the receiving countries usually (albeit not always) register more migrants than the sending countries, as migrants tend to have more incentives to register at the destination than to deregister at home (Kupiszewski, 2002b, p. 106). A detailed analysis of the magnitude of these differences can be assessed by constructing the *double-entry matrices* comparing data concerning the same flows according to the statistics of receiving and sending countries.[1] On that basis, for each flow, the greater of the two values can be taken as the estimate of the real magnitude of migration, following Kupiszewski (2002b, p. 111).[2] However, in order to retain consistency of the definitions applied in the current analysis, the German sources have been universally used for the historical data on all the population flows under study, as the ones that apply the broadest definition of a migrant. Such a solution, despite its clear shortcomings, for example, non-adherence to the United Nations (1998) recommendations in the case of the German definition, allows for the accounting of the underestimation of all international population movements in the Polish migration data.

For the purpose of the research presented in this book the numbers of migrants registered in particular countries by their origin or destination have been used to calculate crude migration rates,[3] according to the definitions provided in Chapter 1. The idea of *net migration*; that is, the balance of immigration less emigration from a given country cannot be applied in a coherent manner – the denominator of the rate should reflect the population at risk, which is entirely different in the

[1] Such matrices were first published in the demographic yearbook of the United Nations (1978) and are often used in the analysis of migration data by origin and destination (Kupiszewska & Nowok, 2005).

[2] Shortcomings of this approach are discussed in Kupiszewska and Nowok (2005). An alternative statistical methodology of estimating the magnitude of migration flows on the basis of two sources is proposed by Brierley, Forster, MacDonald, and Smith (2008), notably also within the Bayesian framework. Their approach is under further development in the research project 'IMEM' ('Integrated Modelling of European Migration'), financed by the NORFACE network. For more details, see www.norface.org/migration12.html (accessed on 6 March 2010).

[3] A discussion of advantages and disadvantages of using migration rates can be found for example in McDonald and Kippen (2002), yet limited to the demographic reality of Australia, in that case visibly differing from Europe.

case of immigration and emigration. Although net migration as such is a widely used analytical category, it has been fundamentally criticised for being an artificial measure of a population residual rather than of migratory processes, obscuring both the magnitude, as well as the age profiles of migration (Rogers, 1990). This implies that each of the flows, from A to B and from B to A, have to be modelled separately.

The definition of population at risk, which is a denominator of migration rates, is strictly linked to the definition of international migration. Particularly, the legal (*de jure*) concepts of long-term migrants and usual residents are interrelated – when a person becomes a migrant from country A to B, (s)he gives up their usual residence at A and takes it up at B. Ideally, this should be reflected in the population statistics of both countries involved in the migration process. However, in practice the yearly population stocks are calculated on the basis of the registers, applying the definitions of migration actually in use in a given country and, if these definitions differ between the countries, the population stock figures may be inconsistent. The stocks are usually corrected only after a population census takes place, which can lead to significant breaks in the time series of the size of population stocks. Poland can be referred to as an example, being notorious for continuous underestimation of emigration flows (GUS, 2003).

In order to take these problems into account, in the presented empirical examples the post-census statistical adjustments have been distributed proportionally throughout the period between the censuses (1998–2002), against the registered balance of migration between Poland and Germany, as registered by German official statistics (after: Bijak & Koryś, 2009). This solution, although simplistic and based on the assumption that both registered and unregistered migration follow the same pattern over time, allows for obtaining population estimates that are closer to reality and for avoiding artificial breaks in time series. More details on the procedure can be found in Annex A. Note that for three other countries (Germany, Italy and Switzerland), similar inter-census corrections were unnecessary, as in these cases the population stocks followed rather stable trends, without administrative adjustments causing substantial breaks in series, as was the case for Poland.

It has to be remembered that despite the above-mentioned corrections, the approach to data proposed in the current study is still inconsistent: migration flows are taken according to the common German definition, while population stocks – according to the definition in use in particular countries of origin of migrants. This discrepancy is visible especially for Poland, where official population estimates follow the definition based on the permanent residence concept. In order to properly re-calculate the Polish population stocks to arrive at a more meaningful definition, demographic developments (including vital events and re-estimated migration flows) should be traced back to the moment, at which the information on population size can be believed to be relatively complete. In Poland this moment is likely to be as early as the population census of December 1950, carried out under a non-democratic regime. Therefore, a thorough inquiry of this problem, as well as other issues of data quality, compatibility and completeness remains beyond the scope of this book and would require a separate, dedicated research project.[4]

Nevertheless, it has to be noted that the above-mentioned problems significantly affect the results of any analysis based on migration data. Since this issue is a serious limitation of all international migration studies, particular caution is therefore needed with respect to the interpretation of the results. The data problems contribute to enhancing the uncertainty of such research tasks as migration forecasting and thus require the development of robust prediction methods that would to some extent compensate for the mentioned deficiencies. More general issues related to uncertainty in migration forecasting and the possible ways to deal with it are addressed in the following section, together with a brief justification for selection of the Bayesian approach as a possible way of dealing with the conditions of prevalent uncertainty.

2.2 Uncertainty, Subjectivity and Judgement in Population and Migration Forecasting

The concept of uncertainty refers to the indeterminism or randomness of the phenomena under study, which cannot be assessed or predicted using the present knowledge. There is no general agreement in science as to whether uncertainty is an inherent feature of the phenomena, as in quantum physics under the Heisenberg uncertainty principle, or is merely a result of imperfect knowledge of the deterministic rules that govern the world, as supported by, for example, A. Einstein (Hawking, 1988/1990, pp. 59–61).[5] The latter is sometimes referred to as 'epistemic' uncertainty, as opposed to the 'aleatory' immanent one (O'Hagan et al., 2006, p. 10). This controversy is also to a large extent present in the social sciences, where both deterministic and non-deterministic explanations of socio-economic phenomena develop in parallel. This philosophical debate is crucial for the current study; the non-deterministic point of view in migration (and more generally, population) forecasting is strongly advocated in this book.

---

[4]Credits to Marek Kupiszewski. Notably, a recent Eurostat-funded research project 'MIMOSA: Modelling of statistical data on migration and migrant populations' (website: mimosa.gedap.be, accessed on 17 November 2009), aims at producing coherent estimates of population stocks and flows for the European Union and EFTA. The final results, concerning the period 2001–2008, were made available in late 2010.

[5]In the current study, 'deterministic' is thus understood as 'not allowing for randomness or uncertainty', in contrast to the dictionary-based definition, describing 'determinism' as 'a theory or doctrine that [...] social phenomena are causally determined by preceding events or natural laws' (Merriam Webster Online Dictionary, www.m-w.com, accessed on 25 April 2006). Notably, determinism in the latter interpretation may refer to stochastic explanations of the phenomena under study, provided that the 'natural laws' involved contain an element of randomness, as, for example, in many areas of contemporary theoretical physics. A discussion on uncertainty and (in-)determinism is also offered further in Chapter 5.

In the field of economics,[6] Knight (1921) made an attempt to systematise various categories of indeterminism, introducing the important differentiation between *uncertainty*, which is immeasurable, and *risk*, which can be measured by a quantitative probability.[7] This distinction is currently being used in economic and social sciences, although it remains to some extent questionable. The main controversy regards the interpretation of probability: whether it is an objective measure of randomness, or a subjective degree of belief. The former interpretation is more associated with the concept of risk, while the latter can be also applied in the situation of uncertainty. Bearing in mind this distinction, the term *uncertainty* will be used throughout the book to depict any type of indeterminism, following the subjectivist viewpoint on probability embodied in the Bayesian statistics. As noted by Dawid (1984, p. 278), from a Bayesian viewpoint, probability is the only concept really needed to express uncertainty. Similarly, Robert (2001, p. 508) observed that the Bayesian approach allows for 'a probabilization of uncertainty: axiomatic reduction from the notion of unknown to the notion of random'.

The objective interpretation of probability is predominantly associated with the relative frequency of the events in question. This approach is applied in the traditional (frequentist, sampling-theory) mathematical statistics, developed since the 1920s mainly on the basis of contributions by R. A. Fisher (1922), and K. Pearson, J. Spława-Neyman, E. S. Pearson and A. Wald (Rao, 1989/1994, pp. 62–63, 68–70). As observed by O'Hagan et al. (2006, p. 11), the frequency-based definition pertains only to the aleatory concept of uncertainty as the one related to intrinsic randomness.

Nevertheless, since the beginning of the twentieth century there have been attempts to create a theory of probability that on one hand would be independent from the frequency of events, and on the other would constitute a logical relation between knowledge (evidence) and rational beliefs. An example of such an approach can be found in Keynes (1921/1973), who interpreted probability as a logical relationship between two sets of alternative propositions. The propositions are either true or false, but our knowledge of them depends on our circumstances. Hence, probability is objective, as it is based on rationality and not on the subjective beliefs of the individuals (*Idem*, pp. 3–8). Noteworthy, such probability may not be easily measurable. Keynes also observed that the nature of statistical inference is inductive, as it allows for generalisations of conclusions drawn on the basis of observed samples, where probability gives grounds for a rational preference of one belief over another (*idem*, p. 359; see also the next section).

[6] The subsequent four paragraphs are inspired by the History of Economic Thought website (cepa.newschool.edu/het, accessed on 3 June 2005), which includes comprehensive essays on various topics concerning the economic theory developments. The website author, Gonçalo L. Fonseca, deserves credit for stimulating ideas and general views on the philosophy of probability, and for providing very useful references to primary sources.

[7] A history of the idea of risk, as well as of the attempts to accommodate it in human life has been provided for example in Bernstein (1996/1997).

The interpretation proposed by Keynes has been criticised by Ramsey (1926), who suggested that probability should be rather a subjective measure of beliefs, linked with the utilities and rational preferences of particular individuals (see Bernardo & Smith, 2000, pp. 83–85). Independently, a similar approach has been taken by De Finetti (1937), who made an attempt to operationalise subjective probability through a betting scheme (a 'Dutch book' argument, *idem*, pp. 86–88). Drawing from their works, as well as from the ideas of expected utility and rational expectations of von Neumann and Morgenstern (1944), Savage (1954) worked out an all-inclusive axiomatic theory of subjective probability and utility, based on the assumption of rational preferences (Gilboa, 2009). The probability measure derived from the axioms of Savage fulfils three 'classical' postulates of probability given by A. Kolmogorov: takes values between 0 and 1, probability of the logical sum of all possible events is 1, and probability of the sum of mutually exclusive events equals the sum of their individual probabilities (Bernardo & Smith, 2000, p. 35).

Another attempt to develop a theory of subjective probability, based on a notion of probability as a measure of a degree of belief, has been initiated by Jeffreys (1939/1961), and elaborated by Zellner (1971) and DeGroot (1970/1981). The latter has designed an axiomatic framework of subjective probability independent from utility, but fulfilling the postulates of Kolmogorov. In all these approaches subjective probability is a fully formalised concept, which can describe both the aleatory and epistemic uncertainty; that is, the intrinsic randomness, as well as imperfect knowledge (O'Hagan et al., 2006, p. 11).

Finally, another objectivist interpretation worth mentioning is Popper's (1990/1996) propensity definition of probability. Propensities are characteristics of various possibilities, measuring, how likely is it that these possibilities will occur. In this way, they can be seen as akin to forces in physics. Propensities are formed in the past and influence the future with a certain degree of probability. The asymmetry between the known (closed) past and the unknown (open) future is here one of the main arguments against the deterministic character of the world, with indeterminism interpreted as the simplest of possibly competing explanations (see also Popper, 1982/1996).[8] The indeterminate character of the future, which is contained neither in the present, nor in the past, implies the 'open universe' of possibilities and the freedom of choice between various alternatives (*idem*).

Different interpretations of probability gave grounds for the parallel development of various philosophies of mathematical statistics. Apart from the traditional inference, based on the notion of frequency of events and repeatable samples, there are also other statistical methodologies that deal with uncertainty in different manners. Thus, the subjectivist approach has been incorporated into Bayesian statistics, which is more thoroughly discussed in the next section. The ideas of Keynes (1921/1973), further developed by Jeffreys (1939/1961, 1946), gave grounds to the notion of 'objective Bayesianism', focusing on statistical inference in the instance

---

[8]Popper (1982/1996, p. 26), quoting a 1954 paper of the British philosopher P. H. Nowell Smith in *Mind*, labelled determinism as a somewhat outdated concept – an 'eighteenth-century bogey'.

of minimum or no information about the events under study prior to obtaining evidence in the course of a statistical experiment. In that respect, it can be even argued (e.g. Robert, 2001, p. 510) that human knowledge in general has a subjective base, which is then enhanced by further observations and experiments. Notably, the evolutionary character of knowledge, which sequentially builds upon the prior knowledge and sensual data (observation), has been acknowledged by theorists of the objectivist approach, such as Popper (1990/1996, p. 47).

The relevance of the debate on different approaches to probability for the social sciences, and demography in particular, is discussed in detail in Courgeau (2004). As the current book takes the Bayesian viewpoint, it is also by nature inherently subjective.

2.2.1 Uncertainty in Demographic and Migration Forecasting

In comparison with two other components of the demographic change – fertility and mortality – migration is clearly lagging behind in terms of the availability of quantitative research (Jennissen, 2004, p. 5). Massey et al. (1994; after Jennissen, *idem*) observed that a vast majority of migration-related literature in North America (and even more so in Europe) is not empirical, often being limited to 'polemic arguments and theoretical discourses' (*idem*), while most of the existing empirical studies are merely descriptive. Besides, in migration research, the uncertainty issue, if at all present, is usually treated much less formally than in the general demographic literature. The notable exceptions are discussed in more detail in Chapter 4.

Therefore, due to the scarcity of relevant literature, the discussion in the current subsection is related to demographic forecasting in general, without going into the specificity of particular population processes, such as migration. However, as migration is probably much less definite and more vague than other components of demographic change (cf. Paradysz, 2006, p. 234), the uncertainty associated with it is very likely much higher. Notwithstanding, the main conclusions of the presented argumentation are sufficiently universal to be also valid in the case of migration forecasts, depending on the methodology used.

The philosophical issues concerning uncertainty have profound consequences for forecasting in general. The major source of uncertainty in all forecasts is that they consider future events, which, depending on the viewpoint, either cannot be precisely determined in the light of the present knowledge, or are inherently uncertain by nature. This is especially important in the social sciences, where the predictions aim at assessing future development paths of the variables, which result from interactions within very complex social systems that are often very difficult to predict in a precise manner. With respect to this, demographic, and in particular international migration forecasting are by no means exceptions – they are 'inherently unpredictable' in terms of a point forecast; for practical reasons their uncertainty can be 'interpreted as degree of forecast inaccuracy and operationalised as forecast error' (Keilman, 1990, pp. 9–10).

With respect to the sources of uncertainty in population (and thus also migration) forecasting, Keilman (*idem*, pp. 19–20) distinguishes seven types of possible forecast errors. Three of them are related to measurement problems: errors in observed trends, errors in jump-off data and errors due to rounding. A separate source of forecast inaccuracy is the randomness of the parameters of a forecasting model itself. Finally, three other types of errors are associated with forecasts of exogenous variables, with sudden and unpredicted shifts in parameters that may occur in the future causing discontinuity in trends, as well as with improper model specification.

According to Rees and Turton (1998, Section 2), there are several possibilities to deal with uncertainty in population and thus also international migration forecasting:

1. To ignore uncertainty and develop a single-variant deterministic projection.
2. To develop multi-variant projections, containing various scenarios (e.g., baseline, high and low), where uncertainty is not quantified in probabilistic terms. This approach is widely used by national statistical offices worldwide, as well as by the international organisations (United Nations, 2005a; Eurostat, 2005).
3. To apply the stochastic approach to quantify uncertainty in terms of probabilities of future events. Formally, a probabilistic or stochastic population forecast is defined by Alho (1999, p. 1) as 'a joint predictive distribution of the future population vectors'.

In the last group, Rees and Turton (1998) distinguished statistical forecasts and the analysis of forecast errors. This distinction, however, does not cover all possibilities of stochastic forecasts, which, according to Keilman (2001, *passim*) and Wilson and Bell (2004), include:

- Extrapolation of time series using mainly the Box and Jenkins (1976) ARIMA models, applied for example in de Beer (1990a), Lee and Tuljapurkar (1994), Keilman, Pham, and Hetland (2001, 2002), as well as Keilman and Pham (2004a, 2004b);
- Probabilistic projections based on expert judgement, developed in the IIASA and presented in the studies of Lutz et al. (1996, 1998, 2000, 2004);
- Extrapolation of time series through propagation of historical forecast errors, applied for example in the studies of Keyfitz (1981), Stoto (1983), Alho and Spencer (1985, 2005), Alho (1990, 1998) and NRC (2000).

A thorough and comprehensive overview of the issues related to dealing with uncertainty in population forecasting is offered by de Beer (2000), as well as Alho and Spencer (2005).

The first of the three possibilities mentioned above does not address the issue of uncertainty at all, ignores the uncertain character of the population-related phenomena, and thus produces a point forecast, which in formal terms has a probability of occurrence equal zero under any continuous probability distribution. According to Rees and Turton (1998, Section 2), 'this can only be characterised as the approach of an ostrich burying its head in the sand'.

From the probabilistic point of view the scenario approach is also methodologically inconsistent. Firstly, there is no information on what are the expected *ex-ante* chances that the forecasted variable will actually fall into the range between low and high scenarios (Lutz et al., 2004, p. 19). Such probabilities can be assessed *ex post*, as in the study of Stoto (1983), who found that in the 1977 US Census Bureau projections, the High-Low spans roughly corresponded to empirical two-thirds confidence intervals. Secondly, the methodology implicitly assumes that there is a single common factor underlying the scenario selection (baseline, high or low) for all variables (fertility, mortality, migration) and regions under study. Therefore, their joint effect or aggregation is examined de facto under the assumption of a perfect correlation between the variables or regions; which is very often not true (NRC, 2000, pp. 191–192).

The advantages of the third, stochastic approach are the straightforward quantification of uncertainty in terms of probability and methodological coherence in that respect. Moreover, there is a rather clear agreement in the mainstream of population forecasting that the future belongs to probabilistic predictions, due to their ability to quantify uncertainty (Lutz & Goldstein, 2004, pp. 3–4). However, this approach also incorporates some problematic issues that remain still unresolved. Firstly, Cohen (1986) noted that uncertainty characterises not only the point forecasts of the variables under study, but also their own uncertainty measures, like the confidence intervals. Secondly, as observed by Clark (2003), demographic models do not account for variability among individuals, which can be another significant source of uncertainty, often overlooked in population forecasting. The latter is a general remark that holds for all macro models without micro-level foundations, not limited to demographic predictions.

Finally, there is no clear agreement on which type of stochastic forecasts (i.e., time series extrapolation, expert-based, or error propagation) produces the 'best' results. A comparison of theoretical properties of variance in the two former model types, offered by Bauer, Feichtinger, Lutz, and Sanderson (1999), did not to lead to universal conclusions in that respect. On the other hand, Tuljapurkar, Lee, and Li (2004) showed that the expert-based random scenarios are very sensitive to the assumptions, but tend to underestimate uncertainty in comparison to the time-series approach.

With respect to the potential reluctance of some forecast users to adopt the stochastic methodology, Booth (2004) observed that in all other cases the users would nonetheless treat multi-variant projections as if they were forecasts sensu stricto, despite the clear differences in definitions mentioned in Chapter 1. On the other hand, Booth (*idem*, p. 10) noted that 'the provision of well-explained probabilistic prediction intervals would in all probability be welcomed by users: at last they would have a range of population futures that are rational, consistent and valid for use'. Besides, if point forecasts are needed, in the stochastic approach they can be simply taken as for example means, medians, or other characteristics from the appropriate predictive distributions (see Chapter 11 for a more detailed discussion).

As noted by de Beer (1997) and also discussed in brief in Chapter 1, international migration is perceived as the component of population change bearing the highest uncertainty. An earlier study of the same author (de Beer, 1990b) showed that in population projections and forecasts for the Netherlands prepared in the late 1970s and in the 1980s the role of international migration was heavily underestimated. According to Kupiszewski (2002b, pp. 123–124), several issues are especially important in the context of international migration forecasting; that may underpin the forecasting difficulties. Such problems, including a lack of comprehensive migration theories, difficulties in operationalisation of the theoretical framework of international migration, the problem of uncertainty of potential explanatory variables themselves and ignoring forced migration and policy elements in the forecasts, are discussed in more detail in Chapters 3 and 4.

Unfortunately, given that much of migration research did not pay enough attention to a formal analysis of uncertainty, the issue is still waiting to be investigated. Therefore, despite the fact that the current study aims to fill this gap at least partially, a more in-depth discussion on the issue of uncertainty in migration forecasting still needs to be addressed in a separate enquiry.

2.2.2 Subjectivity and Judgement in Population and Migration Predictions

Addressing the issue of uncertainty in population and migration models and forecasts requires making use of many subjective assumptions. As remarked in the previous subsection, the scarcity of literature pertaining to migration forecasting as such necessitates reliance on a much more comprehensive discussion concerning the subjectivism of demographic predictions in general. Nonetheless, the presented conclusions also remain valid specifically for international migration forecasts, albeit with a reservation that, given very high uncertainty associated with such predictions, the impact of the subjective elements is likely to be much higher than in the case of other, better researched population phenomena, like fertility or mortality.

In general, Pittenger (1980) noted that population forecasts are based on assumptions, which may be either *extrapolative*; that is derived from the historical data, or *judgemental* – subjective, and independent from the past trends. Judgement itself can be either explicitly stated in the forecast, or concealed among various assumptions used, but nevertheless constitutes an inherent element of the process of selecting the forecasting model and creating the scenarios of demographic change. Nevertheless, incorporation of expert judgement in forecasting has been made explicit only in a handful of studies; with those of Alho and Spencer (1985, 2005), Alho (1990) and the IIASA team (Lutz et al., 1996–2004) as notable examples.

Gjaltema (2001, pp. 4–9) distinguished three levels of population forecasting, where the subjective and judgemental elements are included. The first level, *conditional*, includes all technical issues that influence the forecasting outcome, although not necessarily in a direct way. These elements include: choice of the base year of the forecast, length of the forecast horizon and forecast steps (intervals), as well as specification of the output of the process. The second level, *operational*, contains the issues related to translation of time series into the extrapolative model. This level includes most notably selection of a forecasting model, explanatory variables and methods of their extrapolation, historical data to be used, methods of preliminary data treatment, specification of parameters and constants, decisions on combining the results of various forecasts, and the ways to express uncertainty. Finally, the *contextual* level concerns the actual judgement regarding the future levels of variables under study. The subjective elements on this level include target values or whole trajectories of the variables, assumptions on causal relations with their determinants, judgemental expressions of uncertainty, as well as all mechanical adjustments. A selection of a group of experts who are consulted about the forecast assumptions also belongs to the contextual level.

The factors from all three levels: conditional, operational and contextual, are especially important in forecasting international migration, which requires combining knowledge from various disciplines: demography, economics, sociology, geography, or political science. Using interdisciplinary expert judgement seems to be appropriate to address a complex phenomenon, characterised by a large dose of uncertainty. Nevertheless, subjectivity, which is inevitable in any forecast, should be explicitly visible in the formulation of the model. On the other hand, including additional information in the model should enable the balance between its complexity and efficiency to be kept. Willekens (1994, p. 29) observed that

> [expert] knowledge must be formalized, i.e. represented in a way that can be included in a data base and used in a model, [and] combined with statistical data in order to produce a coherent, consistent and accurate picture of migration that can be used in policy-making and research.

These arguments constitute the basic rationale for selecting Bayesian statistics as a promising framework of migration forecasting, for the reasons briefly addressed in the next section.

2.3 Bayesian Inference in Statistics: Introductory Notes

2.3.1 The Bayesian Paradigm

A way to deal with uncertainty and variability within the real world is to apply the tools of mathematical statistics; which therefore plays a major role in all empirical sciences (Bartholomew, 1995). As Lindley (2000, p. 294) has put it, 'statistics is essentially a study of uncertainty'. However, in the field of mathematical statistics there is no universal agreement with respect to philosophy and methodology of the discipline. Contemporarily, there are two major competing standards of statistical inference: the traditional (frequentist) sampling theory and the Bayesian approach. As both of them are fully comprehensive and self-contained, covering not only theoretical aspects, but also applications in research practice, they both constitute independent scientific paradigms, following the definition of Kuhn (1962; cf. Bernardo and Smith, 2000, p. 93).

The traditional, sampling-theory paradigm in statistical inference is based on the interpretation of probability related to the frequency of phenomena under study. Hence, only events that can occur in repeatable experiments (samples) are considered. This approach rejects the subjectivist notion of probability as a measure of belief concerning chances of occurrence of various states of Nature. For this reason the frequentist approach leads to substantial philosophical problems in analysing the phenomena that are by nature unique and not repeatable, as is often the case in social sciences. Moreover, as shown by Jaynes (1976), this is a source of inconsistency in the interpretation of various important analytical categories, like the confidence intervals. More details concerning the frequentist approach are provided in Chapter 8.

Dogmatic treatment of the frequency interpretation of probability became subject to significant criticism of some mathematic statisticians in the late 1940s and early 1950s (Barnard, 1947, 1949, 1951; Savage, 1954; after: Bernardo & Smith, 2000, pp. 1–10, 83–85). On the basis of the subjectivist notions of probability worked out by Ramsey (1926) and De Finetti (1937), they developed an alternative statistical paradigm (*idem*). The theoretical framework they developed became known as 'Bayesian statistics', from the name of the English Nonconformist clergyman, mathematician and logician from the eighteenth century, Reverend Thomas Bayes (1701–1761). The most important work of Bayes (1763) contains a special case of a theorem allowing for computation of a conditional probability of an event A, given the occurrence of event B, known thereafter as Bayes' theorem. The formula, generalised by Laplace (1812), can be written as:

$$p\left(A|B\right) = \frac{p\left(B|A\right) \cdot p\left(A\right)}{p\left(B\right)}. \tag{2.1}$$

Together with the development of analytical tools and methods, as well as with the advancement of computational techniques and capabilities of computers, Bayesian statistics began to acquire increasingly more supporters in a wide variety of scientific disciplines.

One of the major differences between the sampling theory and Bayesian approach regards including in the latter an additional element of analysis. Apart from the sample space $\mathbf{X}$, parameter space $\boldsymbol{\Theta}$ and the family of all probability distributions over $\mathbf{X}$, $\{p_\theta : \theta \in \Theta\}$, the Bayesian approach assumes the existence of a probability distribution depicting knowledge, intuition or belief of a researcher with respect to the possible values of θ, unconditional on the empirical evidence from data. This distribution, $p(\theta)$, is therefore referred to as the *prior distribution* of θ (Silvey, 1975/1978, p. 201). The prior distribution is thus a probability measure (or at least a σ-finite measure) on the parameter space $\boldsymbol{\Theta}$ (Osiewalski, 2001, pp. 14–15).

Further, let p_θ be defined by a probability density function $p(x\,|\theta\,)$ over $\mathbf{X}$ and let the prior distribution of θ have a density $p(\theta)$ over $\boldsymbol{\Theta}$. The density of the distribution of parameter θ conditional on the observations x can be calculated using the so-called Bayes' formula[9]:

$$p\left(\theta|x\right) = p\left(\theta\right)\frac{p\left(x|\theta\right)}{p\left(x\right)}, \tag{2.2}$$

where $p(x) = \int_{\boldsymbol{\Theta}} p(\theta)\, p(x\,|\theta\,)\, d\theta$ denotes the density of the marginal distribution of the random variable x. The discussion, presented here for univariate θ and x, can be generalised for the case of a multivariate parameter vector $\boldsymbol{\theta}$ and data vector $\mathbf{x}$.

---

[9]To preserve coherence with notation prevailing in the Bayesian literature, although somewhat ambiguously in terms of mathematical precision, $p(.)$ can denote different functions, depending on the argument (θ or x).

The conditional distribution with the density (2.2) reflects the belief about the unknown value of θ, given the results of the experiment x (evidence from data) and is therefore labelled as the *posterior distribution* of θ. The essence of Bayesian inference is therefore a formal transformation of the prior beliefs about various parameter values to the posterior knowledge by incorporating the information provided by the sample of observations (Silvey, 1975/1978, p. 202). It is worth noting that the relationship between the prior and posterior distributions need not be necessarily chronological (although usually is), as the distributions differ only with respect to their conditionality on the observations (Bernardo & Smith, 2000, p. 45). In essence, Bayesian inference consists in appropriate conditioning and marginalising of the joint probability distribution $p(x, \theta)$, also known as *full (Bayesian) probability model*, being a product of the prior distribution $p(\theta)$ and the *likelihood of data*, $p(x\,|\theta)$.[10]

The above-mentioned feature of Bayesian inference, focusing on a posterior distribution $p(\theta\,|x)$ of an unknown parameter θ, is opposite to the frequentist sampling theory. In the latter the main object of the analysis is the distribution of observations conditional on the parameters, $p(x\,|\theta)$, where observations are samples from the random variables, which are assumed to follow certain distributions. In the Bayesian approach also the unknown parameter is treated as a random variable due to the uncertainty about its true value (Bernardo, 2003, p. 2).

This difference between the frequentist sampling-theory approach and Bayesian statistics has profound philosophical consequences. The former one corresponds to the deductive reasoning (a logical inference about a thesis, given the premises are true, or in other words – verification of hypotheses) advocated by such theorists of science as Popper (1935/2003). The latter one refers to the inductive reasoning (generalisation of the observations), promoted in scientific inquiry for example by Russell (1912/2004). Without going into a deeper philosophical discussion, it is worth mentioning that contemporarily both paradigms interfere to some extent. Some theorists call for a need for an eclectic approach in mathematical statistics; for example introducing a notion of a 'pragmatic statistics', using either sampling-theory or Bayesian methods, depending on the context of the analysis (Chatfield, 2002). Also the recent developments in the area of 'objective Bayesianism' outlined in Section 2.2 can indicate that the quest for a unique research paradigm in mathematical statistics is ongoing and is becoming increasingly important with the development of new ideas and analytical possibilities (Bayarri and Berger, 2004).

2.3.2 Decision-Theory, Empirical and Orthodox Approaches

The Bayesian statistical approach in itself is not monolithic. Within the wide framework of the contemporary Bayesian paradigm, there are several methodological approaches: the decision-theory, the empirical, and the orthodox ones. The

[10] As noted by Jacek Osiewalski, in the orthodox Bayesian approach, these elements should be sufficient for any inference, 'for whatsoever is more than these, cometh of evil' (personal communication, English citation after the King James Bible, Mt 5:37, www.kingjamesbibleonline.org, accessed on 6 March 2010).

decision-theory approach has been worked out based on the theory of statistical games, developed by Wald (1950). It includes models of decision making under complete uncertainty, where both probability and utility are equally important in solving the decision problems (Lindley, 2000, p. 312). In this approach, the existence of a certain *loss function* is assumed, $L : \mathbf{D} \times \mathbf{\Theta} \subset \mathbf{R}$, depicting the 'loss' of a statistician making decision $d \in \mathbf{D}$ with respect to the parameter under study, in the case that the true value (state of Nature) of this parameter is θ. The estimation of θ can be therefore narrowed down to minimising the expected posterior loss over all possible decisions $d \in \mathbf{D}$, given data x. In particular, under natural assumptions and with a quadratic loss function $L(d, \theta) = \alpha(\theta - d)^2$, the decision-theory Bayesian point estimate of θ given its prior distribution $p(\theta)$ is the expected value of the posterior distribution of θ, while with an absolute value loss function $L(d, \theta) = \alpha\,|\theta - d|$, the point estimate is a median of the posterior distribution (DeGroot, 1970/1981, pp. 196–202).[11] The relevance of this approach for political decision making is discussed further in Chapter 11.

The 'Empirical Bayes' approach, initiated by von Mises (1943), is in fact a hybrid of the frequentist and Bayesian inferential mechanisms. On one hand, the existence of a prior distribution is assumed and the Bayesian techniques are used to infer on the decision function from the prior distribution. On the other hand, the estimation of $p(\theta)$ and the decision rules is based on the sampling-theory statistical methodology. The prior distributions can therefore have a frequency interpretation and be derived from a sample of observations. A thorough overview of empirical Bayes methods has been provided in Maritz (1970). However, the empirical approach is being criticised for methodological incoherence. Osiewalski (2000, p. 251) presented a simple example showing that if the prior distribution depends on sample data, the posterior distribution corresponds to a different sampling model $p(x\,|\theta)$ than the one originally assumed.

In the orthodox Bayesian statistics, however, both the decision-theory and empirical approaches are criticised: the former for an unnecessary limitation to a particular loss function, the latter for incoherence in applying frequentist methods for estimating prior distributions. The orthodox Bayesians claim that the point estimation applied in the decision-theory approach, as well as a large part of the applied methodology may be useless, as there is no need to optimise the loss function, unless making a particular decision is the explicit goal of the research. There is also no need to calculate the standard errors, or otherwise averaging over the space of observations, **X**. What is only needed for inference is the conditional distribution of the parameters θ given the data x (Lindley, 1991, p. 7; 2000, p. 309).

In the orthodox Bayesianism, two approaches co-exist, which differ with respect to the philosophical aspects of inference, rather than to the formality of the applied

[11] DeGroot (1970/1981) provides a comprehensive discussion on the theory of statistical games and the decision approach. Interestingly, a decision-theory approach to demographic forecasting, involving the presence of a loss function, has been advocated by Lee (1998), as well as by Alho and Spencer (2005) as one of the possibilities of methodological improvements of stochastic population predictions (see also Chapter 11).

mathematical procedures. The subjective approach applies informative prior distributions based on the expertise of the researcher. The objective Bayesianism follows the propositions of Keynes (1921/1973) and Jeffreys (1939/1961, 1946), and focuses on the statistical inference based on non-informative prior distributions. The latter approach is also useful as a reference in all other types of Bayesian analysis, depicting the conditions of minimum prior knowledge and thus may be applied to assess the sensitivity of the results on selection of particular prior distributions. The problem is covered by a separate area of Bayesian inference, referred to as the *robustness analysis*, presented in detail for example by Męczarski (1998), and briefly discussed in Chapter 8.

2.3.3 Bayesian Interval Estimation and Forecasting

In the Bayesian approach, a natural analytical tool is interval estimation, based on the intervals or regions for the unknown parameters θ, analogous to the confidence regions in the frequentist approach. Within the Bayesian framework, the construction of the regions (interchangeably referred to as the *Bayesian [confidence] regions / intervals* or *credible regions / intervals*[12]) is made under the assumption of a particular posterior distribution of θ given the data x, having the density $p(\theta\,|x)$.

Notwithstanding, it has to be noted that Bayesian credible regions are not unique. In fact, any region of the parameter space, which has fixed posterior probability $1-\gamma$ can be called a 'credible region'. The construction of a unique interval (region) estimator for θ, with the posterior probability that θ is actually located within equal $1-\gamma$, can be for example performed by determining such k_γ that would satisfy the condition (Silvey, 1975/1978, p. 203):

$$\int\limits_{\{\theta:p(\theta|x)>k_\gamma\}} p\,(\theta|x)d\theta = 1-\gamma \tag{2.3}$$

The formula (2.3) may concern not only the univariate θ, but also its multivariate generalisation, $\boldsymbol{\theta}$. The interval or, in general cases, region (2.3) contains all a posteriori most likely values of θ, so it is called a *highest posterior density* (HPD) interval (region). Moreover, each value of θ outside (2.3) has smaller probability density than any value from this interval (region). In certain cases, when θ is univariate, and the posterior distribution $p(\theta\,|x)$ is symmetric and unimodal, the quantiles of rank $\gamma/2$ and $1-\gamma/2$ from $p(\theta\,|x)$ can be simply taken as the respective lower and upper limits of the Bayesian interval (Bernardo & Smith, 2000, pp. 259–262). It is worth noting that HPD regions are not invariant under reparameterisation: a transformation $f(\mathbf{C})$ of a HPD region $\mathbf{C}$ may not be HPD itself, although still is a credible region (Bernardo, 2003, p. 20). In any case, owing to the subjective interpretation

[12] After the International Statistical Institute (ISI) Multilingual Glossary of Statistical Terms (isi.cbs.nl/glossary.htm, accessed on 10 June 2005).

of probability, the problems with interpreting confidence regions are avoided, what is an advantage of the Bayesian approach over the frequentist one.

Bayesian statistics also provides straightforward tools for predictions of future values of variables under study. Let $\mathbf{x}$ denote a time series of T sample realisations of a stochastic process in the periods $t = 1, \ldots, T$; $\mathbf{x} = [x_1\ x_2 \ldots x_T]'$. Further, let $\mathbf{x}^{\mathbf{P}}$ denote a vector of q future values of the said process, that are to be predicted: $\mathbf{x}^{\mathbf{P}} = [x_{T+1}\ x_{T+2} \ldots x_{T+q}]'$. In the Bayesian approach, the predictive probability distribution of $\mathbf{x}^{\mathbf{P}}$, conditional on the observations, can be calculated through the posterior distribution of the vector of parameters $\boldsymbol{\theta}$ (Zellner, 1971, p. 29),

$$p\left(\mathbf{x}^{\mathbf{P}}|\mathbf{x}\right) = \int_{\Theta} p\left(\mathbf{x}^{\mathbf{P}},\boldsymbol{\theta}|\mathbf{x}\right) d\boldsymbol{\theta} = \int_{\Theta} p\left(\mathbf{x}^{\mathbf{P}}|\boldsymbol{\theta},\mathbf{x}\right) \cdot p\left(\boldsymbol{\theta}|\mathbf{x}\right) d\boldsymbol{\theta}. \tag{2.4}$$

The predictive distribution (2.4) can be therefore interpreted as an average of the conditional predictive distributions $p(\mathbf{x}^{\mathbf{P}}|\boldsymbol{\theta},\mathbf{x})$, weighted with the values of the posterior density functions of the parameters, $p(\boldsymbol{\theta}|\mathbf{x})$. This distribution can be applied for point and interval estimation of the predicted values of the variable x. The point forecast can be obtained minimising a certain loss function over the full set of possible candidate values, similarly to the approach to estimation presented before, in the discussion of the decision-theory approach in Bayesian inference (see also Chapter 11). Again, with the quadratic loss function, the point forecast is the expected value in the predictive distribution (2.4). Likewise, the interval forecast associated with a given probability $1 - \gamma$ can be determined following the idea presented in (2.3), using the credible or, in particular, HPD regions from the predictive distribution $p(\mathbf{x}^{\mathbf{P}}|\mathbf{x})$.

The brief overview of selected aspects of Bayesian inference presented above is by no means complete, as it has been merely designed to provide some background information for the purpose of the current research. Comprehensive studies in the theory of Bayesian statistics can be found in, among others DeGroot (1970/1981), Box and Tiao (1973), Bernardo and Smith (2000), Robert (2001), and Bernardo (2003). Examples of theoretical and practical aspects of Bayesian econometrics are provided in Zellner (1971) and Osiewalski (2001). Recently, an overview of the methods of Bayesian modelling, with a broad set of sample applications has been provided in Congdon (2001a, 2003, 2005). From the point of view of social sciences, a book definitely worth reading is that of Lynch (2007), which provides a very clear and comprehensive step-by-step introduction to Bayesian analysis, illustrated by a variety of social science applications.

2.4 Markov Chain Monte Carlo (MCMC) Simulations

2.4.1 Numerical Solutions to Bayesian Problems: General Remarks

Bayesian estimation and prediction problems, as briefly introduced in the previous section, in practical applications usually require many thorough computations, involving multi-dimensional integrations over the space of parameters and latent variables (if they are present in the model). Moreover, in many cases, the results cannot be obtained analytically. More complex tasks require the use of numerical algorithms that would enable obtaining the desired results, including posterior distributions of parameters and predictive distributions of variables under study.

There are several classes of numerical algorithms devised to solve numerical problems encountered in the Bayesian inference. Osiewalski (2001, pp. 31–44) lists the *Monte Carlo-Importance Sampling* (*MC-IS*) method (Kloek & van Dijk, 1978; Geweke, 1989), as well as the *Markov chain Monte Carlo* (*MCMC*) techniques. A special case of the latter is the *Gibbs sampling*, originating from the work of Geman and Geman (1984), and Gelfand and Smith (1990) (after: Neal, 1993, p. 47). A thorough and comprehensive overview of various MCMC methods and their modifications is provided for example in the book by Robert and Casella (2005).

Among various estimation methods for Bayesian applications, the Gibbs sampler is characterised by a relative simplicity and automatism of computations (through sampling from full conditional distributions), and has several theoretical advantages for instance over the MC-IS (details in Osiewalski, 2001, p. 31, 44). Hence, the empirical part of the current study is based on calculations made using the Gibbs algorithm, implemented within the WinBUGS 1.4 software developed by Spiegelhalter, Thomas, Best, and Lunn (2003). As all methods have good stochastic properties, selection of the Gibbs sampling also followed a purely practical motivation. Conveniently, the method was readily available in a flexible software environment, which allowed for calculations without a necessity to go in-depth into numerical issues (see also Congdon, 2003).

2.4.2 Simulation of Posterior Distributions Using Gibbs Sampling

After Osiewalski (2001, p. 39, drawing from Casella and George, 1992, p. 168), the Gibbs algorithm of simulating the joint posterior distribution $p(\boldsymbol{\theta}\,|\mathbf{x})$ of the vector of parameters $\boldsymbol{\theta}$, split into d blocks, $\boldsymbol{\theta} = [\theta_1, \ldots, \theta_d]'$, where $d > 1$, consists of the following steps:

1. The vector of initial values, $\boldsymbol{\theta}^{(0)} = [\theta_1^{(0)}, \ldots \theta_d^{(0)}]'$, is taken arbitrarily, e.g. at random.
2. In the q-th cycle of the algorithm, the realisation $\boldsymbol{\theta}^{(q)} = [\theta_1^{(q)}, \ldots, \theta_d^{(q)}]'$ is generated in sequential stages from the following d full conditional distributions:

$$\left\{\begin{array}{l} \theta_1^{(q)} \sim p\left(\theta_1|\theta_2 = \theta_2^{(q-1)}, \theta_3 = \theta_3^{(q-1)}, \ldots, \theta_d = \theta_d^{(q-1)}, \mathbf{x}\right) \\ \theta_2^{(q)} \sim p\left(\theta_2|\theta_1 = \theta_1^{(q)}, \theta_3 = \theta_3^{(q-1)}, \ldots, \theta_d = \theta_d^{(q-1)}, \mathbf{x}\right) \\ \qquad \vdots \\ \theta_d^{(q)} \sim p\left(\theta_d|\theta_1 = \theta_1^{(q)}, \theta_2 = \theta_2^{(q)}, \ldots, \theta_{d-1} = \theta_{d-1}^{(q)}, \mathbf{x}\right) \end{array}\right\} \tag{2.5}$$

In the *burn-in phase*, the second step is repeated S times, until the algorithm converges to the stationary distribution of the Markov chain defined by (2.5). Thereafter, M more iterations of step 2 are generated, which ultimately serve as a sample from the posterior distribution, used for an approximation of the required characteristics thereof: moments, quantiles, etc. (Osiewalski, *idem*). In prediction problems, on the basis of (2.5) the future values of the variable under study, $\mathbf{x}^{\mathbf{P}}$, can be generated, given the data $\mathbf{x}$.

It can be shown that the distribution of the random vector $\boldsymbol{\theta}^{(q)}$ generated using (2.5) converges to $p(\boldsymbol{\theta}\,|\mathbf{x})$, as $q\rightarrow\infty$. Formal proof is offered and discussed for example in Tierney (1994), whereas stationarity is demonstrated in Osiewalski (2001, pp. 40–42) using a series of heavy integrations of the density of the vector $\boldsymbol{\theta}^{(q)}$ drawn from a Gibbs sampler.

As noted by Congdon (2003, p. 5), the basic Gibbs algorithm works well for sampling from standard probability distributions (e.g., Normal, Gamma or Student's *t*). In other cases, the method may be modified by additionally implementing the Metropolis-Hastings algorithm (Metropolis, Rosenbluth, Rosenbluth, Teller, & Teller, 1953; Hastings, 1970; for description see e.g. Neal, 1993, pp. 54–64), or the *adaptive rejection sampling* technique (Gilks & Wild, 1992), the latter especially important in hierarchical Bayesian models. Especially for dealing with complex and multi-dimensional problems of Bayesian model selection, the *Reversible Jump MCMC* algorithm (*RJMCMC*) has been developed (Green, 1995), alongside the Carlin and Chib (1995) procedure applied in the current study (for a more detailed description, see Chapter 5).

Many of these possibilities, including the Metropolis-Hastings method, and rejection sampling, as well as several other modifications of the standard Gibbs algorithm, have been applied in the WinBUGS software (Spiegelhalter et al., 2003) used in the current study. The software itself offers very flexible possibilities of conducting various types of Bayesian analysis, many examples of which can be found in the book by Congdon (2003). Besides, as noted by Lynch (2007, p. xi), an important advantage of WinBUGS is that the users need not derive the form of full conditional distributions (2.5) themselves, as the calculations are performed automatically within the programme. This feature may be especially important for more complex, hierarchical models. For these reasons, as well as due to simplicity of coding statistical models in WinBUGS, the current study employs this software (version 1.4) for the purpose of numerical computations. A more detailed discussion of WinBUGS and selected other practical possibilities for conducting Bayesian analysis using other available software is provided in Chapter 9.

Part II
Explaining and Forecasting Migration

Chapter 3
Explaining Migration: Brief Overview of Selected Theories

Attempts to synthesise knowledge on population movements in the form of a coherent theory date back to *migration laws* of Ravenstein (1885; 1889). Ravenstein generalised observations on internal migration in Great Britain and found that the intensity of the process was influenced by several factors: distance, population size of the origin and destination regions, absorption capacities of the latter, sex of migrants, etc. Although he found the rural-urban flows to be dominant, he also acknowledged the presence of return migration. Concerning the twentieth century followers of Ravenstein, the efforts to develop a theoretical framework of migratory phenomena have been made by the representatives of various disciplines of science, including sociology, economics, and human geography. The current chapter[1] attempts to outline selected theoretical fundamentals of international migration forecasting. The discussion firstly offers a brief overview of selected migration theories (in Section 3.1), followed by an evaluation of their potential applicability in forecasting international population flows (Section 3.2). The systematics of the theories presented in the current chapter largely follows, with slight modifications, the one proposed by Zlotnik (1998), and adopted by Kupiszewski (2002b, pp. 118–122).

3.1 Different Theoretical Perspectives on Migration Flows

3.1.1 Sociological Theories

According to Zlotnik (1998, p. 9), sociological theories of migration date back to the *intervening opportunities* idea of Stouffer (1940, 1960). In his approach, migration is proportional to the number of attracting opportunities (e.g., jobs) available for migrants at the destination, and inversely proportional to the number of such opportunities existing closer to the place of origin. It is worth stressing that this concept does not relate migration directly to spatial distance (and cannot be thus

[1] An earlier version of Chapters 3 and 4 (Bijak, 2006) is available at: www.cefmr.pan.pl (as of 1 May 2007).

J. Bijak, *Forecasting International Migration in Europe: A Bayesian View*, The Springer Series on Demographic Methods and Population Analysis 24,
DOI 10.1007/978-90-481-8897-0_3,

classified as a purely geographical theory), but to the interplay between distance and opportunities available for prospective migrants in various locations.

The notion of opportunities gave grounds for the synthesising *push-and-pull factors* approach of Lee (1966). His theory explains that migration is determined by the presence of attracting (*pull*) factors at destination, and repelling (*push*) factors at origin, which prevail over the positives at origin and negatives at destination, and are further moderated by intervening obstacles and personal factors. For international migration these factors can be further divided into *hard* and *soft* ones (Öberg, 1996). The former group includes dramatic circumstances like humanitarian crises, armed conflicts, environmental catastrophes, etc., while the latter – less critical problems, such as poverty, social exclusion, or unemployment. The dominance of particular factors determines to some extent the characteristics of the migrating population: the favourable pull factors at destination tend to attract migrants who are positively selected in terms of human capital or motivation. This is not the case, when the unfavourable push factors at origin play a crucial role in instigating the migration process. Besides, the 'soft' factors can be intuitively seen as corresponding to long-term migration trends, which are easier to forecast, whereas the 'hard' ones mainly concern extraordinary circumstances instigating large migration waves, which are very difficult, if at all possible to predict. As suggested by Arango (2000), the 'push-pull' framework can be further extended to include the 'retain' and 'repel' factors, in order to help answer not only the question, why people move, but also why so few of them do so.

Taylor (1986) noted that a very important pull factor is the presence of interpersonal *migrant networks* in the receiving country, which are comprised of people sharing kinship, friendship or origin (after: Massey et al., 1993, p. 448). Such networks of people interconnected by family or acquaintance ties assist subsequent migrants in many aspects of everyday life. The main role of networks is to diminish various costs (not only monetary, but also psychological, and other) and risks associated with migration, and to facilitate the flow of migrants between the origin and destination countries. Population flows are thus characterised by a large degree of inertia: once started, they are difficult to control by the authorities of the receiving country, and become more and more independent from the factors that originally caused them. The network hypothesis has been verified for example for historical migratory outflows from Europe in the period 1850–1914 by Hatton and Williamson (1998), who used foreign population stocks as proxy variables for the size of particular networks. Nevertheless, as noted by Orrell (2007, p. 245) the existence of a positive feedback loop due to the presence of migrant networks in general increases complexity and decreases predictability of the whole migration process under study (for more on complexity, see Chapters 10 and 11 of this book).

More recently, the idea of networks has been generalised within the theory of *transnational social spaces* (Pries, 1999; Faist, 2000), providing a meso-level analytical framework for international migration studies.[2] The theory acknowledges the

[2] I am very grateful to Izabela Koryś for drawing my attention to this notion.

presence of various migration-related cross-border linkages between individuals and groups. As defined by Faist (2000, p. 199),

> transnational social spaces consist of combinations of social and symbolic ties, their contents, positions in networks and organizations, and networks of organizations that can be found in multiple states. These spaces denote dynamic processes, not static notions of ties and positions.

Apart from the networks theory, this approach builds on the concept of *social capital*, seen as the

> resources that help people or groups to achieve their goals in ties and the assets inherent in patterned social and symbolic ties that allow actors to cooperate in networks and organizations, serving as a mechanism to integrate groups and symbolic communities (*idem*, p. 102).

The social capital, embodied in social and symbolic ties between individuals and groups, such as the norms of reciprocity, solidarity, and mutual obligations, is a set of local assets. Hence, social capital may not be easily transferable across borders. This fact, overlooked by the networks theory, provides justification for the relative immobility of certain social groups. On the other hand, once migration starts, the social capital facilitates both the adaptation to the host society, as well as the maintenance of ties with the country of origin. The transnational social spaces that are formed in this way can be thus seen as bridges between the source and destination countries of migrants (Faist, 2000, pp. 195–241). The notion of transnational social spaces is sufficiently general to explain the phenomena of chain migration, return migration, perpetuation of migratory processes, and the 'saturation' of population flows at a certain level. However, despite the potential attractiveness of this concept, its development is relatively recent, and it is still very far from possible operationalisation for a practical use in migration forecasting.

The important role of institutions in facilitating the migration process is acknowledged in the *institutional theory* of migration sketched by Massey et al. (1993), concerning both regular (e.g., active job recruitment, counselling), as well as irregular (e.g., human smuggling and trafficking) aspects of population flows. This notion is complementary to the network theory, extending the set of subjects facilitating migration to various institutions, from for-profit entities and enterprises (legal or illegal), to humanitarian organisations, NGOs, etc. Also in the light of this theory, the process of institutionalisation of migratory flows seems to be to some extent self-perpetuating, independent from the initially-dominant migration factors, and increasingly difficult to regulate (Massey et al., 1993, p. 451). With respect to the importance of institutions, these observations can be also linked with the institutional theory in economics, thus de facto forming a hybrid of sociological and economic perspectives.

Another example of a cross-cutting, interdisciplinary approach is the *cumulative causation hypothesis* (Massey, 1990), according to the author elaborating on the economic theories of Veblen (1898) and Myrdal (1957). The theoretical construction is based on the proposition that migration is an evolutionary process that contributes to institutional and socio-economic change both at origin and destination, through various feedback mechanisms. Examples of the latter include the redistribution of income of households involved in migration, both in absolute and relative terms, as well as the redistribution of land and capital. These processes ultimately result in changes in the social hierarchy (Massey et al., 1993, pp. 451–454). As one of the consequences, due to the fact that return migrants usually possess more human capital than the immobile population, the non-migrant group increasingly aspires to a privileged position in the community. It is worth noting that the principle of cumulative causation is also an important element of the transnational social spaces theory (Faist, 2000, pp. 129–132).

The factors listed above instigate a 'migrant culture', in which migration is perceived as an activity with expected positive gains, which in turn contributes to changes in the human capital distribution of the source regions. In such a way, the 'brain drain' of a sending country can actually turn into the 'brain gain'. This theoretical possibility has been further corroborated by Stark and Wang (2001) within the microeconomic analytical framework. Their analysis provides arguments that expected positive returns from migration are a source of externalities in the form of an overall increase in the human capital in the region of origin, although a very important necessary condition for such an outcome is a rather limited size of the 'emigration of brains'.

3.1.2 Macroeconomic Theories

The economic theories of labour migration, as all economic theories in general, usually follow either the macro- or micro-level perspective, and reflect a diversity of paradigms of theoretical economics that either dominated, or co-existed in various periods. To begin with, the neoclassical macroeconomic migration theory (Lewis, 1954; after: Massey et al., 1993, p. 433) explains that – given wage differentials between two capitalist economies, one characterised by a surplus of labour (unemployment) and the other by a surplus of capital – migration and capital movements occur. The flows of both production factors in opposite directions, and the related convergence of wage levels are a way of return to the economic equilibrium, where the wage differential reflects only the costs of undertaking migration. It is worth noting that in the whole neoclassical approach, migration is a disequilibrium phenomenon, which ceases as soon as the equilibrium is reached (see for example Harris & Todaro, 1970, p. 129).

Nevertheless, the presented approach is very simplistic and does not explain for example return migration, or population flows in the absence of wage differentials, as indicated by Stark (2003). Additionally, Jennissen (2004, p. 46) points to the fact that there exists an alternative Keynesian view on migration-induced labour market adjustments towards the economic equilibrium, through the elimination of differences in unemployment, not in wages (see for example Hart, 1975; *idem*). What follows is the pro-cyclical nature of migration, as observed for example for Canada by Milne (1993), who showed that net migration rates of particular regions go along with the business cycle, although with a time lag.

A similar framework is explored in the neoclassical macroeconomic migration theory of Harris and Todaro (1970). They formulated their conceptual model in terms of *expected* income (wages), taking into account the probability that labour migrants from the 'agricultural' sector (A) in the 'rural' region r, find jobs in the 'manufacturing' sector (M) of the 'urban' economy (u), the latter characterised by minimum wages (W_M^*) and unemployment. The equilibrium condition, to which the system should optimally converge, is defined as (Harris & Todaro, 1970, p. 129):

$$\frac{dN_u}{dt} = \psi\left(\frac{W_M^* \cdot N_M}{N_u} - P \cdot \frac{dX_A}{dN_A}\right), \tag{3.1}$$

where N_u is the total urban labour supply (original population and migrants), N_M is the number of employed in the manufacturing sector, and N_A – in the agricultural sector. Further, $X_A = X_A(N_A)$ is the increasing and concave production function of the agricultural sector, and P denotes the price of the agricultural output defined in terms of the goods produced by the manufacturing sector (the terms of trade). The expected urban wages are equal to the minimum wages adjusted for the chances of being employed, $W_M^* \cdot N_M/N_u$. The function ψ is such that $\psi' > 0$ and $\psi(\alpha) = 0$ for some $\alpha \in \mathbf{R}$, which under (3.1) implies that migration stops, when the expected wage differential equals α (in particular α=0 can be assumed).

Harris and Todaro additionally tested the impact of various policy instruments (among others, minimum wages and migration restrictions) on the economic equilibrium. They found that in the absence of full wage flexibility 'either a limited wage-subsidy or a migration-restriction policy will lead to a welfare improvement [at the destination]' (*idem*, p. 137). On the other hand, the outcome of both of these policies will be sub-optimal from the point of view of the economy as a whole system of the sending and receiving regions.

Apart from the purely neoclassical and Keynesian approaches, there have been many further attempts to explore the economic motives of migrants and the nature of migratory processes. The *dual labour markets theory* (Piore, 1979; after: Massey et al., 1993, pp. 441–443) is based on an observation that migration flows are to a large extent determined by labour demand characteristics at the destination. This notion puts forward that immigrant labour is necessary for the economies of the developed countries, and hence an immanent and permanent demand for foreign workforce. An important explanation is that wages are not only the price of labour, but also a proxy measure of the position of the employee in the occupational and social hierarchy. If there are labour shortages at the bottom of the hierarchy, the entrepreneurs would prefer to hire immigrant workers without aspirations to a higher social status than to raise wages in order to attract the local labour force. The latter option would require proportional wage increases in the whole sector to preserve the whole job ladder, and would ultimately lead to a structural inflation.

The dual labour markets theory justifies the segmentation of labour markets – the local population moves to more attractive professions, while immigrants take up the '3D' (dirty, dangerous and difficult) jobs. On the top of that, this division is intensified by the very nature of various occupations – the attractive ones are usually capital-intensive, while the 3D are far more labour-intensive. Moreover, immigrant workforce is a more flexible production factor than the local one, protected by various institutions (trade unions, regulations of work conditions, etc.), which is another factor petrifying the labour market dualism. This also gives reasons for the dominance of women and young people among migrants – they are more willing to accept poorly-paid jobs in unfavourable conditions than other groups (*idem*). Thus, the dual labour markets theory provides useful explanations for many features of migratory phenomena, but nevertheless seems difficult to operationalise to suit macro-level migration predictions. Another drawback it has lies in its concentration exclusively on the labour market of the receiving country in explaining migration (Arango, 2002).

In an attempt to generalise the macroeconomic perspective, the *world systems theory* (Wallerstein, 1974) assumes that international migration is associated with the advances of the capitalist system and global markets, not only in the world economic 'core', but also in semi-peripheral and peripheral regions. A flow of goods and capital from the core to the peripheries 'in search of land, raw materials, labour, and new consumer markets' (Massey et al., 1993, pp. 445) is counterbalanced by the flow of labour in the opposite direction. There are many driving forces behind this process. On one hand, there is an increasing demand for low-skilled labour in the core regions, where the jobs in the manufacturing sector become less and less desirable for the local population, following the shift towards the service-based economy. On the other hand, the commercialisation of agricultural production in the peripheries caused by the capital flows leads to an increased productivity and the resulting reduction of demand for the local labour. These factors instigate migration of the surplus of agricultural workforce, which is 'uprooted' from the peripheral regions in the search for either formal or informal job opportunities in the low-paid segments of the manufacturing and services sectors in the core (Zlotnik, 1998, pp. 7–8).

The world systems theory also acknowledges the presence of many other links (not only material, but also historical, cultural, linguistic, etc.) between the origin and destination countries, which influence migratory flows. These elements are the reason why the whole theory cannot be simply classified as 'macroeconomic'; rather, it offers another cross-cutting perspective, combining economic and sociological explanations. As noted by Massey et al. (1993, p. 448), in the world systems approach, 'international migration ultimately has little to do with wage rates or employment differentials between countries; it follows from the dynamics of market creation and the structure of the global economy'. Special attention is paid to the asymmetrical relationships between the former colonies and their former metropolis, the latter having more favourable terms of trade. In that respect, Jennissen (2004, pp. 53–54) noted that although this theory gives grounds to the philosophy of alter-globalisation movement, its view on international trade can be seen as controversial, as free trade is contemporarily thought to reduce income and employment disparities (and thus also migration). Nonetheless, the world systems theory is articulated only verbally, not in formal mathematical terms, and is too general in order to serve as a direct reference for migration forecasting.

3.1.3 Microeconomic Theories

The individual-level migration decisions are explained for example by the neoclassical microeconomic theory of Sjaastad (1962), which treats migration as an investment in human capital, and a result of a rational cost-benefit analysis. According to this theory, prospective migrants choose the destinations that are maximising the net present value of their expected future income, minus the sum of direct and indirect costs of migration. In a more comprehensive framework, this can be formalised as (Massey et al., 1993, p. 435, also referring to Borjas, 1990):

$$ER(0) = \int_0^n [p_1(t)p_2(t)Y_d(t) - p_3(t)Y_o(t)] \cdot e^{-rt}dt - C(0). \tag{3.2}$$

In the above equation, $ER(0)$ denotes the expected returns from migration at the moment 0, n is the time horizon of the decision making process, p_1 is the probability of not being deported $p_1 < 1$ for irregular migrants), Y_o and Y_d are earnings at the origin and destination, while p_2 and p_3 denote the respective probabilities of finding a job. Further, r is the discount rate, and $C(0)$ is the sum of all costs of migration, both economic and psychological (*idem*), although the latter seem to be very difficult to operationalise and measure.

A generalisation of the microeconomic decision framework is the *value-expectancy* concept of DeJong and Fawcett (1981; after: Faist, 2000, pp. 36–37). The underlying formal model defines the individual motivation to migrate (*MM*), subject to maximisation, as:

$$MM = \sum_i P_i \cdot E_i. \tag{3.3}$$

The values of P_i refer to the preferred outcomes of migration, and E_i – to the 'expectations' of their realisation through migration which are held by a potential migrant, and thus can be interpreted as subjective probabilities (*sic!*). The index i denotes the 'values' or 'desires' of an individual; that is, various dimensions of the decision making problem. The presented theory is very comprehensive and may cover different aspects of human decisions in the migration context (different values of i), not only limited to the economic ones (unemployment, wages), but also potentially taking into account the social and psychological spheres of life. On the other hand, Faist (*idem*) points to the fact that the value-expectancy theory would be difficult to apply for prediction-making, as people tend to rationalise their actions *ex post* rather than to reason them *ex ante*, as it is suggested by the formulation of (3.3). Moreover, many of the dimensions of the migratory decision problem may be, again, almost impossible to measure.

The *new economic theory of migration* (Stark & Bloom, 1985; Stark, 1991; after: Kupiszewski, 2002b, p. 119) suggests that migration-related decisions are made by households rather than by individuals. This finding coincides with the observations that migratory processes are characterised by visible family patterns, as noted both by economists (Mincer, 1978) and demographers (Castro & Rogers, 1983). From this point of view, diversified migration strategies of particular household members are an instrument of risk management at the household level, rather than a simplistic maximisation of the expected income. In this approach, wage differentials between origin and destination countries are not a prerequisite for migration. One of the possible extensions of this perspective is the life-cycle theoretical model of Dustmann (1997), who formulated a stochastic framework for the analysis of migration and return migration under the condition of uncertainty, with focus on precautionary savings of the individuals. Savings can be seen as another element of risk management at the household level, related to remittances – monetary transfers from migrants to their families in the country of origin.

Other theoretical examples explaining the presence of migratory flows in spite of the absence of differences in income levels have been summarised by Stark (2003). He stresses an important role of individual preferences ('taste') in favour of migration, as well as of the difference in the purchasing power of savings generated by migrants between the regions of origin and destination, which justifies the presence of return migration. Another very important migration factor is the community context, which can be associated with the *relative deprivation* concept (Stark & Taylor, 1989). This approach proposes that migration is fuelled not by absolute, but by relative income differentials towards the reference group of potential migrants. This notion provides a link between economy and psychology, the latter with respect to the sense of resentment towards the reference group. The relative deprivation RD of a household with income y is defined as (Stark & Taylor, 1989; after: Massey et al., 1993, p. 439):

$$RD(y) = \int_{y}^{y_{max}} h[1 - F(z)]dz\,, \tag{3.4}$$

where $F(y)$ denotes the cumulated income distribution in a given community, y_{max} is the maximum income observed in this group, and h is a monotonously increasing function, measuring the dissatisfaction with the relative position of the household income in the reference community.

Recently, Quinn (2006) extended the definition of relative deprivation to other durables, such as land and housing, in addition to income. In his empirical analysis of migration within Mexico, as well as from Mexico to the United States, the data supported a hypothesis about a combined effect of absolute wage differentials and relative deprivation on migration flows. Such a synthesising approach may provide a partial solution to the problem noticed by Massey et al. (1993, p. 440); that the neoclassical and new economic theories of migration 'lead to divergent conclusions about the origins and nature of international migration'.

Further attempts to reconcile the two paradigms (neoclassical and 'new economic') may potentially be aimed at combining the two levels of decision making (individual or household), their distinct goal functions (maximising income or minimising risk), or presence of the social context (e.g., whether income is perceived in absolute terms, or in relation to the income distribution in a reference group). Such a comprehensive micro-level model explaining migratory decisions would have to be based on the multi-criteria programming with many parameters and constraints. Problems of this type, although potentially very complex, may either have relatively simple analytical solutions, or be solved using numerical methods, and thus be able to provide useful input for practical forecasting applications. Nevertheless, the issue of constructing a multi-level, multi-criteria, micro-model of migration remains far beyond the scope of this book.

3.1.4 Geographical Theories

In the tradition of human geography, the theories of migration are focused on the role of distance in explaining spatial movements. Distance is viewed as a factor moderating the *spatial interactions* between regions, which include population flows. For example, the *gravity theory* of migration (Stewart, 1941; Zipf, 1946; after: Isard, 1960/1965), analogous to Newton's law of gravity, assumes that migration between regions i and j, m_{ij} is proportional to the product of population sizes in the origin and destination regions (P_i and P_j), corresponding to masses in the Newtonian model, and inversely proportional to the b-th power of distance between the two regions, d_{ij}, which is a discounting factor (Isard, 1960/1965, p. 350):

$$m_{ij} = G \cdot \frac{P_i \cdot P_j}{d_{ij}^b}. \tag{3.5}$$

There have been various hypotheses concerning the values of b that would model the role of distance in determining the number of spatial interactions. The proposition of Stewart (1941), assuming $b=2$, defines (3.5) as the 'demographic force' in a full analogy of Newton's law, while the hypothesis of Zipf (1946) is based on the assumption of $b=1$ and on a logarithmic transformation of the right-hand side of (3.5) (after: Isard, 1960/1965, pp. 352–356).

Isard (*idem*, pp. 357–358) noted that in empirical research on spatial interactions, the notions of mass and distance can be defined in many different ways. Instead of population sizes, such economic measures as employment or income can be used as masses, while distance can be measured according to a different metric: either Euclidean (crow-fly), or taking into account the structure of the existing transport network, time, or cost of transportation. Various mass factors can be also considered jointly, like for example in the model of Lowry (1966), who built a gravity model relating migration to unemployment rates, wage levels, and numbers of persons in the civilian labour force (non-agriculture), as well as in the armed forces, both at the origin and at the destination (after: Morrison, 1973, pp. 132–133). Alecke, Huber, and Untiedt (2001) observed that the notion of gravity is also used in many econometric models, where income (GDP) differentials per capita are most commonly applied as masses, instead of population sizes.

In addition to the gravity framework, there have been several attempts to utilise advanced mathematical tools to find the patterns of spatial interactions. Wilson (1967, 1970; after: Mazurkiewicz, 1986, pp. 25–34) proposed that interactions between regions i and j, including migration (m_{ij}), maximise the *entropy* of the whole system of regions under study:

$$\Sigma_{ij}\, m_{ij} \ln(m_{ij}) \rightarrow \max\,. \tag{3.6}$$

The maximisation is constrained by the costs of such interactions ($x_{i,j}$), which are related to the distance between i and j ($\Sigma_{ij}\, m_{ij}\, x_{ij} = c$).

In addition to the entropy concept, the same author also undertook an effort to describe spatial interactions within the framework set by the catastrophe and bifurcation theory (Wilson, 1981), where the dynamic systems under study may undergo substantial qualitative changes as a consequence of very small modifications of some of their parameters.

Another geographical theory, the one of the *mobility transition* (Zelinsky, 1971), attempts to explain changes in spatial mobility by a hypothesis akin to the 'demographic transition' concept. According to Zelinsky (*idem*; after: Gawryszewski, 1989, pp. 11–19), social modernisation caused an increase and a continuous diversification of human mobility patterns. During a transition from a pre-modern to a modern society, in the eighteenth and nineteenth centuries, migratory movements were primarily undertaken towards the national borders, as well as to other countries. Along with the industrialisation processes, migration from rural to urban areas was on the rise – it began to decline only in the advanced societies in the second half of the twentieth century. In these advanced societies, in turn, migration between and within urban areas have been increasing, as have the short-term circulatory movements (commuting, business trips, tourism, etc.). Recently, circulation has absorbed more and more mobility of other types, and has itself been substituted by an increasing role of communication systems.

The mobility transition theory has been designed as a comprehensive framework describing human mobility, although Kupiszewski (2002b, p. 122) noted that it ignores the phenomena of suburbanisation and counter-urbanisation, characteristic of the advanced societies. Nevertheless, the hypothesis of substitution between spatial mobility and other means of communication seems increasingly important in the contemporary world of the Internet and telecommuting. The whole theory, however, would be very difficult to apply directly to migration forecasting, given the limited availability of data on various types of mobility and communication.

3.1.5 Unifying Perspectives

Apart from the discipline-specific theories of migration, there have also been several attempts to propose a unified explanation for population flows. The *migration systems theory* (Kritz, Lim, & Zlotnik, 1992; following the pioneering work by Mabogunje, 1970; after: Zlotnik, 1998, pp. 12–13) distinguishes migration systems comprised of various sending and receiving countries characterised by considerable migratory links. In such a dynamic system, migration is in a continuous interplay with historical, economic, cultural and political linkages between the countries, both at the the micro and macro levels. The presence of feedback effects makes population flows both a cause *and* an outcome of other interactions. Despite the clear advantages of such a synthesising and multi-perspective approach, it is at the moment too complex to be applied in practice, especially given the problems with availability and quality of internationally-comparable migration statistics (Zlotnik, 1998).

Another attempt to create a theoretical synthesis of international migration has been recently undertaken by Massey (2002). His proposition combines economic, political, sociological and psychological determinants with the notion of migration transition (similar to the one proposed by Zelinsky, 1971), and with the role of duration-of-stay effects. In general, Massey (2002) perceives international migration in the post-industrial countries as an outcome of socio-economic development and integration processes. Nonetheless, at the current stage of development, these ideas are a long way from constituting an all-inclusive theory of migration, and it seems hardly possible that they could be operationalised in practical applications.

3.2 Theory in Migration Forecasting: A Global Outlook

3.2.1 Migration Theories: General Remarks

According to Chojnicki (1977), one of the approaches in socio-economic forecasting, referred to as *nomothetic*, is to derive predictions about the future directly on the basis of theories or laws governing the phenomena under study. However, such theories should be universal and robust enough in order to be straightforwardly concretised in the forecasts. Nevertheless, even if a direct application of the existing theories is not possible, they can provide suggestions for the construction of more specific forecasting models. For these reasons, the current chapter presents a selective discussion of theories that could be most relevant for that purpose, followed by a brief assessment of their applicability in the macro level prediction-making. The aim is thus not to offer a comprehensive overview of migration theories, but rather to indicate and evaluate approaches that can be potentially useful for the construction of forecasting models. A brief insight into the structure of the discussion, reflecting a distinction between sociological, economic, geographical, and unifying (cross-disciplinary) theoretical perspectives of human population flows, presented before, is shown in Fig. 3.1.

It has to be noted that the survey of migration theories presented in the current chapter is by no means complete, and that much more comprehensive discussion is offered in Massey et al. (1993), Zlotnik (1998), and Jennissen (2004, pp. 31–57). For example, the distinction between theories explaining the initiation of international migration, and the ones related to the perpetuation of population flows, introduced by Massey et al. (1993), is not considered in the current study. Such a

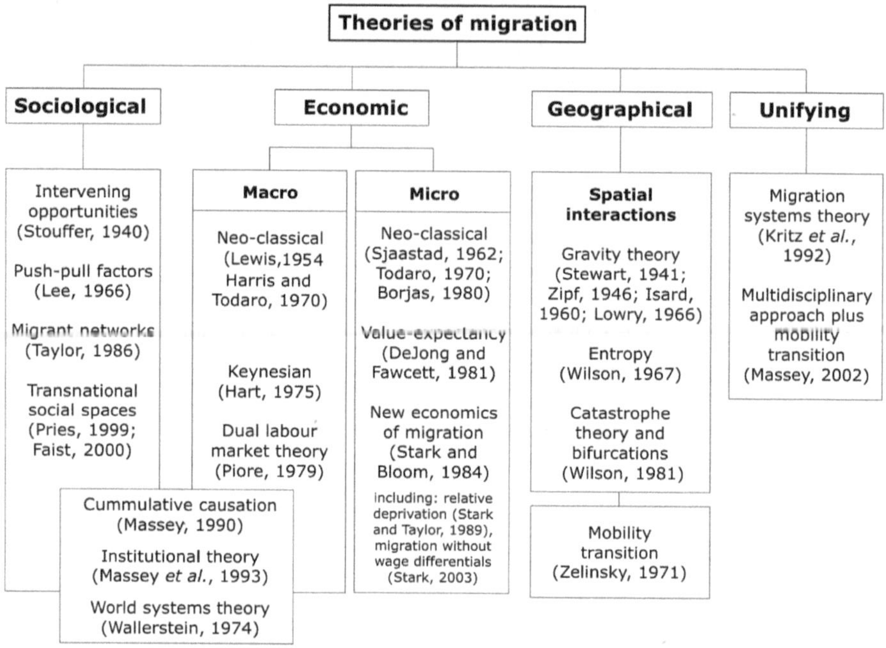

Fig. 3.1 Selected migration theories offered by various disciplines of science. Source: Zlotnik (1998) and Kupiszewski (2002b), own elaboration

division might be potentially useful, if a forecaster would make an attempt to distinguish migration continuing under unchanging external circumstances, from the ones related to system shocks (political, economic, etc.); the latter explained by the theories of initiation of population movements. In the current study, however, the issue of discontinuity is treated in a relatively simplified fashion, as described further in Chapter 7.

As remarked in Chapter 1, there is a clear distinction between *internal* and *international* migration. Although this book focuses on the latter, involving crossing the national borders, many theories attempting to explain population movements refer to internal migration. Nevertheless, contemporarily, given the globalisation processes, the complexity and dynamics of migratory phenomena increases (cf. IOM, 2008), while the diversity between the mechanisms driving internal and international migration becomes less and less obvious. Such convergence of both types of migration processes would allow applying internal migration theories more or less directly to the studies of international population flows.

The empirical evidence either in favour of or against the convergence hypothesis remains inconclusive. For example, the Polish case study of interplay between internal and international migration prepared by Korcelli (1994) suggests that in the present-day world both types of population flows may become more and more substituting, and that prospective migrants would migrate either internally or externally, making decisions on the basis of a cost-benefit analysis. This hypothesis has been recently cautiously confirmed in a study of Polish migration prepared by Kupiszewski (2006). However, in an analysis of 132 European regions from seven countries, prepared by Stillwell, Eyre, and Rees (1999), very little statistical evidence has been found for the existence of direct linkages between regional international migration and interregional migration.

Generally speaking, in contemporary Europe the differentiation between internal and international migration seems to be rendered less relevant by the process of European integration with respect to the freedom of movement of persons and labour force, which directly applies to the four countries under study: Germany, Italy, Poland and Switzerland. This notion is supported in a study of Janicki (2005, pp. 13–17), who also observed that in many border regions within the European Union, international migration flows dominate over the internal ones. This is despite the continuing presence of various socio-cultural barriers, which remain in place even after the administrative ones are gradually being removed.

On the other hand, it may be also argued that especially in the situation of Poland and other Central European countries after the EU accession in 2004, the freedom of movement intensifies international migratory processes (as already indicated by the preliminary Eurostat data for 2005), rather than the internal ones.[3] This line of reasoning is consistent with the suggestions presented before: given the partial dissolution of institutional barriers, prospective migrants rationally choose their destinations, and reveal their 'true migration preferences', rather than limit themselves to what is available inland. However, it has to be noted that in the case of Poland, the substitution of internal and international migration is influenced by migration policies, as well as legal and administrative solutions adopted by particular destination countries, which can significantly vary in that respect.

Given the above, as the hypothesis of convergence of internal and international migration is difficult to validate, while the existence of such a relationship cannot be entirely excluded, the current overview presents theories pertaining to both types of population flows. Nevertheless, this distinction should be borne in mind, especially with respect to the importance of formal, administrative or political barriers, which are crucial in interpreting international movements, but neglected in the internal migration theories.

---

[3] Although at the time of writing this chapter, Switzerland did not participate in the free movement of workers in Europe, relevant regulations were gradually liberalised by 2007 for the relations with the 'old' EU-15 countries (what actually happened on 1 June 2007). Subsequently, the same is going to happen with respect to the new EU member states by 2011, but not later than in 2014 (*Agreement between the European Community and its Member States, of the one part, and the Swiss Confederation, of the other, on the free movement of persons*, Official Journal of the European Communities L 114, 30 April 2002, and *Protocol to the Agreement between the European Community and its Member States, of the one part, and the Swiss Confederation, of the other, on the free movement of persons regarding the participation, as contracting parties, of the Czech Republic, the Republic of Estonia, the Republic of Cyprus, the Republic of Latvia, the Republic of Lithuania, the Republic of Hungary, the Republic of Malta, the Republic of Poland, the Republic of Slovenia and the Slovak Republic pursuant to their accession to the European Union*, Official Journal of the European Communities L 89, 28 March 2002).

3.2.2 Use of Theories for Migration Predictions

With respect to the evaluation of migration theories, Öberg and Wils (1992) observed that each of the existing ones explains the actual phenomena only partially and therefore has a limited use in the forecasting process. They stressed that the geographical theories are more suitable for internal migration, as they do not include such elements as institutional barriers (state borders, visa requirements, etc.), which are inherent in the case of international flows (Zolberg, 1989). Also Willekens (1994, pp. 33–34) noted that the existing theories of international migration do not substantially differ from those of internal migration, although contemporarily this may be less relevant due to the reasons discussed before, such as globalisation and integration processes in Europe. In addition, the existing theories ignore forced migration and migration policy factors, which are crucial for the actual magnitudes of observed population flows. Also the economic theories do not adequately explain migration during system shocks, like for example the socio-economic transformation in the post-socialist Europe in the late 1980s and early 1990s.

According to Öberg and Wils (1992, pp. 6–7), although all theories – geographical, economic and sociological – are useful in explaining migration *ex post*, their forecasting potential is very limited. These objections are shared by Kupiszewski (2002b, pp. 122–124), who argues that the existing theories of migration can be hardly used in a direct manner to forecast international population flows. None of the theories is comprehensive and self-contained, while migration is too complex a phenomenon to be explained by a single, narrow theory. In contrast, the wide-ranging theories, as for example, the mobility transition theory of Zelinsky (1971), the world systems theory of Wallerstein (1974), the migration systems theory of Kritz et al. (1992), and the unifying perspective of Massey (2002) are difficult to operationalise, because they are not sufficiently formal in terms of the mathematical expressions applied. Moreover, the imperfect availability and quality of data renders such comprehensive approaches hardly useful in practical applications. Other theories, although potentially transformable into the forecast input through a range of proxy variables, listed for example by Jennissen (2004, p. 57), can have a limited explanatory capacity. Due to incompleteness and various deficiencies of macro-level statistics on migration, as indicated in Chapter 2, the reliability of any empirical research aimed at verifying particular theories would be only partial and could be easily questioned.

An even more substantial criticism of the existing migration theories was provided by Arango (2002, pp. 283, 294), who stated that they have a built-in 'epistemological frailty', being in reality 'models, analytical frameworks, conceptual approaches, empirical generalisations, simple notions, and only seldom real theories'. Although providing several useful explanations *ex post*, the existing theories can hardly justify the observed magnitudes of migration, differences in that respect between otherwise similar countries, or answer, why some people do not migrate. Besides, the up-to-date efforts in order to create a migration theory have not been cumulative: the presented approaches are mutually not interrelated, and do not represent a sequential development of knowledge (*idem*).

For these reasons, the nomothetic forecasting of migration, based directly on the laws or theories of population movements, is not an option, as the existing laws and theories are not universal enough to allow for the practical application of this approach. Therefore, migration forecasting should ideally be model-based, rather than follow any particular theory. An exception may be to include selected push and pull factors as explanatory variables in forecasting models, which would partially refer to the theory of Lee (1966), to the extent possible given the data availability. Notwithstanding this general conclusion, there may be some situations where the theory-based forecasting can still be useful, such as in the absence of reliable data for modelling. Specifically with such applications in mind, an attempt to reconcile the data-based and theory-based forecasting can be undertaken, a sketch of which is provided in Chapter 6 on the example of multivariate models of migration and its determinants. Some extensions of the proposed ideas are also provided in Chapter 10.

Chapter 4
Forecasting Migration: Selected Models and Methods

In the current chapter, a survey of various models and methods used in migration predictions to date is offered. The rationale is that socio-economic predictions can be based not only on general, well-grounded laws and theories, but also on descriptive models designed to suit specific research questions. The presented overview follows a distinction between deterministic and probabilistic approaches, presented respectively in Sections 4.1 and 4.2, depending on the way the uncertainty issue is treated. The presented models and methods are finally compared and evaluated from the point of view of their usefulness for the purpose of the current and possible future studies.

4.1 Deterministic Methods of Predicting Migration

4.1.1 Judgemental Migration Scenarios

Judgemental scenarios used in demographic forecasting describe possible future trajectories of particular components of population change (fertility, mortality, and migration). They are constructed on the basis of qualitative and quantitative argumentation about what is considered to be the plausible development of the variables in question. The scenarios have to be coherent with the underlying judgement and assumptions behind them,[1] and show the demographic consequences of the latter using a 'what-if' approach. Scenarios usually serve as input for deterministic population projections, showing the outcome of various variants of change in the demographic parameters: usually, the baseline, high, and low (see Chapter 2).

Probably the first judgemental scenario of expected future migration flows set up in terms of quantitative demography can be found in the 'political arithmetick' (*sic!*) study of Sir William Petty (1682) concerning the future growth of the City of London. Although internal migration from rural England to London is hidden in different assumptions regarding the future demographic growth for both areas,

[1] Informally referred to as 'the underlying story' (a term likely coined by Dutch demographer Harri Cruijsen).

J. Bijak, *Forecasting International Migration in Europe: A Bayesian View*, The Springer Series on Demographic Methods and Population Analysis 24, DOI 10.1007/978-90-481-8897-0_4,

population inflows to the British capital are explicitly mentioned as the causes of the expansion of the city.

According to the knowledge of the author,[2] deterministic scenarios of international migration are contemporarily used in the official population projections in a vast majority of developed countries, at least in Europe. Apart from the national forecasters, judgemental scenarios of population flows are also widely applied in the supra-national demographic projections prepared by international organisations and agencies (for example, United Nations, 2005a; Eurostat, 2005[3]), as well as by individual researchers (e.g., Bijak et al., 2007).

With respect to the construction of scenarios, Kupiszewski (1998) pointed out that in some cases they can take into account analogies of the projected migration flows to the ones that have already occurred in the past in other countries, under similar conditions. An example is the enlargement of the European Union in 2004. An analysis of similarities to the previous extensions of the then-EEC (to include Ireland in 1973, Greece in 1981, Portugal and Spain in 1986) would allow for assuming a reasonable scenario, in which the post-accession increase of migration from Central to Western Europe will likely be temporary and rather moderate in size (*idem*). Such an analysis may be also performed within the formal framework of econometric modelling, which is further discussed in Chapter 7.

An example of a scenario-based study dealing with the forecast of the East-West migration in Europe after the EU enlargement is the research of Layard, Blanchard, Dornbusch, and Krugman (1992), who extrapolated earlier empirical findings concerning migration from Southern to Northern Europe in the 1950s and 1960s, as well as from Mexico to the United States in the 1970s and 1980s. Their estimate of a gross inflow of Central and Eastern Europeans to the West over 15 years totalled less than 3% of the population of the sending countries (Layard et al., 1992; after: Alvarez-Plata, Brücker, & Siliverstovs, 2003, p. 11).

A similarly straightforward calculation has been made by Franzmeyer and Brücker (1997), who built a gravity model of net migration between ten Central and Eastern European countries, i, and the 'old' European Union, EU-15. The model was based exclusively on the difference of the GDP per capita (PPP-adjusted), $Y_i / Y_{\text{EU-15}}$, and it was calibrated on the basis of the earlier empirical findings of Barro and Sala-i-Martin (1995; *idem*). The key assumption was made on the elasticity of migration on income differentials, so that an income gap of 10% was presumed to drive between 0.08 and 0.16% of the population of the worse-off country out of

[2]The mentioned issues have been discussed for example during the 'Meeting of the Working Group on Population Projections' in Eurostat (Luxembourg, 15–16 July 2004), and the 'Joint Eurostat – UN ECE Work Sessions on Demographic Projections' (Vienna, 21–23 September 2005 and Bucharest, 10–12 October 2007).

[3]In Eurostat (2005), the methodology differed for the 'old' 15 EU member states and for the 12 accession and candidate countries as of 2004. For the latter, the projections were purely based on judgemental scenarios, while for the former they involved averaging of forecasts yielded using three methods: extrapolation of trends, and econometric analysis of migration determinants, as well as national forecasts (Lanzieri, 2004).

the place of origin. This supposition is the main reason for extremely high migration forecasts yielded by the model: population flows from Central and Eastern Europe to the EU-15 have been estimated as 590–1,180 thousand persons a year, depending on the pace of income convergence.

The example of the study of Franzmeyer and Brücker (1997) shows that the judgemental scenarios should not only be consistent with the underlying 'stories', but also controlled with respect to the results they produce. Neither reasonable assumptions, nor the use of a particular theory (here: gravity with respect to income differentials) alone guarantee that the outcome will be plausible from a demographic point of view.

4.1.2 The Delphi Method and Surveys Among Experts

Expert judgement used in setting the scenarios can be either made by the authors of forecasts themselves, or derived from a survey carried out among a larger group of specialists from various countries and fields of expertise. The latter approach may include obtaining migration scenarios using a Delphi method, designed to 'elicit and refine group judgement' in an anonymous and interactive communication process among experts in the field (Dalkey, 1969, p. v). The exchange of knowledge is made in subsequent rounds, between which all feedbacks from the participants are controlled, and the final output is formed from the aggregation of all individual opinions. A potential drawback of the method lies in the pressure for conformity among the experts (e.g. Orrell, 2007, p. 240). Some other issues related to the elicitation of knowledge from experts are also briefly addressed in Chapter 10.

An example of an application of the approach in migration forecasting is the study of Drbohlav (1996), who used a two-round Delphi method on a sample of 70 experts in the first round and 39 in the second one. He obtained rough estimates of the magnitude and timing of the expected East-West migration flows in Europe, as well as the envisaged directions of migration policy changes in the future. Another migration-related Delphi study was conducted in 1991 in Russia by V. Tichonov. It produced an estimate of emigration from the former USSR in the period 1992–1997, whereby about half of 30 experts predicted between two and four million emigrants (Vishnevsky & Zayonchkovskaya, 1994; after: Willekens, 1994, p. 28).[4] In turn, immigration flows to selected European countries were forecasted by Bijak and Wiśniowski (2010) using Bayesian methods coupled with a Delphi survey, as described in Chapter 10.

Also Bauer and Zimmermann (1999) made an attempt to gather information on future migration flows from Central to Western Europe from a survey conducted

[4] Apparently, the majority of population flows concerning the ex-USSR in the period 1992–1998 proved to be internal migration, and migration between the former republics. For example, in Russia alone, over 20.5 million people migrated internally in that period, 5.5 million immigrated from abroad, and 2.6 million emigrated (Wegren & Cooper Drury, 2001, pp. 16, 39). It is, however, worth noting that the two last figures relate to a large extent to the population exchange with the other republics of the former Soviet Union (*idem*).

among experts, and use it as a supplementary source of information for their econometric model. Despite the very low response rate (20 questionnaires received out of 446 sent), the authors found the results consistent with the IOM (1998) survey aimed at estimating the 'migration potential', which is briefly discussed in the next subsection.

4.1.3 'Migration Potential' Assessment Surveys

Other survey-based studies that are used to assess the future international migration flows are the analyses of 'migration potential'. Examples of such research with respect to the East-West flows in Europe are presented in the reports of Fassmann and Hintermann (1997), and of C. Wallace (IOM, 1998). Such studies are typically based on questionnaires filled in by a representative random sample of respondents, who are asked questions on whether they consider undertaking migration, for what reasons, under which circumstances, etc.

As noted by Kupiszewski (2002a, pp. 633–637), the main problem with the survey-based studies is that categories used in such analyses are defined very vaguely. For example, the commonly-used 'migration potential' can be defined differently in various studies, depending on what is asked in the questionnaire. In this way, the formulation of questions may heavily influence the results of a survey. There are, no doubt, different numbers of 'potential migrants' giving a positive answer to the questions: whether they intend to migrate any time in the future, or whether they have already actively searched for any real possibilities of settling abroad.

Besides, the results of surveys are usually presented in a manner that does not address the issue of uncertainty. In fact, although the authors of such studies usually do not refer to their outcomes as forecasts *per se*, the results they present are often interpreted in this way by the users. For example, in summary studies dealing with the expected future migration flows, the survey-based estimates are often treated on a similar footing with other forecasts. Moreover, in some of such studies the differentiation between 'migration potential' and the actual forecast is not made explicit. Importantly, this concerns several analyses prepared specifically for the policy makers, such as the research of Alvarez-Plata et al. (2003), or report of the Netherlands Bureau for Economic Policy Analysis (CPB, 2004).

Moreover, migration *intentions* declared by the respondents cannot be directly transformed into the actual behaviour, so as to become a proxy measure of the magnitude of population flows in the future. In that respect, either the declared intentions can reflect 'wishful thinking' rather than reality, or additional factors may get involved in the decision making process between the moment of the survey and the expected time of migration. The problem is thus not with survey-based research as such, but with interpreting its results as migration forecasts, irrespective of clear disclaimers and warnings provided by the authors of such studies (e.g., IOM, 1998, p. 11). On the top of that, due to high costs of conducting a survey, the sample sizes are usually too small to allow for obtaining significant results involving breakdowns by sex, age, regions, motives, etc.

For illustration: Fassmann and Hintermann (1997) surveyed 4,392 persons in the Czech Republic, Hungary, Poland and the Slovak Republic. Depending on a country, they identified between 17 and 30% of respondents, expressing an overall willingness to leave their country, as a 'general migration potential', between 6 and 18% as the 'probable potential', concerning people who started to gather information about the possible destination, and between 1 and 2% as the 'real potential' – those who have actually undertaken any particular steps to move. The preferred destinations of prospective migration were Germany and Austria, which is not surprising in the light of their geographic proximity to the countries under study, and of the presence of strong migrant networks. Notably, only the 'real potential' category corresponds to the magnitudes of migrants that can be seen as plausible from the demographic point of view, whereas two other ones reflect rather 'wishful thinking' of the respondents. An additional problem of the study of Fassmann and Hintermann was the sample size – too small to disaggregate the results jointly by sex, age groups, regions, motives, education levels, etc.

In the study of Wallace (IOM, 1998), between 1,000 and 1,200 respondents in each of the eleven selected Central and Eastern European countries were asked six multiple-choice questions. In particular, the survey aimed at identifying, whether the respondents were going to migrate, for how long, why (or why not), do they have relatives or friends abroad, and have they already taken any preparatory steps. Depending on the type of migration potential distinguished by the author, the country-specific propensity to migrate ranged between 7 and 26% for permanent emigration, between 18 and 57% for long-term temporary labour migration, and between 13 and 68% for short-term labour migration (*idem*, p. 11). The preferred destinations were: the USA for permanent emigration, and Germany and Austria for labour migration.

The magnitude of the 'migration potential' assessment shows that the IOM (1998) survey identified dissatisfaction of the respondents with the state of affairs in the home country, rather than real migration perspectives. This conclusion can be supported by the fact that the highest percentages of 'potential migrants' have been found in Croatia and the then-Federal Republic of Yugoslavia, at that time both coping with a tense political situation. Very high estimates have been also obtained for Romania, the country to become, upon its accession to the European Union, the poorest EU Member State in terms of GDP per capita.

The IOM (1998) survey also contains some unexpected findings. For example, in Poland as many as 29% respondents identified 'ethnic tensions' as an important push factor out of the country (*idem*, p. 48). In an almost ethnically homogenous society this is a very surprising outcome, which likely indicates some problems with the execution of the survey, or with the translation of a questionnaire in such a multi-country study.[5]

All the problems mentioned above support the conclusion that the results of the survey should be interpreted with caution. Another general problem relates to

---

[5] I am grateful to Izabela Koryś for drawing my attention to this.

the uncertainty issue: although the sample-based surveys as such have potential in addressing it in a statistical manner, this possibility is not addressed in the 'migration potential' assessment studies presented above.

4.1.4 Macro-Level Mathematical Models in Demography

With respect to mathematical models of migration, Kupiszewski (2002b, pp. 16–17) noted that they predominantly stem from two different disciplines: demography and human geography. The former approach focuses predominantly on population distributions by sex and age, as well as on the impact of migration on the overall demographic dynamics (and *vice versa*), while the latter – on the spatial outcomes of the redistribution of migrants. Although both methodologies apply mathematical tools to model and forecast migratory flows within their framework, the classical demographic models of population dynamics use multi-dimensional linear algebra and calculus in a deterministic fashion. In turn, many geographical models are based on the Markov chains, thus using a stochastic approach, and as such are discussed in the next section.[6]

Another important distinction is the one between macro-level (aggregate) and micro-level models.[7] In general, the macro-level demographic models of population change stem from the cohort-component approach, pioneered by Bernardelli (1941) and Leslie (1945) (after: Jóźwiak, 1992, p. 20). The cohort-component model is usually employed to forecast population size and age structure on the basis of judgementally-assumed scenarios of change in the particular components of demographic dynamics (originally, births and deaths) concerning the country or region under study. On their basis, the survivorship of particular birth cohorts is calculated in order to yield population size and age structure in subsequent projection steps.[8]

Despite the continuous development of a cohort-component model, population forecasts until the 1970s either ignored the migration component or treated it simplistically. According to Rogers (1975, pp. 1–2), net migration has been for the first time included in a comprehensive manner in population modelling in the report of the United Nations (1970).[9] Since then, migration has been incorporated in an

[6]For a systematic overview of various model classes, see also the typology introduced in Section 4.4 (Figure 4.1).

[7]A detailed overview of selected micro- and macro-level migration models, with special focus on the latter (mainly Poisson regression, gravity models and spatial interactions), is also offered in Stillwell and Congdon (1991).

[8]The single-region mathematical demography, including the cohort-component approach, is presented in details for example in Keyfitz (1968), Rogers (1975, pp. 7–55), and Jóźwiak (1992, pp. 21–50).

[9]Paradysz (2004, pp. 130) noted that an earlier study addressing the same problem was due to Korčak-Čepurkivs'kij (1970), which, however, was published only posthumously. Its author was persecuted under the Stalinist regime and spend 18 years in a Gulag forced labour camp (cf. the interview with A. Višnevskij in *Demoscope weekly* 197–198, 4–17 April 2005; www.demoscope.ru/weekly/2005/0197/analit01.php, accessed on 25 August 2006).

increasing number of official population projections or forecasts. A survey on the methodology of such studies carried out in 30 developed countries in the early 1990s (Keilman & Cruijsen, 1992) showed that in a vast majority of cases, deterministic scenarios of all components of demographic change have been used as input. For international migration, Cruijsen and Keilman (1992, pp. 20–22) found that at that time six countries did not include the migratory component in the projection models. Among the remaining ones, only Belgium used a regression model to extrapolate migration, while the other relied on assumptions on political plans or targets, 'visual' interpolation of curves, or simply assumed constancy of migration rates throughout the forecast horizon.

Kupiszewski (2002b, pp. 39–49) noted that the basic cohort-component model has ultimately evolved to include migration, following various methodological approaches. In the *migrant pool* method, applied for example in the 1970s and 1980s in the works of the US Bureau of the Census, the numbers of migrants from all regions under study are projected, merged, and subsequently distributed among the receiving regions using an allocation algorithm. The *population accounting models* (Rees & Wilson, 1973; Rees & Willekens, 1986) are based on the rates of transitions or movements of people between different regions. In this approach, the balance of changes in population size on one hand, and births, deaths, in- and out-migrations on the other hand, should hold both at the aggregate level, as well as for each age group specifically. In turn, the *multi-regional model* (Rogers, 1975) is a generalisation of the cohort-component approach, based on the concept of a multi-dimensional life table, and treating simultaneously populations of a system of regions, between which migration can occur. The projected numbers of migrants in particular age groups are thus an outcome of an analysis of demographic interdependencies involving all components of demographic change, including initial assumptions on migration propensities. The multi-regional approach is in turn generalised in the multi-state models (Rogers, 1980), where regions can be substituted by any other 'states' that an individual can 'occupy', e.g., marital status, economic activity, educational level, etc.[10] Examples of further extensions include non-linear multidimensional models, like the exponential LIPRO (LIfestyle PROjection) model developed by van Imhoff (1990), and van Imhoff and Keilman (1991). A very important feature of all multi-regional and multi-state models (as well as of the Markov models, presented in more detail in Section 4.2) is that they treat population dynamics as a system of regions or states, with migration (respectively, transitions) reflecting interactions between the elements of such system.

Another attempt to synthesise the multi-regional and population-accounting approaches has been undertaken in the multi-level models,[11] aimed at a coherent

[10]For a mathematical discussion of multi-regional and multi-state models, see Jóźwiak (1992, pp. 51–94, 95–111).

[11]Such models are also sometimes referred to as 'hierarchical' due to the presence of a hierarchy of regions, although in the current study only the term 'multi-level' is used, in order to avoid confusion with hierarchical models in a statistical sense (i.e., models with random parameters or with latent variables, see also Chapter 12).

analysis of migration flows on various levels of spatial disaggregation. According to Kupiszewski (2002b, p. 144), the idea dates back to the work of Rees, Stillwell, and Convey (1992) and Rees (1996). Recently, Kupiszewska and Kupiszewski (2005) constructed the multi-level model Multipoles (Multistate Population model for multi-Level Systems), treating migration on three different geographic levels (between regions, between the countries under study, and the population exchange with the rest of the world). Models like the Multipoles offer an internally-coherent possibility of an analysis and forecasting of migratory phenomena on a very detailed level of disaggregation (by regions, sex and age), taking into account other demographic features of the population under study.

The deterministic nature of the presented mathematical models of population dynamics stems from their algebraic formulation, as well as from the dominant forecasting practice in demography. As it has been noted at the beginning of this section, the cohort-component or multi-regional models are until now usually fed with judgemental scenarios of particular components of demographic change, including migration. However, there exist probabilistic exceptions to this tendency, which are presented in Section 4.2.[12]

4.1.5 Demo-Economic Modelling Attempts

In addition to purely demographic forecasts, several authors made efforts to create models that would combine population and economic aspects of social development. According to Kupiszewski (2002b, p. 38), the demo-economic approach to modelling dates back at least to the late 1970s and the development of such a model ('REGINA') in France (Courbis, 1979; *idem*).

From the point of view of migration forecasting, an interesting example is the recent study of Fachin and Venanzoni (2002). Their model for Italy ('IDEM' – *Integrated Demographic and Economic Model*) combines a multi-regional cohort-component model of population dynamics with an economic input-output analysis (Leontieff's table). Migration between regions is one of the key intermediary modules in IDEM, linking demography with economic aspects: labour supply and productivity. As the analysis is limited to a single country, the focus is on internal rather than international migration, the latter flows being treated very simplistically, as constant yearly inflows (in terms of numbers) and outflows (in terms of rates). In this way, international migration is exogenous, as are two other components of demographic change – mortality and fertility. Nonetheless, one can easily imagine an extension of the model to a multi-national economic system, where migration between particular countries would also play an important role. In turn, internal migration is modelled and forecasted in terms of origin-and-destination-specific rates, disaggregated by age according to the patterns observed in particular regions. Rates and other exogenous and intermediary variables are

[12]A discussion on the stochastic versions of population dynamics models is offered by Jóźwiak (1992, pp. 113–121).

estimated and predicted using econometric models (in the case of migration, the logit regression). However, despite the stochastic character of the forecasting tools for particular model components, the very construction of IDEM is algebraic and deterministic, similarly as in the macro-level models of mathematical demography.

As regards the future research on possible ways to introduce a stochastic framework in demo-economic, as well as in multi-regional or multi-state demographic models, it is definitely worth making reference to over 50 years of experience of the theory of econometrics. Systems of linear or non-linear equations, as involved in the deterministic models presented above, can be made probabilistic in a relatively simple manner, following several methodological options widely used by econometricians. There have been a variety of research paradigms in econometrics over the last half of the century, changing from the simultaneous equations models, through the 'atheoretical' vector autoregression (VAR) models, to the structural modelling within the latter, involving, among other issues, the 'from general to specific approach' and the analysis of cointegration.[13] These developments offer very interesting paths of potential methodological advancements of the demographic and demo-economic forecasting in the future.

4.2 Probabilistic Migration Forecasts: Assessing Uncertainty

4.2.1 Markovian and Related Models of Aggregate Population Flows

As noted in the previous section, one important group of stochastic models of migration stems from the tradition of human geography, with focus on spatial redistribution of population through migration. In particular, there are numerous examples of models applying the *Markov chains*, originally due to A. A. Markov (1906).[14] The basic assumption of a Markov chain – a stochastic process X_t, which in a given discrete moment of time $t \in \{0, 1, \ldots\}$ can be in any of m states, $x_t \in \{1, \ldots, m\}$ – is the *Markov property* in its simplest form:

$$p(X_t = x_t \,|X_{t-1} = x_{t-1}, \ldots, X_0 = x_0) = p(X_t = x_t \,|X_{t-1} = x_{t-1}). \tag{4.1}$$

The Markov property assumes that the process X_t 'lacks memory' from its past, so that the probability of being in a state x_t in the moment t depends only on the state occupied in the preceding period, $t-1$. This allows for formulating a transition

[13] Cf. Charemza and Deadman (1992/1997, pp. 13–22, 151–166); some issues are also addressed in Chapters 6 and 10.

[14] See 'Markov chain' and 'Markov process' in the Springer online *Encyclopaedia of Mathematics* (eom.springer.de/M/m062350.htm, eom.springer.de/M/mm062490.htm, accessed on 25 August 2006), and the review of probabilistic literature of the Kolmogorov Library (www.kolmogorov.pms.ru/uspensky-predvarenie.html, accessed on 25 August 2006), albeit the latter quoting the 1907 edition of the seminal work of Markov (1906).

matrix $\mathbf{P_t} = [p_{ijt}]_{m\times m}$, whereby $p_{ijt} = p(X_t = j\,|X_{t-1} = i)$. If all p_{ijt} are time-invariant (i.e. for every t, $\mathbf{P_t} = \mathbf{P}$), the Markov chain is called *homogenous*. In such cases, given the probability distribution of the process over the space of states (in migration studies: of regions) at the time t, $\mathbf{x_t}$, the model for use in forecasting the future distribution of population n periods ahead, $\mathbf{x_{t+n}}$, can be written as:

$$\mathbf{x}'_{\mathbf{t+n}} = \mathbf{x}'_{\mathbf{t}}\,\mathbf{P}^n. \tag{4.2}$$

In general, the pioneering works in the applications of Markov chains to modelling social mobility[15] are due to Prais (1955) for movements of people between income classes, as well as Blumen, Kogan, and McCarthy, (1955) for changes of jobs. Also in migration studies, population flows between regions (states of a Markov chain) have been modelled in various ways, originating from chains with homogenous transition matrices. Specifically for inter-regional migration, the early examples are the studies of Rogers (1966), Brown (1970), and Joseph (1975). The pioneering work of Blumen et al. (1955) is also important, for it includes the notion of heterogeneity of the population under study. The population is divided into 'movers' and 'stayers'; only the former ones changing the states they belong to. This approach has been later generalised in a model with different transition matrices for various subpopulations (Goodman, 1961). The population heterogeneity has been also formally examined by Spilerman (1972), who applied a regression analysis based on several exogenous variables within the Markov chain framework.

The homogeneity of a Markov chain and the related stationarity of the process, meaning in particular that the conditional distribution of $\mathbf{x_t}$ given $\mathbf{x_0}$ stabilises as t increases to infinity, is desired from a mathematical point of view, although rarely observed in real-life migration modelling problems (see for example Huff & Clark, 1978). On the other hand, Brown (1970, p. 401) noted that certain properties of Markov chains, like mean times of passage between states, time of recurrence to a given state, or stationary distributions of a chain, are very useful for interpreting the properties of the system under study. For these reasons, there have been many attempts to retain certain assumptions of the model and lift some other ones, leading for instance to migration models using heterogeneous Markov chains with non-stationary transition matrices. An application of this approach can be found for example in models characterised by the 'cumulative inertia' property (McGinnis, Myers, & Pilger, 1963). This notion links the transition probabilities with the duration of stay in particular states, assuming that the longer the stochastic process is in a given state (a person does not migrate), the less likely it will move to another state (region) in the future.

---

[15] Overview presented after Kupiszewski (2002b, pp. 28–34). Paradysz (2006, pp. 232) labels such mobility as 'vertical', as opposed to 'horizontal' (moves in geographic space). Whereas the former can also concern mobility within other social structures (educational, occupational, national, ethnic, etc.), the latter explicitly refers to migration.

Characteristics of the 'cumulative inertia' and 'mover-stayer' approaches are combined in the semi-Markov processes proposed by Ginsberg (1971). These models are a generalisation of Markov chains, and relate the probability of leaving a state to the time previously spent in it, as well as to the state entered next (destination). Although for such processes the Markov property does not hold, according to Ginsberg (*idem*), including duration-of-stay effects is a more suitable premise for modelling social processes than the Markovian 'lack of memory'.

Further to semi-Markov extensions, an attempt to construct a stochastic demo-economic model for forecasting interregional migration has been made by Plane and Rogerson (1985). They combined a Markov chain with an economic gravity model (see Chapter 3), in which changes in the spatial distribution of economic opportunities modify the spatial distribution of the population. Their basic model has the following form:

$$p_{ij}^{t} = \frac{p_{ij}^{t-1}(A_j^t/A_j^{t-1})^{\gamma}}{\sum_k p_{ik}^{t-1}(A_k^t/A_k^{t-1})^{\gamma}}, \qquad (4.3)$$

where p_{ij}^t is the probability of migration from i to j in the period $(t,\ t+1)$, while A_j^t is the 'attractiveness' of the j-th region (Plane & Rogerson, 1985; after Rogerson, 1984 [*sic!*], p. 115). The variable A_j^t was defined here as $A_j^t = (\Delta E_j^{t-1} + s_j^{t-1} E_j^t)/(U_j^{t-1} + s_j^{t-1} E_j^t)$, with E_j^t and U_j^t respectively denoting employment and unemployment in the j-th region at the time t, and s_j^t being a rate, at which employees separate from their jobs in the period (t, t+1). The parameter γ approximates the elasticity of migration probability with respect to changes in A_j^t (*idem*). Alternatively to (4.3), Rogerson (1984) formulated a logit model, additionally including the job application and competition process, through the region-specific rates of the numbers of job searchers to job vacancies.

Recently, Constant and Zimmermann (2003) combined a Markov chain model for the moves of migrants between Germany and their countries of origin, with an estimation of transition probabilities using logit models on the basis of micro-level data. For that particular example the authors found that 'a Markov chain is an appropriate representation of the structure of the behavioral process of repeat migrants' (*idem*, p. 22). Further extensions of the approach in migration research can include for example the idea of Hidden Markov Models introduced by Baum and Petrie (1966), whereby the outcomes of an unobserved (hidden) Markov chain are observable only after filtering through state-specific probability distributions or neural networks.[16]

With respect to the practical applicability of Markovian and related models in migration studies, Kupiszewski (2002b, pp. 34–35) noted that they form a very

---

[16]Many thanks go to Katarzyna Bijak for drawing my attention to Hidden Markov Models, and for a brief tutorial.

general and elegant tool for modelling any transitions involving human populations, including the migratory ones. Moreover, Lindsay and Barr (1972) have presented Monte Carlo simulations based on dynamic transition probabilities, which would change over time according to pre-defined deterministic patterns. In these Monte Carlo simulations, the probabilities were derived using the gravity model, additionally acknowledging the presence of various barriers between the regions.

Despite their advantages, Markovian models are contemporarily of a rather marginal use in geographical studies, regardless of their relative popularity in the 1960s and 1970s. There are several reasons for that. Firstly, the assumption that a stochastic process lacks memory longer than from the preceding period is strong, and sometimes very far from the reality of migratory phenomena, for example with respect to return migration. Secondly, the 'ideal' properties of Markov chains, like the homogeneity of populations under study and stationarity of the stochastic process, are also rather artificial in the real-world modelling. Moreover, the above-listed attempts to overcome these problems, although very well-designed from a mathematical point of view, require very detailed empirical data, that are usually not available (Kupiszewski, 2002b, pp. 34–35).

4.2.2 Micro-Level Methods: Event-History Analysis and Ethnosurvey

In the micro-level event-history analysis, migration can be one of many possible demographic events that may happen to an individual. In the event-history modelling, people can move not only between geographic regions, but also between other socio-economic and demographic 'states', like marital status, type of economic activity, level of education, etc., where the moves are modelled on the basis of estimated transition probabilities. The methodology is thus a natural analytical framework for use in multi-state demographic models (Rogers, 1980).

In human geography, the origins of the event-history approach date back to the studies of migration and residential histories, or even to the 'time geography' approach, linking the human activity in time and space, following by Hägerstrand (1970). Among many examples of applications of event histories, Ginsberg (1978) used a probabilistic framework (a semi-Markov process mentioned in the previous subsection) to analyse residential changes of individuals, focusing on times between particular moves. In demography, the event-history framework in its wider sense, encompassing many possible aspects of human life, had numerous applications up to date. Among the first ones there was a study of interrelations between migration, family formation and stage of career in the individual life course (Courgeau, 1985).[17]

[17] Methodology, and an overview of possible demographic applications of the event-history analysis is provided in Courgeau and Lelièvre (1992), while a detailed analysis of migration dynamics is discussed in the life-course context by Mulder (1993).

The event-history models can be formulated for either continuous or discrete time. In the former case, following Courgeau (1995, p. 23), let the duration of stay of an individual a be denoted by T_k^a, where $k-1$ is the number of migrations (transitions) undertaken previously by that person. Further, let the respective region of residence be described by a random variable I_k^a. Additionally, let various characteristics of the individual a, $x_i^a(t)$, as well as sets of information this person has about the destination region j, $y_i^a(t)$, be given. The equation linking the instantaneous migration rate $m_{ijk}(t)$ of the individual a with the remaining quantities is:

$$m_{ijk}^a(t|x_i^a(t),\ y_j^a(t)) = \lim_{dt\to 0} \frac{P(T_k^a < t+dt,\ I_{k+1}^a = j|T_k^a \geq t,\ I_k^a = i, x_i^a(t),\ y_j^a(t))}{dt}. \tag{4.4}$$

In an analogous model with discrete time, the probabilities of migration p_{ijkt} would be modelled instead of instantaneous rates.

Courgeau (*idem*) observed that (4.4) is 'a multivariate model for failure-time data with competing risks', which may take into account not only variables related to migration as such, as duration of stay, sequence of migrations, or preferences of regions, but also different personal characteristics of migrants (sex, age, etc.). Additionally, other important events from the individual biographies, related for example to family formation, childbearing, and changes in education level, as well as to employment status, can be considered in the model, provided that appropriate data are available.

The instantaneous rates of migration between particular regions, given in (4.4), can be estimated for example from representative retrospective surveys. Especially in the case of international migration, these rates should be ideally calculated on the basis of large comparable cross-country studies, which are nevertheless very costly and thus performed very seldom.[18] An alternative approach to prepare forecasts of migration on the basis of an event-history model is based on Monte Carlo micro-simulations, which are comprehensively presented for example by van Imhoff and Post (1998). As noted by Courgeau (1995, p. 24), the micro-simulation methods in migration forecasting, allowing 'to obtain an estimation of future regional population, under certain hypotheses', date back to the pioneering work of Hägerstrand (1957). The hypotheses predominantly concern

[18]One well-known example is the Fertility and Family Survey (FFS), carried out in the 1990s in 20 European countries under the auspices of the Population Activities Unit (PAU) of the UN Economic Commission for Europe, following a common research design. Migration-related questions have been included in one of the optional modules of the original FFS. According to the knowledge of the author, up till the time of writing of this book, the FFS has been repeated only once in a handful of originally-participating countries. Although the PAU also conducts another European-wide survey on *Generations and Gender*, it does not include any information about migration. More detailed information is available on the PAU website: www.unece.org/pau (accessed on 9 August 2006).

probabilities of making a migratory move within a given time after the previous one, and can be extended to include the impact of other socio-demographic characteristics. On the basis of such probabilities, the migratory behaviour of each individual from a population under study can be simulated using the Monte Carlo method. The individual-level trajectories can be subsequently aggregated in order to yield the required migration forecast. An important element assumed here, not always realistically, is the population homogeneity (Courgeau, 1995, p. 24).

Another micro-level research framework designed specifically for studying international migration is an ethnosurvey (Massey, 1987), combining the features of a quantitative survey with ethnographic methods. As traditional analytical techniques do not properly deal with undocumented migration, circular movements, and dynamic character of migratory processes, there was a need for a research design that would allow capturing these phenomena. The qualitative part of the ethnosurvey methodology involves semi-structured interviews, participant observations and case studies, while the quantitative one – gathering data on individuals, their life-event histories, as well as on households and communities, by the means of parallel sampling in the origin and destination areas. The main drawbacks of the ethnosurvey approach are twofold. Firstly, it is cost- and labour-consuming, and secondly, it is very difficult to obtain representative results from the samples used (*idem*, p. 1515).

Among other applications, the results of an ethnosurvey can be directly used to predict international migration flows. For example, Massey and Zenteno (1999) prepared a forecast of a Mexican migration to the United States, using a system of the following dynamic equations for the probability of immigration of an individual i in the year t, $p(Mig_{it})$, as well as for the probability of return, $p(Ret_{it})$, depending on the personal characteristics (*idem*, p. 5330):

$$p(Mig_{it}) = f(Age_{i,\,t},\ Sex_i,\ Itrips_{i,\,t-1},\ Iexp_{i,\,t-1},\ Ctrips_{t-1},\ Cexp_{t-1}), \qquad (4.5a)$$

$$p(Ret_{it}) = f(Age_{i,\,t},\ Sex_i,\ Itrips_{i,\,t},\ Iexp_{i,\,t-1},\ Ctrips_{t-1},\ Cexp_{t-1}). \qquad (4.5b)$$

In the above equations, the variables *Age* and *Sex* are self-explanatory, $Itrips_{i,t}$ denotes the number of migratory moves made by the i-th person in the year t, while $Ctrips_t$ – the number of such moves made by the other members of the community of origin. Further, $Iexp_{i,t}$ and $Cexp_t$ depict the migratory experience, measured by the number of moves, which was cumulated respectively by the i-th person, and by other community members in the year t. The probabilities (4.5a) and (4.5b) can be estimated within the event-history framework using a logistic regression, in order to obtain the rates of out- and return migration. Notably, although the quantitative part of an ethnosurvey is defined in probabilistic terms, there are also qualitative components of the research, not discussed in this book, which make the whole methodology de facto a stochastic-descriptive (hence, deterministic) hybrid.

4.2.3 Selected Attempts to Bridge the Micro and Macro Perspectives

With respect to the general discussion on combining the advantages of micro- and macro-level approaches in population modelling, Courgeau (2003) observed that both methodologies are complementary. In a comprehensive multi-level research framework, both individual and aggregate characteristics should be introduced into the model, as it has been applied for example in a study of inter-regional migration in Norway presented by Courgeau and Baccaïni (1998). In such a way, the macro-level variables provide an exogenous context for the analysis of individual biographies within the event-history approach. Interestingly for the current research, Courgeau (2003, pp. 81–82) sees large methodological potential in applying the Bayesian paradigm and subjective probability in multi-level studies, involving hierarchical Bayesian modelling.

Another framework for migration modelling and forecasting, using the tools of theoretical physics has been proposed by Weidlich and Haag (1988) within an approach labelled as 'sociodynamics'. The authors developed a model that aimed at linking micro-level migratory decisions of the individuals with their macro-level outcomes for inter-regional population flows. On the micro level, migratory processes are described in terms of dynamic utility and mobility functions, applying the *master equation* method used in statistical physics.

Master equations are first-order differential equations defined for probability distributions of the process over a vector of N states, $\mathbf{n} = [n_i]_{1\times N}$. In migration studies, they may refer to the population distribution over the space of N regions, with n_i denoting the number of people in the i-th one. The probabilities are related to the rates of transition from one state (region) to another, w_{ij}, according to the formula (Weidlich & Haag, *idem*, pp. 9–10, 23):

$$\frac{dp(\mathbf{n},t)}{dt} = \sum_{i\neq j} w_{ij}\left(\mathbf{n}^{(ij)},t\right)\cdot p(\mathbf{n}^{(ij)},t) - \sum_{i\neq j} w_{ij}\left(\mathbf{n},t\right)\cdot p(\mathbf{n},t), \tag{4.6}$$

where $\mathbf{n}^{(ij)}$ depicts the state of the process after a single move of one person from the i-th state (region) to the j-th one, $\mathbf{n}^{(ij)} = [n_1, \ldots, (n_i - 1), \ldots, (n_j + 1), \ldots, n_N]'$.

On a macro level, the transition rates w_{ji} are derived using a regression analysis with several socio-economic and geographic (distance) explanatory variables, which constitute the background of migration processes. In Weidlich and Haag (*idem*), the model was tested on the examples of selected European countries: West Germany, France, Italy, and Sweden, as well as Israel and Canada. Nevertheless, despite the potential usefulness of the model, and the level of detail and mathematical precision of a dynamic analysis it offers, its complexity rendered it hardly exploited in practical applications up to date.

At the time of writing this chapter, several leading European demographic research institutes have been making a joint effort in order to develop a unified framework for demographic predictions within the EU-sponsored project 'Bridging the micro-macro gap in population forecasting' (MicMac). The analysis aimed at producing a forecasting model that would 'offer a bridge between aggregate projections of cohorts [...] and projection of the life courses of individual cohort members [...]' (van der Gaag et al., 2005, p. 3). The outcome would combine the features of multi-regional models with the event-history analysis, paying special attention to the uncertainty issue. Unfortunately for migration predictions, this component of population change has been treated in MicMac much less comprehensively than mortality and fertility. Nevertheless, such a micro-macro approach has no doubt considerable potential also for modelling and forecasting of population flows, subject to caveats mentioned in Chapter 10.4.[19]

4.2.4 Econometric Forecasts of International Migration

Econometric models are a natural tool not only to predict migration, but also to verify particular economic theories on the basis of empirical data. The recent boom in using econometric models to forecast international migration in Europe dates back to the 1990s, and focuses almost universally on population flows from Central and Eastern European countries to the West after the expected enlargement of the European Union. Many such studies have been published in Austria and Germany, both located very close to the potential sending countries, and already accommodating large groups of immigrants from Central and Eastern Europe. Due to the many research studies that havealready been undetaken, the current subsection presents only a handful of models in more detail, focusing on different methodological approaches to econometric modelling of migration, and, in most cases, referring to the European circumstances.[20] The current overview, rather than being chronological, follows the modelling approaches from the most simple to more structured.

To start with, a very simple model has been proposed by Fertig and Schmidt (2000), who estimated immigration rates to Germany, m, from the four EU candidate countries at the time of preparing the study (the Czech Republic, Estonia, Hungary, and Poland). Their model covers the country-specific, time-specific, and cross-sectional effects, in addition to the overall mean migration rate, and can be written in a simple form:

$$m_{it} = \mu + \varepsilon_i + \varepsilon_t + \varepsilon_{it}. \tag{4.7}$$

Denoting the country of origin by i, the model (4.7) assumes that $\varepsilon_i \sim$ N(0, σ_i^2), $\varepsilon_{it} \sim$ N(0, σ_{it}^2), and ε_t is a Gaussian autoregressive process AR(1). The forecast produced on its basis yielded that in the period 1998–2017 the average

[19] A brief description of the project, its rationale, basic assumptions, and methodology are offered for example by van der Gaag, de Beer, and Willekens (2005). More details can be found on the project website: www.micmac-projections.org (accessed on 30 April 2006).

[20] More complete and detailed literature surveys of existing econometric forecasts are presented for example in Alvarez-Plata et al. (2003), in a report of the CPB (2004), as well as in Brücker and Siliverstovs (2005).

annual population inflow to Germany would range between 15 and 57 thousand immigrants a year (*idem*, p. 37).

The idea of Fertig and Schmidt (2000) has been further elaborated, among others, by Dustmann, Casanova, Fertig, Preston, and Schmidt (2003), who also forecasted European migration after the EU enlargement. In the latter study, the error term was decomposed into origin-specific, destination-specific, time-specific, and various cross-sectional components, additionally involving relative income per capita, and dummies for particular countries. Their approach, however, assumed stationarity of the series under study, and thus did not take into account the expected lifting of restrictions on the labour market access for the new EU citizens. As an outcome, their model produced rather moderate forecasts of post-EU-enlargement migration flows. Consequently, given the magnitude of migration after 2004, especially in the case of the United Kingdom, the misspecification of the forecasting model has led to large *ex-post* errors already in the first years of the forecast horizon, with real migration flows higher by more than one order of magnitude (IPPR, 2008; see Chapter 10; for figures on migration from Poland, see also Kaczmarczyk & Okólski, 2008).

In a comprehensive modelling framework, Sinn et al. (2001) made a forecast of foreign population stocks in Germany (B), considering the five largest EU candidate countries of that time: Poland, Romania, the Czech Republic, Hungary and the Slovak Republic. The authors applied a partial adjustments model conditionally on a set of exogenous variables:

$$B_t = \lambda \, [\alpha_0 + \alpha_1 \, YV_t + \alpha_2 \, G_t + \alpha_3 \, EU_t + \alpha_4 \, FR_t + (1/\lambda - 1 + \alpha_5) \, B_{t-1}] + \varepsilon_t. \quad (4.8)$$

In (4.8), λ satisfies the long-term relationship: $B_t = B_{t-1} + \lambda \, (B_t^* - B_{t-1})$, and B_t^* is the equilibrium trajectory of foreign population stocks under study, $B_t^* = \alpha_0 + \alpha_1 \, YV_t + \alpha_2 \, \mathrm{G}_t + \alpha_3 \cdot EU_t + \alpha_4 \, FR_t + \alpha_5 \, B_{t-1}$. A reference in B_t^* to the number of migrants in the preceding period, B_{t-1}, reflects the 'migrant network' hypothesis (*idem*, pp. 8–9, see also Chapter 3). The exogenous predictors used in the model are: YV – fraction of PPP-adjusted GDP per capita in the sending country to the German one, G – the output gap in Germany, EU – a dummy variable regarding the EU membership, and FR – a dummy related to the freedom of movement of the labour force. The model has been estimated on the basis of empirical data on German population stocks originating from Greece, Italy, Portugal, Spain, and Turkey, and applied for the simulations concerning the 'new' prospective Member States. As a result, the forecast yields that by 2015 the B_t would increase from the initial 459 thousand to 3.2–4.1 million people (*idem*, pp. 18–21). It has to be stressed that, as these results concern in fact conditional forecasts (see also Chapter 6), their results heavily depend on the assumed future values of exogenous variables.

Another forecast of post-enlargement migration to the EU-15 from ten countries of Central and Eastern Europe has been prepared by Alvarez-Plata et al. (2003). Their study explains the share of migrants from country i residing in country j, expressed as a percentage of the total population of the latter ($ms_{i,j}$). Also conditionally on the set of predictors, treated as given and thus non-stochastic, the dependent variable has been modelled as:

$$ms_{i,\,j,\,t} = \alpha + (1-\delta)\, ms_{i,\,j,\,t-1} + \beta_1 \ln(w_{j,\,t}/w_{i,\,t}) + \beta_2 \ln(w_{i,\,t}) + \beta_3 \ln(e_{i,t}) + \beta_4 \ln(e_{j,t}) + \beta_5 \ln(P_{i,t}) + Z_{i,j} + u_{i,j,t}, \quad (4.9)$$

where $u_{i,j,t} = u_{i,j} + v_{i,j,t}$ and $v_{i,j,t}$ denotes the Gaussian white noise. The remaining explanatory variables are: w – real income levels, e – employment rates, P – population sizes, and Z – dummy variables denoting geographic and cultural proximity of particular countries. Additionally, the forecast assumed a long-term convergence of economic explanatory variables (in particular, w), which concern ten Central and Eastern European countries, to the average EU-15 levels. As a result, net migration from the countries under study was forecasted to decline exponentially from 367 thousand persons a year, shortly after the introducing the freedom of labour force movement in the enlarged EU, to below zero by 2030 (*idem*, p. 60). It is worth noting that in model (4.9) population size is exogenous, taken from the predictions of the World Bank, and hence is not internally consistent with the forecasted migration numbers.

The features of several types of the models presented above, (4.7), (4.8), and (4.9), have been combined in the study of Brücker and Siliverstovs (2006), who made an attempt to explain the share of migrant stocks from the country i in the country j, expressed as a percentage of the total population of the latter (thus, defined as in Alvarez-Plata et al., 2003). In addition to including exogenous variables (real income levels and fractions, employment rates, etc.), the model has been prepared using the partial adjustments framework, where destination-specific shares of immigrants change stepwise towards their long-term equilibrium levels, and the error term is decomposed into the country-specific effect and the white noise. Their study also examines the impact of heterogeneity across countries on the estimation of migration models, as well as on the forecasts obtained.

Similar models, although with slightly varying sets of explanatory variables, have been proposed also by other authors. In addition to income (or GDP) per capita differences between the sending and receiving countries, present in virtually all models, Hatton and Williamson (1998) added a share of employment in agriculture, a birth rate of a sending country (lagged 20 years), migrant stock at the destination, and country-specific dummies. In their model, the dependent variables were emigration rates related to the population of the source country. Orłowski (2000) added population size and unemployment rates of the destination region, as well as geographic distance, in order to explain emigration rates. Boeri and Brücker (2001) found a significant effect of such variables as: employment rates, institutional restrictions to migrate, presence of migrant networks, and cultural (language) proximity, as well as of standard of living, on the stocks of Central and Eastern European migrants residing in the 'old' EU-15 countries. Alecke et al. (2001) used unemployment rates in both the sending and receiving countries, alongside country-specific dummies to model the origin-destination emigration rates. Several models involving various macroeconomic determinants of country-specific net international migration rates in Europe have been also tested by Jennissen (2004), whose work is discussed in Chapter 6, in the part devoted to problems with the selection of predictors in migration forecasting models.

Another option of applying econometric techniques in migration forecasting is based on generalised linear models (GLM). Willekens and Baydar (1986) suggested that in such models, there is a need to separate effects that are specific to the regions of origin, destination and their interactions, as well as the effects of the 'environment' (exogenous socio-economic, political, cultural, and other variables). They observed that overall migration can be decomposed into the *level effect* concerning the number of migrants, the *generation effect* related to the shares of particular sending areas in all out-migrations, and the *distribution effect* concerning the share of these out-migrations among all possible destinations. Letting N_t stand for the total number of migrants, $w_{i,t}$ for the probability that a migration originates from region i in the year t, and $p_{i,j,t}$ for the probability that a migration originating from i ends in j during t, the model for migration from i to j in time t ($m_{i,j,t}$) can be thus written as (*idem*, p. 207):

$$m_{i,j,t} = N_t \cdot w_{i,t} \cdot p_{i,j,t}, \tag{4.10a}$$

or, for a system of regions, in matrix notation ($\mathbf{M_t} = [m_{i,j,t}]_{n\times n}$, diagonal $\mathbf{W_t} = [w_{i,t}]_{n\times n}$, $\mathbf{P_t} = [p_{i,j,t}]_{n\times n}$):

$$\mathbf{M_t} = N_t\, \mathbf{W_t P_t}. \tag{4.10b}$$

For the elements of $\mathbf{W_t}$ and $\mathbf{P_t}$, additional explanatory analysis can be performed. Willekens and Baydar (*idem*, p. 225) suggested decomposing the logits of $w_{i,t}$ and $p_{i,j,t}$ into components related to regions, time, and their interrelations, possibly applying certain restrictions on the variables under study.

Apart from studies dealing specifically with migration in Europe, there are some other, important models, which call for separate attention. For example, the GLM approach for forecasting migration in a closed system of countries has been recently proposed by Cohen, Roig, Reuman and GoGwilt (2008), based on the notion of gravity of population sizes of both origin and destination countries. The model equation for migration flow from country i to j in the year $t(M_{i,j,t})$ is:

$$\log(M_{i,j,t}) = c + a\ln(P_{i,t}) + b\ln(P_{j,t}) + c\ln(A_i) + d\ln(A_j) + f\ln(D_{i,j}) + \\ + \sum_{h\in S_0} g_h z_h + \sum_{h\in R_0} h_z z_h + \sum_{h\in S_0} i_h z_h s_h + \sum_{h\in R_0} j_h z_h s_h + k\cdot(t-1985) + \varepsilon_{i,j,t}. \tag{4.11}$$

In the above model, $P_{i,t}$ denotes the population of the i-th country in the year t, A_i is the area of country i (in square kilometres) and $D_{i,j}$ is the distance between countries i and j (here, being the great-circle distance between the capital cities). Further, S_0 and R_0 are the subsets of origins and destinations, for which detailed country-specific data are available. For such countries, z_i denote dummy variables equal 1 for i being the countries of origin or destination, while s_i are additional dummies equal 1 for countries i being a source of data (adapted from Cohen et al., 2008, p. 15270).

Cohen et al. (2008) argue that their model has two features that are crucial in forecasting migration for a closed system of countries, such as all regions of the world, and thus can provide a consistent input for global population projections or forecasts. Firstly, the formulation of the model in terms of origin-destination flows ensures that the system-wide balance of migration is zero, and, secondly, in such a model, population of any country cannot be depleted exclusively due to migration under reasonable model parameters. Such a model would ideally fit into the multi-regional framework developed by Rogers (1975), as discussed in Section 4.1.

Another general category is formed by the simultaneous-equations econometric models of migration, the examples of which, however, are less numerous than of single-equation models. Already in the 1960s, Okun (1968) modelled interrelations between inter-state migration in the USA and the inequality of income levels per capita among the states. Later, Greenwood (1973) presented a complex model with seven equations and seven endogenous (jointly dependent) variables: the civilian labour force (*CLF*) out-migration (*OM*) and in-migration (*IM*), income change (ΔInc), employment change (ΔEmp), unemployment change ($\Delta Unemp$), as well as *CLF* change (ΔCLF) and natural increase (*NatInc*). Equations for the two last variables are identities ($\Delta CLF = \Delta Emp + \Delta Unemp$; $NatInc \equiv \Delta CLF + IM - OM$), while the remaining ones are structural, involving several exogenous variables (proxies of age and education structures, government expenditures, and several dummies), and the stochastic component (Greenwood, 1973, pp. 92–95). The simultaneous-equations models can be potentially very useful for forecasting migration within larger supra-national systems, yet the weak international comparability of migration data forms a clear practical limitation in that respect.

4.2.5 Limitations of Econometric Models

Despite the clear advantage of econometric studies in the context of uncertainty, they are often being criticised for the shortcomings of model specification, especially with respect to demographic variables and country-specific effects, which are missing in most of them (Fertig & Schmidt, 2000; Alecke et al., 2001). Limiting the possible explanations of population flows to the economic aspects is also challenged by Kupiszewski (2001), as often leading to very high forecast errors. Ideally, the forecasting model should control at least the basic demographic characteristics, as the size and age structure of the population. In that respect, Kupiszewski (2002a, pp. 637–641) argues that if crude numbers of migrants are forecasted instead of the occurrence-exposure rates and there are no demographic constraints on migration, this may lead to extreme and unlikely results, like the ones obtained in the judgemental scenarios of Franzmeyer and Brücker (1997).

To avoid extremities, a reasonable level of control of the limiting factors should be exerted, depending on various dimensions (demographic, social, economic, environmental, and cultural) of the forecasting context, as well as on the scale of the predictions (Cohen, 1998). Examples of applications of models successfully taking into account various constraints of the socio-demographic setting include Massey and Zenteno (1999), as well as Cohen et al. (2008). A possible analysis of the impact of the age composition of the population on migration may follow the hypothesis of

Plane (1993), who asserted that age-specific internal mobility rates of young people (especially aged 20–24 years) are visibly lower in the 'baby-boom' generations than in the other ('baby-bust') ones. However, such hypotheses still need to be carefully verified on the basis of empirical data, before being applied in the migration forecasting practice.

When migration rates or shares of migrants in the total population are forecasted within a model which uses population size as one of the explanatory variables, as for example in Alvarez-Plata et al. (2003), the movements of people occur de facto outside the model. In a coherent framework, migration flows would increase the population of the destination country and diminish in the source country by the same number, changing also the denominators of migration rates or shares of migrants between subsequent forecast periods. Otherwise, treating population size as exogenous is another source of bias, as in the study of Alvarez-Plata et al. (2003), who used the population stocks projected by the World Bank as predictors. The consequences of this omission can be very serious especially for regions with rapid population growth or decline.

The advantages of various modelling approaches have been combined in a study prepared recently for the UK Department for Communities and Local Government by Barrell, Gottschalk, Kirby, and Orazgani (2009). The aim of their study was to assess the impact of the 2008–2009 economic crisis on immigration to the United Kingdom. Their model, too complex to be presented here in all its detail, consisted of a migration equation similar to the one proposed by Alvarez-Plata et al. (2003), additionally equipped in autoregressive distributed lag features (ADL). Unlike in the previously discussed models, in the study of Barrell et al. (2009) population flows and stocks were combined in a coherent manner, although still not linked with the general population dynamics, for example via a cohort-component mechanism. The model was then fed with macroeconomic inputs from the neo-Keynesian general equilibrium model 'NiGEM' to produce conditional forecasts. The advantage of the model presented by Barrell et al. (2009) is the explicit presentation of uncertainty bands in the form of predictive intervals. The problem here is that, similarly as in the study of Dustmann et al. (2003) one of the underlying assumptions of the model was the one on stationarity, resulting in relatively narrow predictive intervals. These assumptions need not hold especially in the turbulent times of global economic crisis, when uncertainty can be larger rather than smaller.

A way to overcome the incoherence caused by population moves in such models is to examine the forecasted rates or shares in a broader model of demographic dynamics. In such a way, the rates or shares would be transformed into numbers of migrants according to the changes of the population size predicted within the same model. For this purpose, a multi-regional framework of Rogers (1975) can be used, or its extensions, like the exponential LIPRO model (van Imhoff, 1990; van Imhoff & Keilman, 1991), the MUDEA (MUlti-level DEmographic Analysis) model (Willekens, 1995), which combines the features of the population accounting and the multi-regional methodologies, or the generalised multi-level MULTIPOLES approach (Kupiszewska & Kupiszewski, 2005). In order to include a stochastic component in a multi-regional setting, the GLM framework suggested by Cohen et al. (2008) for forecasting migration in a consistent manner can be also applied. Another possibility is to use a statistically-coherent framework proposed within the 'Uncertain Population of Europe' project (Alho et al., 2005), taking into account cross-country correlations of particular migratory processes under study. The latter approach is, among others, discussed in more detail in the next subsection.

4.2.6 Stochastic Forecasts of Migration Time Series

In addition to econometric models involving additional explanatory variables, and thus leading to conditional forecasts, another important group of migration forecasting approaches is based on the analysis and extrapolation of time series. The most common methodology, originally due to Box and Jenkins (1976), applies autoregressive integrated moving average (ARIMA) models, predominantly within the framework of traditional, frequentist statistics. In general terms, the ARIMA(p,d,q) model can be formulated as (e.g., Greene, 2000, pp. 776–777):

$$\left(1-\sum_{i=1}^{p}\phi_i L^i\right)\left[(1-L)^d x_t\right] = c + \left(1-\sum_{j=1}^{q}\theta_j L^j\right)\varepsilon_t, \qquad (4.12)$$

where x_t is the variable under study, ε_t is the error term (usually assumed to be a Gaussian white noise), L denotes the lag operator ($L^k x_t = x_{t-k}$), and $(1-L)^d x_t$ is the d-th difference of x_t.

There are several examples of migration studies based on ARIMA models with various parameters p, d and q. For instance, a forecast of migration to and from the Netherlands prepared by de Beer (1997) uses an AR(1) autoregressive process $x_t = c + \varphi\, x_{t-1} + \varepsilon_t$, both for the total volume of emigration and immigration (x_t). In the same study, a moving average process MA(1), $x_t = c + \varepsilon_t - \theta\, \varepsilon_{t-1}$, has been found best fitting the data in the case of net migration (after: Keilman et al., 2001, p. 62). In predictions for Finland, Alho (1998) applied ARIMA(0, 1, 1) models $x_t = c + x_{t-1} + \varepsilon_t - \theta\, \varepsilon_{t-1}$ for logarithms of immigration and emigration (*idem*). Keilman et al. (2001) prepared a Norwegian forecast using an ARMA(1,1) model for the logarithm of immigration, with $x_t = c + \varphi\, x_{t-1} + \varepsilon_t - \theta\, \varepsilon_{t-1}$, and an ARIMA(0,1,0) model; that is a random walk with drift, with $x_t = c + x_{t-1} + \varepsilon_t$, for the logarithm of emigration. Further to the presented studies, the most important classes of stochastic models for forecasting demographic rates (random walks, linear stationary processes, and ARIMA) are also discussed in Alho and Spencer (2005, pp. 198–225), together with the possibilities of their extension by allowing for example for heteroskedasticity of the error term, ε_t.

The time series models presented above have their multivariate generalisations, allowing for including other variables in addition to the key one under study (migration). An example of this approach is the vector autoregressive (VAR) modelling, originally due to Litterman (1979) and Sims (1980) (after: Greene, 2000, p. 741, see also Chapter 6), which has been used in the forecast of migration between Australia and New Zealand prepared by Gorbey, James, and Poot (1999). Their VAR(4) model, based on quarterly data, has the form:

$$\mathbf{X_t} = \mathbf{C_0} + (\mathbf{C_1}\, L + \mathbf{C_2}\, L^2 + \mathbf{C_3}\, L^3 + \mathbf{C_4}\, L^4)\, \mathbf{X_t} + \boldsymbol{\varepsilon}_\mathbf{t}. \tag{4.13}$$

where $\mathbf{C_i}$ are the matrices of cross-variable coefficients, L denotes the lag operator, and $\boldsymbol{\varepsilon}_\mathbf{t}$ is a multi-dimensional Gaussian white noise. The choice of variables in the VAR models has been based on the formal tests of the unit root, and the Granger-Sims tests of causality (*idem*, pp. 78–84), yielding the following sets of interdependent variables, $\mathbf{X_t}$, applied in the forecasts:

$$\mathbf{X_t} = [NMR_t,\ YRCGQ_t]', \tag{4.13a}$$

$$\mathbf{X_t} = [NMR_t,\ YRGQ_t,\ AUGQ_t]', \tag{4.13b}$$

$$\mathbf{X_t} = [D4NMR_t,\ YRCGQ_t,\ ZUGQ_t]', \tag{4.13c}$$

$$\mathbf{X_t} = [NMR_t,\ YRGQ_t,\ UDIFQ_t,\ ERGQ_t]'. \tag{4.13d}$$

In (4.13a) – (4.13d), the following abbreviations of particular variables are used: *NMR* – net migration rate, *D4NMR* – yearly difference of *NMR*, *YRGQ* – growth of the real GDP ratio for the two countries, *YRCGQ* – growth of the real GDP ratio per capita, *UDIFQ* – differences in unemployment rates, *AUGQ* and *ZUGQ* – country-specific unemployment growth indices for Australia and New Zealand, and *ERGQ* – growth of the earnings ratio for the two countries. Although Gorbey et al. (1999) observed that migration between Australia and New Zealand is largely visa-free and resembles internal population flows, the same modelling framework can be also tested for typical international migration. The possible modifications include adjustments taking into consideration exogenous, not modelled, variables related to migration policies and to the freedom of movement of persons.

Apart from migration models based on the analysis and extrapolation of time series, there is an example of a forecasting framework that makes an attempt to partially depart from the traditional statistical paradigm. Lutz et al. (1996, 1998, 2000, 2004) developed a concept of 'expert-based probabilistic population projections', in which subjective expert judgement is applied to prepare stochastic forecasts. In general, let υ_t denote the phenomenon under study, for example international migration rates. Within the expert-based framework, the forecasting model has the form $\upsilon_t = \bar{\upsilon}_t + \varepsilon_t$, where $\bar{\upsilon}_t$ is the average trajectory of the process, assumed a priori by specialists in the field, and ε_t follows a chosen stochastic process, e.g., AR(p) or MA(q).

Lutz et al. (2004) applied $\varepsilon_t \sim$ MA(30) for yearly data, assuming additionally that the standard deviation of ε_t, $\sigma(\varepsilon_t)$, is equal to a pre-defined value $\sigma^*(\varepsilon_t)$, also pre-set on the basis of subjective expert opinion. For international migration forecasts, the authors assumed that the average of the process is time-invariant ($\bar{\upsilon}_t = \bar{\upsilon}$), while $\sigma^*(\varepsilon_t)$ was calculated in such a way that 80% of the density of the probability distribution of υ_t was concentrated between zero and the judgementally-chosen value $\upsilon_{\max}$. As subjectivity is explicitly expressed in the 'expert-based' forecasting approach, it can be seen as a hybrid between the frequentist and some Bayesian methods. On one hand, the probabilities (weights) attached to the trajectories of forecasted variables are subjective, but on the other hand, no observations (data) are used to change the prior beliefs into the posterior knowledge, and thus the essence of the Bayesian inferential mechanism is not involved. Notably, a partial Bayesian interpretation of 'expert-based probabilistic projections' has been given by Tuljapurkar (1997, p. 760) in terms of attaching 'a priori probabilities to each static or dynamic scenario' in the forecasts presented by Lutz, Sanderson, and Scherbov (1997).

The features of various approaches to probabilistic population forecasting have been also synthesised within the framework of the EU-financed research project 'Uncertain Population of Europe' (UPE) (Alho et al., 2005). The UPE predictions combine the cohort-component model of population dynamics with probabilistic forecasts of fertility, mortality, and migration, based on the analysis of time series, historical forecast errors, and on expert opinion. The results have been calculated on the basis of 3,000 simulations. One specific innovation of the UPE Project is a comprehensive empirical analysis of correlations between forecast errors for various components of population change (including international migration[21]), as well as between the countries under study (Alho et al., 2005, p. 2). With respect to migration, the balance of flows for particular countries was modelled using linear trends, random walk models or autoregressive models, with dummy variables for some country-specific years. Among the mentioned models, the AR(1) has been ultimately chosen for forecasting net migration among the 18 countries of the European Economic Area (Keilman and Pham, 2004b).[22]

[21]Some authors underline the association of migration with family formation and dissolution processes, concerning marriages, cohabitations, divorces, and, indirectly, childbearing (Paradysz, 2006, pp. 235). The direct impact of (internal) migration on fertility has been assessed for example by Kulu (2005), who found empirical support for a hypothesis that migrants might adjust their fertility to the levels observed in the host community, rather than preserve levels from their regions of origin. In the case of international migration, however, the situation can be more complex, for example due to the presence of ethnic enclaves ('ghettos') which do not facilitate the integration of immigrants with the host society. Moreover, a similar notion of convergence applies to mortality: it can be argued that migrants 'adjust' their mortality patterns to the ones prevailing at the destination, via two channels: access to the same health care services and exposure to the same environmental hazards as the host population (e.g., Bijak et al., 2007).

[22]More information is available from the project website: www.stat.fi/tup/euupe (accessed on 5 May 2006).

4.3 Bayesian Approach in Migration Studies and Demography

4.3.1 Bayesian Models and Forecasts of Population Flows

The existing examples of using Bayesian methods to model and forecast international migration are scarce. Therefore, the current overview applies a somewhat broader perspective, including additionally selected models of non-migratory population flows.

To begin with, Gorbey et al. (1999) estimated their VAR(4) models of migration between Australia and New Zealand defined in (4.13) also within the Bayesian framework. The coefficients of the models have been assumed to follow *Minnesota priors* of Litterman (1979). For the parameters on the first lags of the same, i-th variable, Normal distributions N(1, $\sigma_{i,i,1}{}^2$) have been assumed a priori. In all other cases, for interrelations between the i-th variable and the k-th lag of the j-th variable, that is for $k > 1$ or $k = 1$ and $i \neq j$, Normal prior distributions N(0, $\sigma_{i,j,k}{}^2$) have been used. In this way, the time series of each variable is a priori believed to be most likely generated independently by a random walk process. For the specific variances of the Gaussian priors, it has been assumed that $\sigma_{i,j,k} = \gamma g(k) f(i,j) s_i/s_j$, with $\gamma = 0.4$, $g(k) = k^{-1}$, $f(i,i) = 1$, $f(i,j) < 1$ for $i \neq j$, and s_i denoting a standard error in the autoregressive model for the i-th variable.

The *ex-post* comparison of various frequentist and Bayesian models yielded that the best-performing ones were the ones given by (4.13b), with $\mathbf{X_t}$ comprised of net migration rates, growth of the real GDP ratio for the two countries, and quarterly unemployment growth in Australia. Among them, the Bayesian model produced slightly greater *ex-post* forecast errors than the corresponding traditional VAR (Gorbey et al., 1999, p. 90), one possible explanation being the likely disagreement between the prior distributions and the data sample.

Nevertheless, from a strictly Bayesian point of view, there is one problem with the coherence of such approach. As the s_i values are estimated from the same sample as the model itself, the priors are data-based. Therefore, the presented approach is not fully Bayesian, as in that case the prior distributions should be specified independently from the observations.[23]

The research of Brücker and Siliverstovs (2006), mentioned earlier in this section, also compares the results of estimation of migration models in the sampling-theory and Bayesian approaches. In their comparison of various estimation methods, the hierarchical Bayes estimator (likely the mean from the appropriate posterior distribution) and the frequentist fixed-effects estimator, performed best in terms of *ex-post* prediction errors. The problem is, however, that the authors consider the Bayesian framework merely as an alternative methodology of estimation. The study lacks both the discussion of the prior distributions used, as well as the a posteriori uncertainty assessment, which are inherent in the Bayesian analysis.

---

[23] I am very grateful to Jacek Osiewalski for drawing my attention to this problem.

In addition to the studies of 'pure' migratory movements, examples of Bayesian gravity models based on the Poisson regression, applied to forecasting other types of population flows, have been offered by Congdon (2000, 2001b).[24] It should be noted, that although his studies focus on the flows of patients to hospitals, the models are sufficiently general to cover other types of spatial movements in a similar manner. The common assumption is that the number of patients coming to the hospital j from the region i follows a Poisson distribution with a mean $\mu_{i,j}$, for which various alternative models can be built, as for example (Congdon, 2001b, pp. 297–298):

$$\ln(\mu_{i,j}) = k + \alpha_0 \ln(P_i) + \delta \ln(R_{i,j}) + \xi_1 E_{1,j} + \xi_2 E_{2,j} + \varphi S_{i,j}, \text{ or:} \tag{4.14a}$$

$$\begin{aligned} \ln(\mu_{i,j}) &= k + \alpha_0 \ln(P_i) + \alpha_1 YAN_i + \alpha_2 Aged_i + \delta_i \ln(R_{i,j}) \\ &+ \xi_1 E_{1,j} + \xi_2 E_{2,j} + \varphi S_{i,j}. \end{aligned} \tag{4.14b}$$

The basic set of explanatory variables is comprised of P_i – population size of the i-th region, and $R_{i,j}$ – a proxy of the medical services supply, calculated as the number of beds in the j-th hospital, B_j, weighted by an average distance (crow-fly or car-time), from the i-th region to the j-th hospital, $d_{i,j}$. The further exogenous predictors are: YAN_i – an index of demand for the health-care services, $Aged_i$ – a fraction of population aged 65 years or more, $E_{1,j}$, $E_{2,j}$ – dummy variables for two selected hospitals, and $S_{i,j}$ – a dummy indicator, whether the j-th hospital is located in the i-th region. The prior distributions are assumed to be Gaussian, diffuse for the constant, $k \sim N(1, 100^2)$, and carrying more information in the case of remaining parameters: α_i, $\delta_i \sim N(1, \sqrt{10}^2)$, and ξ_1, ξ_2, $\varphi \sim N(10, \sqrt{10}^2)$. The model (4.14b) based on car-time distance has been found fitting the data best.

Finally, a set of simple AR(1) and VAR(1) models for forecasting rates of migration between Poland and Germany is proposed by Bijak (2008a). However, as the last-mentioned study merely constitutes a prototype for models introduced in the current one, at the same time being less sophisticated from a methodological point of view, its detailed discussion is omitted.

4.3.2 Bayesian Methods in Demography: A Concise Survey

It is worth noting that the scarcity of Bayesian forecasts of international population movements resembles the situation in the other fields of demographic analysis. The rare existing examples include forecasts of fertility (Tuljapurkar & Boe, 1999),

[24] A summary of various probability models that can be used for migration forecasts is offered in Willekens (2008), who argues for using Poisson regression for counts of migrants, logit or logistic models for proportions of migrants in a given population, and Poisson models with offset for occurrence-exposure rates. The discussion further distinguishes models for state occupancies, transition probabilities, and transition rates.

mortality (Girosi & King, 2008),[25] or the whole population size in the case of limited information, as in the study of the Iraqi Kurdish population by Daponte, Kadane, and Wolfson (1999). Population change has also been analysed in the Bayesian framework by Bernardo and Muñoz (1993), and Clark (2003), the latter from an ecological perspective, not limiting the analysis to the human species. Likewise, Wilson and Rannala (2003) estimated the rates of migration and ancestry patterns for thistle plants (*Centaurea corymbosa*) and grey wolves (*Canis lupus*) using Bayesian methods, on the basis of the analysis of the available genotype data.

Interestingly, according to Alho (1999, p. 1), one of the first attempts to formulate the population prediction problem in probabilistic terms instead of producing the baseline-high-low intervals has been made by Törnqvist (1949) within the Bayesian framework. More contemporarily, a complex Bayesian analysis of demographic change using a cohort approach has been proposed by Nakamura (1986), while the application of Bayesian inference in the studies on the Lexis diagram has been shown by Berzuini, Clayton, and Bernardinelli (1993). The latter concept has been recently extended in a form of specialised software 'BAMP' designed for the age-period-cohort modelling and forecasting (Schmid & Knorr-Held, 2001).

Some other examples of Bayesian models that can be used in demography and related fields are given by Courgeau (2004), and include multi-level or hierarchical models (Draper, 1995; Goldstein, 2003), as well as survival models (Ibrahim, Chen, & Sinha, 2001; Gustafson, Aeschliman, & Levy, 2003). In that respect, this book aims at contributing to the further development of the field of Bayesian population predictions, both in terms of the subject (migration), and of the applied methodology. In that respect, the book aims to extend the Alho and Spencer's (2005) volume, going more deeply into the issues of international migration and Bayesian forecasting.

4.4 From Migration Theories to Model-Based Forecasting

4.4.1 Migration Forecasting Methods and Models: State of the Art and Typology

The nomothetic, theory-based, predictions of socio-economic phenomena are contrasted by Chojnicki (1977) with the *heuristic-model* approach, contemporarily prevailing in the forecasting practice. In order to reconstruct the events under study, and analyse their consequences, various descriptive *models*[26] can be built. Unlike

[25] Despite a very general title ('*Demographic forecasting*'), the book by Girosi and King (2008) explicitly deals in very much detail with various methods for Bayesian forecasting mortality rates by age, time, cause of death, etc.

[26] 'Model' can be defined as 'a system of postulates, data, and inferences presented as a mathematical description of an entity or state of affairs' (the Merriam-Webster Online Dictionary; www.m-w.com, accessed on 25 April 2006).

theories, which should be by nature very general, models relate to a particular reality, and are characterised by a large degree of flexibility. Chojnicki (*idem*) noted that in some cases it may be difficult to distinguish between theories and models. This observation is especially important, given that models often serve as formal tools for verification of hypotheses, which are posed by theories, on the basis of measurable quantities. In addition to models, there exist heuristic methods (not model-based) that also can be applied in forecasting (e.g., the Delphi approach).

One group of methods used in preparing forecasts of international migration is based on deterministic mathematical models, as well as other techniques, which do not explicitly address the uncertainty of migratory phenomena. It should be noted that the crucial distinction between deterministic and probabilistic methods or models, applied in the current study, is to some extent arbitrary, and can be related to the common practice in the existing research, rather than to the potential of methods in addressing the uncertainty issues.

Unlike the deterministic methods and models, the stochastic (probabilistic) tools of migration forecasting are rooted in the probability theory, and explicitly address, at least potentially, the issue of uncertainty. With respect to the typology of forecasting methods presented by Chojnicki (1977), they are usually model-based, with a few of them including additional heuristic elements (e.g., qualitative research in the ethnosurvey). Although some of the models make explicit reference to particular theories of migration, like for example econometric models to economic theories, they do not attempt to construct a general theoretical framework of population flows in probabilistic terms. It should be remarked that, as the scope of the current study is strongly related to the uncertainty issue, the stochastic models discussed in this chapter are presented in relatively more detail than the deterministic ones.

As mentioned before, the theories of migration are too fragmented and too general to be used directly as a forecasting tool within the nomothetic approach. The alternative option to obtain plausible predictions of population flows is the use of models or heuristic methods. There are several such methods and models, both deterministic and probabilistic, which can be potentially interlinked, with the outcome of simpler methods used as input in the more general ones. An attempt to present a synthesis of the discussion about various models and methods, as well as the possible relationships between them, is undertaken in Fig. 4.1.

In general, the distinction between deterministic and stochastic models follows the way they are presented in this chapter, with some additional remarks. On one hand, as it has been mentioned before, some of the deterministic macro-level demographic and demo-economic models can acquire stochastic features, when they get probabilistic input. On the other hand, stochastic forecasts can (and often are, as it is discussed further) be interpreted in a deterministic fashion, disregarding the whole context of uncertainty and forecast errors. Moreover, some methods use research techniques of different scientific disciplines in parallel. An example is the ethnosurvey, the ultimate results of which rely both on the probabilistic (survey-based) and deterministic (qualitative) analysis of migration flows. Furthermore, econometric and time series models can be prepared alternatively within the frequentist or

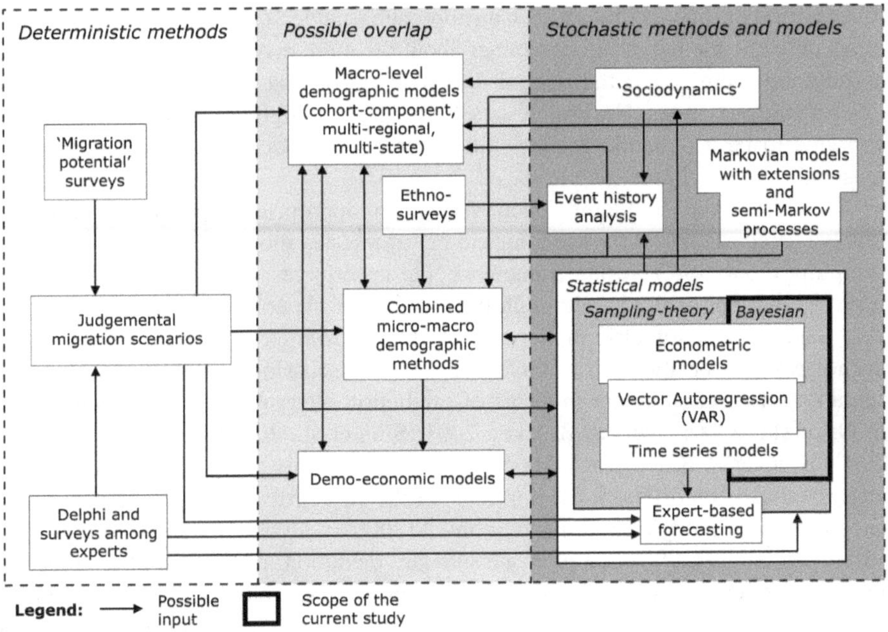

Fig. 4.1 Typology of selected migration forecasting methods and models.
Source: own elaboration

Bayesian statistical paradigm. In the current study, the main methodological scope covers selected Bayesian types of probabilistic models, as indicated by a bold frame in Fig. 4.1.

4.4.2 Deterministic Character of Many Existing Predictions

As noted in Section 4.1, most of the practical applications of the cohort-component and multi-regional demographic models are based on judgemental scenarios of particular components of demographic change, including migratory flows. Although the input for the demographic models could easily consist of stochastic forecasts of these variables, a majority of demographic applications stick to the multi-variant projections, where uncertainty is not properly quantifiable (see Chapter 2). The exceptions are based on the event-history analysis, in which micro-level simulations are used to estimate probabilities of transition between particular states, thus operating within a probabilistic framework (Section 4.2). Also many geographical models, despite being based on stochastic processes (notably, Markov chains), and thus having large potential of simulation-based uncertainty assessment, do not seem widely explored in this context up to date.

Moreover, the judgemental scenarios that often form the input to the models of population dynamics often assume the constancy of migratory flows starting from a given period of time. Although such an approach is understandable, as given the lack of specific knowledge of the researcher about the more distant future, the constancy assumption seems to be the neutral option; it is also likely to generate very high *ex-post* forecast errors. Nevertheless, as international migration will most probably continue to be a dynamic process, this approach cannot be expected to produce reasonable results in the long term.

To address the problems listed above, an appropriate option would be to involve forecasts based on econometric or time series models, applying either the sampling-theory or Bayesian framework, but nevertheless bearing in mind several reservations. For example, although the analysis of uncertainty is embodied in the very nature of econometric models, many studies devoted to migration forecasting do not pay proper attention to this issue. In the discussion of the results, the presentation of forecast errors in terms of prediction intervals is often missing (e.g., in Orłowski, 2000; Boeri & Brücker, 2001; Sinn et al., 2001; Alvarez-Plata et al., 2003; Brücker & Siliverstovs, 2006). In most cases, this is replaced by a scenario analysis, based on different assumptions on the GDP growth and convergence. An exception is the study of Fertig and Schmidt (2000), who based their high-migration variant on the mean forecast plus one standard deviation. In contrast to the practice in econometric forecasting of migration, uncertainty is explicitly addressed in predictions based on pure time series models (e.g., de Beer, 1997; Alho, 1998; Keilman et al., 2001; Lutz et al., 2004; Alho et al., 2005).

All remarks made with respect to the frequentist econometric and stochastic forecasting models remain in force for their Bayesian equivalents. The latter address the issue of uncertainty in a form of whole predictive distributions, as defined in (2.4), from which the HPD or other credible intervals can be derived, in order to assess the uncertainty span. In addition, the Bayesian inference allows for a formal incorporation of expert judgement into the model, with respect to the characteristics of the processes and interactions between variables. This is possible through assuming appropriate informative prior distributions of the model parameters. The expert opinion can be obtained for example from surveys among experts or Delphi studies, similar to the ones presented in Section 4.1, or subjectively assumed by the forecaster.

4.4.3 Notes on Including Theory in Population and Migration Forecasts

The question of including theories in population and migration forecasts, and the related one of complexity and sophistication of tools used for forecasting purposes remains one of the key issues surrounding forecast-making, not only in demography, but across the social science disciplines. There have been various attempts to address this problem, both theoretical and empirical. In general terms, augmenting demographic forecasts with theoretical knowledge was discussed in the seminal paper by Keyfitz (1982). His main conclusion on a limited usefulness of theories in the work of the forecaster was based on the following arguments. Firstly, theories are often designed for explaining the past in static terms rather than for making dynamic, unconditional, *ceteris paribus* extrapolations into the future. Secondly, even if a theory holds, its effect can still be countered by other forces in play. Moreover, the theories may not be specific about the timing of various processes and the potentially dynamic interactions between the forecasted variable and its potential predictors, which themselves may be difficult to forecast. Finally, the theories may hold sufficiently well in a time horizon which is too short for demographic predictions (Keyfitz, 1982, pp. 745–746).

As noted by Keyfitz (1982), the forecasting task differs from the explanatory analysis, aimed at the improved understanding of the processes under study. More precisely, predictions should satisfy different validity criteria than the studies of causal mechanisms and processes. First and foremost, forecasts should be thus unconditional statements about the likely future, which would answer the particular needs of the forecast users. Secondly, forecasts can still remain valid despite not identifying the causal patterns driving the predicted variables. Thirdly, forecasters should explicitly acknowledge the forecast error rather than pretending that they can accurately foresee the future. Finally, demographer's judgement and intuition, 'sharpened by the exposure to literature' (*idem*, p. 747) is a very valuable element, which can enter the forecasting process in various informal and indirect ways.

On the empirical side, there were numerous attempts to verify the usefulness of forecasting methods with a varying degree of complexity. With respect to demographic predictions, another paper by Keyfitz (1981) showed on the basis of over 1,100 forecasts for various countries published between 1939 and 1969 that their accuracy was gradually improving over time. This achievement was attributed to two factors: increasing use of tools and techniques of demographic analysis (such as the cohort-component method) and increasing reliance on demographers' judgement. However, the paper also contains an important caveat that the future developments in reducing forecast errors remain uncertain and too high expectations in that respect 'would seem imprudent' (*idem*, p. 585).

In general terms, efforts to evaluate various forecasting methods, simple and complex alike, were undertaken in a series of 'M-Competitions' organised by S. Makridakis and associates since the late 1970s (for an overview, see Makridakis & Hibon, 2000). These competitions, consisting of comparing the efficiency of various forecasts according to their *ex-post* errors, indicated that the more complex methods do not necessarily outperform the simple ones. Moreover, the results varied depending on the applied measure of accuracy (see also Section 8.2) and forecast horizon, and on whether single or averaged forecasts were considered (*idem*, p. 452; see also Chapter 5). For example, in the M3-Competition, the 413 demographic series studied (out of the total of 3,003) were best reproduced either by simple methods, such as exponential smoothing with trend, or by rule-based expert systems, selecting *ad hoc* an appropriate technique from a menu of simple models, including random walks, linear trends, exponential smoothing, ARIMA processes or other (*idem*, pp. 463–464, 456). These results call for creating an expert-based mechanism for selecting models best suited for particular tasks. An attempt to construct such a framework is undertaken in Chapter 5 using well-established Bayesian methods of model selection and forecast averaging.

Furthermore, for the reasons already mentioned above, and given the complexity of demographic processes and their determinants, and also given the difficulties in predicting the structural changes, Keyfitz (1982) tentatively suggested relying on statistical and mathematical extrapolation models in population forecasting. Such methods should be constantly improved and fine-tuned in order to reduce forecast errors, acknowledging at the same time that the error is an inevitable feature of every forecast. Moreover, demographic judgement, stemming from the experience of forecasters, should be used in full. Elaborating on these postulates, the current study aims at proposing simple statistical methods for forecasting migration, in which at least some judgemental elements would be incorporated in a formal and explicit manner. Within this framework, there still exist possible ways to incorporate theories into forecasting, which might be especially important whenever the data are missing or conflicting. Further discussion is offered in Chapters 6 and 10 on the example of vector autoregression models, where migration is forecasted alongside some of its important determinants.

4.4.4 Implications for the Current and Future Studies

With respect to various methodological options of stochastic modelling and forecasting of demographic phenomena, presented in Chapter 2, Lee (1998) suggested that the choice of a 'proper' method should depend on the availability of good-quality data sets. Longer series of observations enable using the time series approach, while for shorter samples or less reliable data, the expert-based probabilistic scenarios are recommended. In order to synthesise these methodologies, Tuljapurkar et al. (2004) suggested that the new directions in demographic forecasting should include hybrid models, combining time series and the expert-based scenarios. The current study aims at exploring this path in the Bayesian framework, where prior knowledge can to some extent compensate for the shortness of data series.

Importantly in the context of data availability, Salt and Singleton (1995) and de Beer (2008) suggested that population flows should be predicted separately for different categories of migrants. Van der Gaag and van Wissen (1999) tested this idea on several models for immigration into five EU countries, distinguishing nationals, EU-foreigners and non-EU-foreigners, as well as examining several migration motives (economic migration, asylum-seeking, elderly migration,. . .). However, in the current work, the analysis of different categories of migrants is limited to the origin-specific registered flows comprised mainly of labour and family migrants. Asylum seekers are excluded, as migrants between the four countries under study (Germany, Italy, Poland, and Switzerland) by definition do not fall into this category. A further exploration of the possibility of modelling and forecasting migration decomposed by motive and/or by nationality is also restricted by a lack of comparable data for all four countries, which especially concerns Poland. An additional difficulty lies in obtaining relevant judgements to be used in predictive models for all categories of migrants. As summarised by Lawrence, Goodwin, O'Connor and Önkal (2006, p. 508), 'decomposition is not guaranteed to improve accuracy and may actually reduce it when the decomposed judgements are psychologically more complex or less familiar than holistic judgements [...]'.

As a consequence of the above argumentation, examples presented in the current book focus on applying several tailor-made econometric and time series models to forecast specific international migration among the countries under study. Given the limited applicability of theories in migration forecasting, as discussed before, the model-based approach is followed without aiming for a far-reaching generalisation of the obtained results. As stated by Arango (2002, p. 295), 'migration is hard to define, difficult to measure, multifaceted and multiform, and resistant to theory-building'. As this applies also to any formalised modelling of general migration features, the current study is strictly devoted to providing insights into methodological issues related to the potential usefulness of selected forecasting frameworks under the Bayesian philosophy, rather than to verifying any particular theory.

Among the available methods presented in the current chapter, the practical application of deterministic methods is not subject to investigation due to their disregard for the uncertainty issue, as in the case of judgemental scenarios. The survey-based approach is also not considered, as it does not qualify as a proper forecasting method in the light of the discussion offered before. Nevertheless, a survey among experts or a Delphi study concerning either the parameters of a forecasting model, or migratory flows and their uncertainty assessment (advocated for example by Lutz et al., 1996, 2004) as an interesting extension of the analysis are discussed in Chapter 10.

A detailed discussion of the ways of incorporating judgement in migration forecasts, including insights into the Bayesian approach, has been provided by Willekens (1994), whose remarks may serve as a point of reference for an expert opinion-based study. Although such possibilities are generally very broad, and remain beyond the scope of the current study, they are clearly worth addressing in separate research. Other probabilistic methods and models (Markov chains, the event-history analysis, or the 'sociodynamic' approach) are also not explored in the current work, as they are quite complex and require much more detailed statistical information than there is available for international population flows in Europe. An investigation into some of these options, although very promising, like in the case of the event-history approach bridged with macro-level studies of demographic dynamics, would require a separate, very extensive (and expensive) research project.

In general, there is still a large potential in applying alternative forecasting methods, which have not been used in the migration context up to date. A full account of methodological possibilities, which has been provided by Armstrong and Green (2007), is illustrated in Fig. 4.2, together with the indication of the position of methods already used in migration forecasting. Thus, in this typology, the analysis of 'intentions/expectations' covers for example 'migration potential' assessment

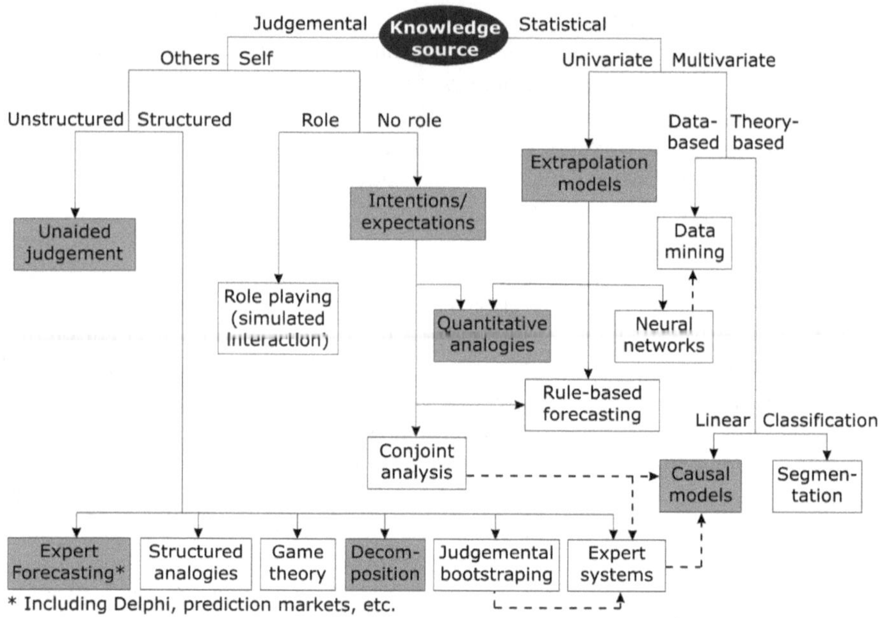

Fig. 4.2 Migration forecasting methods in the general methodological framework.
Notes: Grey shading denotes types of methods used in migration forecasting, straight lines – the relationships between various methods, while dashed lines – possible relationships.
Source: Armstrong and Green (2007), reprinted by kind permission. Own elaboration on migration forecasting

surveys. Judgemental migration scenarios can be classified either as instances of 'unaided judgement', or 'quantitative analogies', depending on whether or not the forecasts are supported by formal methods, and use available information on analogous situations from the past, respectively from other countries. Among the judgement-based methods in migration studies, there are also examples of the use of the Delphi approach, as indicated in Section 4.1. The same applies to judgemental decomposition of migration flows into various groups, which are further analysed based on available data (van der Gaag & van Wissen, 1999).

At the same time, the vast majority of the existing studies of migration either fall into the class of 'causal models', involving econometric analyses, or 'extrapolation models', including deterministic demographic and demo-economic forecasts, and predictions based on stochastic processes: Markov chains and time series analysis. 'Extrapolation' also covers the micro-macro bridging perspectives, as well as the outcomes of event-history studies or the quantitative parts of ethnosurveys. As a consequence, this category refers to a very broad group of methods, covering deterministic and probabilistic approaches alike, according to their potential of addressing the uncertainty issue in a coherent quantitative manner.

Specifically in the probabilistic approach, further paths of methodological developments could possibly include the investigation of data mining techniques based on various sources ('letting the data speak for themselves'), stochastic versions of neural networks, rule-based forecasting, or generalised versions of expert systems, addressing the uncertainty issue. In particular, in the rule-based forecasting, as defined by Armstrong and Green (2007, *passim*),

> expert domain knowledge and statistical techniques are combined using an expert system to extrapolate time series. Most series features are identified by automated analysis, but experts identify some factors. In particular they identify the causal forces acting on trends.

The rule-based approach to forecasting seems therefore especially promising in the context of such a multi-dimensional phenomenon as migration, where the judgemental element is very important. A simple attempt to address this challenge is offered in Chapter 5. Nonetheless, all the mentioned options require conducting detailed research in the future, dedicated particularly to the examination of the applicability of specific methods in international migration forecasting.

Part III
Examples of Bayesian Migration Predictions

Chapter 5
Bayesian Model Selection and Forecast Averaging

This chapter presents the first out of four perspectives proposed for designing a modelling framework for Bayesian forecasting of international migration. In particular, it explores the Bayesian model selection and forecast averaging techniques, based on the posterior odds criterion. Theoretical foundations are laid in Section 5.1, together with a framework proposed for migration predictions on the basis of simple, non-nested stochastic processes. The computational details of the Carlin–Chib procedure used for model selection are also provided. Section 5.2 presents empirical results of forecasts of international migration among selected European countries for 2005–2015, yielded by various models. The Chapter focuses on individual and averaged forecasts from simple stochastic processes.

5.1 Selection and Averaging Problems: Simple Stochastic Processes

5.1.1 Methodological Foundations of Bayesian Model Selection

Within the Bayesian paradigm, the problem of formal model selection on the basis of posterior model probabilities has been paid attention by many authors. One of such approaches is made within the decision-theory framework, through defining and minimising the expected value of a loss function over the space of possible models (e.g., Zellner, 1971, pp. 291–317; Palm & Zellner, 1992; Min & Zellner, 1993; see also Chapters 2 and 11). However, Osiewalski (2001, p. 21) noted that the decision approach is not an essence of the Bayesian inference. Instead, one can restrain the analysis to a comparison of the *posterior odds* of different models, given the data, in order to maximally utilise information from the sample of observations (e.g., Raftery, 1995; Hoeting, Madigan, Raftery, & Volinsky, 1999; Osiewalski, 2001). In the current study, the latter option has been followed.

After Osiewalski and Steel (1993a), let $M_1, \ldots, M_m$, $m \in \mathbf{N}$, be mutually exclusive, non-nested, models of the phenomenon under study, adding up to the whole (finite) space of possible models, $\mathbf{M}$. Assuming prior probabilities $p(M_1), \ldots, p(M_m)$ for the respective models, Bayes' theorem yields their posterior probabilities, given the data x:

J. Bijak, *Forecasting International Migration in Europe: A Bayesian View*,
The Springer Series on Demographic Methods and Population Analysis 24,
DOI 10.1007/978-90-481-8897-0_5,

$$p(M_i \mid x) = \frac{p(M_i) \cdot p(x \,|M_i\,)}{\sum_{k \in \mathbf{M}} p(M_k) \cdot p(x \,|M_k\,)}. \tag{5.1}$$

In (5.1), $p(x\,|M_i\,)$ denotes the marginal density of data in the i-th model, corresponding to $p(x)$ in equation (2.2), in Chapter 2.

The posterior probabilities of alternative models i and j can be used for their direct comparison, based on the calculation of the *posterior odds ratio*, R_{ij}:

$$R_{ij} = \frac{p(M_i \mid x)}{p(M_j \mid x)} = \frac{p(M_i) \cdot p(x \mid M_i\,)}{p(M_j) \cdot p(x \mid M_j\,)}. \tag{5.2}$$

Values of $R_{ij} > 1$ indicate that, given the data x and priors $p(M_i)$ and $p(M_j)$, the model i is more likely to accurately describe the phenomenon in question than the model j, while $R_{ij} \in [0,\ 1)$ gives preference to the model j over i. Some authors argue that it is useful to set thresholds for R_{ij}, denoted O_L and O_R ($O_L < O_R$), so that $R_{ij} < O_L$ provides 'strong evidence' in favour of the model j, $R_{ij} > O_R$ for the model i, while $R_{ij} \in [O_L,\ O_R]$ is inconclusive (Hoeting et al., 1999, p. 385).[1] From (5.2), the posterior odds ratio can be multiplicatively decomposed into the *prior odds ratio*, $p(M_i)/p(M_j)$, and the *Bayes factor*, $BF_{ij} = p(x \mid M_i)/p(x \mid M_j)$ (e.g., Raftery, 1995, p. 129).[2]

Osiewalski (2001) noted that the posterior odds ratio may depend on the priors selected for particular models, both with respect to model-specific parameters, as well as to the models themselves. It has to be noted that $P(x\,|M_i\,)$ is usually not the same as the likelihood function in the traditional sampling-theory statistics; except for very simple cases, with one-element parameter space $\mathbf{\Theta}_i = \{\theta_i^0\}$. In $p(x\,|M_i\,)$, the likelihood is namely weighted by prior densities of the parameters as shown in (2.2), in Chapter 2. Hence, despite a visible similarity under a flat prior to the frequentist likelihood ratio test (see also Chapter 8), in a general case the two procedures need not be equivalent.

---

[1] Historically, the first attempt to set such thresholds was the Jeffreys scale (Jeffreys, 1961). The commonly-used variant thereof treats $R_{ij} < 1/10$ as 'strong evidence for M_j', $1/10 < R_{ij} < 1/3$ as 'moderate evidence for M_j', and $1/3 < R_{ij} < 1$ as 'weak evidence for M_j'. Symmetrically, $R_{ij} > 10$ depicts 'strong evidence for M_i', $3 < R_{ij} < 10$ 'moderate evidence for M_i', and $1 < R_{ij} < 3$ 'weak evidence for M_i' (after: Wasserman, 1997, p. 7).

[2] Another way to compare models is to apply *information criteria* (cf. Congdon, 2003, pp. 32–33). The Akaike (1973; *idem*) Information Criterion (AIC), suggests selecting a model with the smallest value of $AIC = -2\ln L(\theta\,|x\,) + 2p$, where $L(\theta\,|x)$ is the likelihood of θ given x, and p is the number of parameters. An extension, the Bayesian Information Criterion (BIC; Schwarz, 1978; *idem*) strongly penalises models with more parameters, considering also the sample size n ($BIC = -2\ln L(\theta\,|x\,) + p\ln n$). Defining *deviance* as $D(\theta\,|x\,) = -2\ln L(\theta\,|x\,)$, the Deviance Information Criterion (DIC; Spiegelhalter, Best, Carlin, & van der Linde, 2002) implies minimising $DIC = D^*(\theta\,|x\,) + p_D$. Here, D^* is the mean posterior deviance of θ given x, and p_D is the 'effective number of parameters', equal to D^* minus the deviance of the posterior means of θ estimates (Spiegelhalter et al., 2002, pp. 587, 603). Among these criteria, the BIC has an asymptotic relationship with the Bayes factors: for models M_i and $BIC(M_i) - BIC(M_j) \approx -2\ln(BF_{ji})$ (Raftery, 1995, p. 134).

In model selection problems it is often assumed that all models have equal prior probabilities, especially when there are no arguments for the opposite. Alternatively, it may be argued that the formal selection criteria should favour simpler (more straightforward) explanations of the phenomena under study, according to *Occam's razor* principle.[3] The number of the parameters in the i-th model, l_i, can be taken as the measure of complexity, and the prior probabilities can be set in such a way that $p(M_i)$ is proportional to 2^{-l_i} (Osiewalski, 2001, p. 21). Naturally, there many other ways of assuming the priors on the basis of the expert judgement, which for some reasons may favour a particular model or model class.

In general, it is worth noting that the formal model selection technique satisfies the definition of rule-based forecasting, defined in Section 4.4 after Armstrong and Green (2007). The Bayesian methodology presented above acts as an expert system, selecting the model for extrapolating the time series, where the experts can input their judgement through prior probabilities over the model space. In this way, the proposed framework combines the 'good forecasts' postulates of Keyfitz (1982) mentioned in Section 4.4 with the mechanism of choosing the best model out of a presented variety, rather than relying on a single one, which may not be appropriate for all potential applications.

5.1.2 Bayesian Forecast Averaging (Inference Pooling)

With respect to demographic forecasting, Ahlburg (1995) and Smith (1997) noted that there is no firm evidence as to whether simple models perform better (or worse) than more complex ones. They also criticised the quest for the single 'best' forecasting model and suggested that the accuracy of the outcome can be improved by combining various forecasts.[4] A similar conclusion was reached in the series of 'M-Competitions' of forecasting methods in terms of their *ex-post* predictive performance (Makridakis & Hibon, 2000, see Section 4.4).

In the traditional, sampling-theory forecasting, the idea of averaging forecasts dates back to the seminal paper of Bates and Granger (1969) (after: Palm & Zellner, 1992, p. 687). In the Bayesian approach, this possibility has been explored within the framework of *inference pooling*, currently known as *Bayesian model averaging*, which allows for merging the features of various predictive models in order to account for the uncertainty of model specification (Hoeting et al., 1999). Within the Bayesian context, the averaging problem is closely related to model selection, sketched in the previous subsection.

---

[3]According to the Merriam-Webster Online Dictionary, 'Occam's razor' [or 'Ockham's razor'] is 'a scientific and philosophic rule that entities should not be multiplied unnecessarily, which is interpreted as requiring that the simplest of competing theories be preferred to the more complex, or that explanations of unknown phenomena be sought first in terms of known quantities' (www.m-w.com, accessed on 31 March 2006). A thorough discussion of the 'Occam's razor' notion in the Bayesian context is offered for example by [W. H.] Jeffreys and Berger (1992).

[4]For a discussion of various possibilities of acknowledging the model error in demographic forecasting, see Alho and Spencer (2005, pp. 240–242).

In migration studies, averaging of forecasts has been applied only in a handful of cases up to date. Harker (1986) proposed the application of the *analytic hierarchy process* (AHP), as an alternative way to combine expert judgement with quantitative forecasts. In this approach, subjective weights are assigned to a set of alternatives (objects), on the basis of a ratio scale of pair-wise preferences. The alternatives themselves can be composed of further elements, which constitute subsequent levels of an analytical hierarchy. Formally, the proposed model has the form (Harker, 1986, pp. 71–72):

$$\mathbf{T} = x^D\,\mathbf{T^D} + x^E\,\mathbf{T^E} + x^{QOL}\,\mathbf{T}^{\mathrm{QOL}}, \tag{5.3}$$

whereby $\mathbf{T}$ is the forecasted matrix of interregional migration flows, $\mathbf{T^D}, \mathbf{T^E}$ and $\mathbf{T^{QOL}}$ denote forecasts made based on the analysis of distance, economic factors, and the quality of life (QOL), while x^D, x^E and x^{QOL} are respective weights summing up to unity ($x^D + x^E + x^{QOL} = 1$).

Although from the subjectivist viewpoint the AHP approach looks promising, Willekens (1994, p. 35) noted that the idea has been criticised for incoherence in determination of weights on various hierarchy levels. To the knowledge of the author, the AHP does not seem to be used in migration forecasting practice after Harker (1986). Instead, more straightforward weighted averaging of various frequentist model-based forecasts of net international migration flows has been recently applied for net migration in the 15 countries of the 'old' European Union in the population predictions of the Eurostat (2005), based on the methodology of Lanzieri (2004).

Following the notation used in (2.4) in Chapter 2, let $\mathbf{x^P}$ denote the vector of future values of the variable(s) under study. Then, the conditional density of the averaged predictive distribution of $\mathbf{x^P}$ over the model space $\mathbf{M}$, given the data $\mathbf{x}$, denoted by $\bar{p}(\mathbf{x^P}\,|\mathbf{x})$, can be calculated as follows (Hoeting et al., 1999, p. 383; Osiewalski, 2001, p. 24):

$$\bar{p}(\mathbf{x^P}\,|\mathbf{x}) = \sum\nolimits_{i \in \mathbf{M}} p(M_i \mid \mathbf{x}) \cdot p(\mathbf{x^P}\,|\mathbf{x}, M_i). \tag{5.4}$$

The element $p(\mathbf{x^P}\,|\mathbf{x},\ M_i)$ corresponds to the predictive distribution (2.4) in the i-th model, while $p(M_i\,|\mathbf{x})$ is the posterior probability of the i-th model, defined in (5.1).

The pre-selection of models for the purpose of their averaging can be also done on the basis of posterior odds, according to 'Occam's razor' principle. In this context, Hoeting et al. (1999, p. 385) elaborated the idea of the threshold values O_L and O_R for the posterior odds ratio, R_{ij}, as defined in the previous subsection. They labelled the interval $[O_L,\ O_R]$ as *Occam's window* and suggested that models that have strong evidence against them (such models i, for which $R_{ij} < O_L$) should be disregarded in the pre-selection process.

In order to apply Occam's principle, the values of O_L and O_R should be set in favour of simpler models. For example, if the i-th model has a simpler structure (less parameters) than the j-th model, then $O_L = 1/20$ and $O_R = 1$ strongly support the i-th model in the pre-selection procedure, as advocated by Madigan and Raftery (1994) (after: Hoeting et al., 1999, p. 385). Alternatively, Raftery et al. (1996) proposed setting $Q_L = 1/20$ and $O_R = 20$ (*idem*). They found that these values lead to an improved performance of the averaged predictions made on the basis of the ultimately selected models, yet without an explicit reference to Occam's principle. Needless to say, such values of O_L and O_R are to a large extent arbitrary and have been selected for the sake of analogy with the conventional 0.05 significance level in the traditional, frequentist statistics.

In general, most authors agree that under certain conditions, including the unbiasedness of forecasts, the averaged ones have several advantages: minimise the total error rate (sum of probabilities of *Type I* and *Type II errors*[5]), minimise the mean squared error of estimation, and offer better out-of-sample forecasting performance than competing methods (Raftery & Zheng, 2003, pp. 931–932). Nevertheless, averaging requires caution: Palm and Zellner (1992) argue that combining a good model with a biased one yields worse results than the good model alone. In such cases, the 'strength borrowing' between various forecasts can provide an illusion of greater accuracy. A meachanistic averaging can thus mask the real differences between the forecasts, ultimately leading to inappropriate conclusions.

5.1.3 Empirical Application: Specification of Forecasting Models

Let $\mu_{i-j}(t) = \ln(m_{i-j}(t))$ denote logarithms of emigration rates from country i to j per 1,000 population of the country of origin, in year t, where $i, j \in \{\text{CH, DE, IT, PL}\}$ for Switzerland, Germany, Italy, and Poland, respectively. The logarithmic transformation has been used, as emigration rates are by definition positive numbers.[6]

For each of the variables $\mu_{i-j}(t)$, let the following five simple types of models be defined, belonging to the wide class of the autoregressive integrated moving average (ARIMA) models, systematised by Box and Jenkins (1976):

$$M_1 : \mu_{i-j}(t) = c_1 + \varepsilon_1(t) \quad \text{- random deviations from constant} \tag{5.5a}$$

$$M_2 : \mu_{i-j}(t) = c_2 + \mu_{i-j}(t-1) + \varepsilon_2(t) \text{ - random walk with drift} \tag{5.5b}$$

$$M_3 : \mu_{i-j}(t) = c_3 + \varphi_3\mu_{i-j}(t-1) + \varepsilon_3(t); \varphi_3 \neq 0,\ \varphi_3 \neq 1 \text{ - autoregressive process AR(1)} \tag{5.5c}$$

[5] *Type I error* in statistics is related to rejecting a true hypothesis and *Type II error* to accepting a false one.

[6] Alternatively to logarithms, other transformations can be used, such as logit or logistic (e.g. Willekens, 2008), in order to include upper bounds for the modelled variables. However, as migration rates per 1,000 population are much closer to the lower bound than to the upper one (in the current examples by three or four orders of magnitude, cf. Annex A), log-transformations were universally applied for the sake of simplicity. Besides, as rates are not probabilities, but rather intensity measures, they technically need not be bound from above.

$$M_4 : \mu_{i-j}(t) = c_4 + \varepsilon_4(t) - \theta_4\varepsilon_4(t-1); \theta_4 \neq 0 \text{ - moving average process MA(1)} \quad (5.5d)$$

$$M_5 : \mu_{i-j}(t) = c_5 + \varphi_5\mu_{i-j}(t-1) + \varepsilon_5(t) - \theta_5\varepsilon_5(t-1); \varphi_5 \neq 0, \theta_5 \neq 0 \text{ - ARMA(1, 1)} \quad (5.5e)$$

In all models $M_1, \ldots, M_5$, $\varepsilon_i(t)$ denotes the Gaussian white noise, $\varepsilon_i(t) \sim$ iid N(0, σ_i^2). Normality of the random component, instead of applying the Poisson distribution with offset for modelling migration rates, as suggested for example by Congdon (2000, 2001b) and Willekens (2008), has been assumed for two reasons. Firstly, the numbers of migrants under study are sufficiently large to allow for a very close Normal approximation of the Poisson distribution. Secondly, by assuming Normality, the problems with possible over- or underdispersion of a Poisson regression are avoided.

The model class defined in (5.5), denoted $\mathbf{M_1}$, covers all autoregressive moving average ARMA(1,1) processes, in general terms specified as $\mu_{i-j}(t) = c + \varphi\mu_{i-j}(t-1) + \varepsilon(t) + \theta\varepsilon(t-1)$. The mathematical formulation of particular models $M_1, \ldots, M_5$ depends on the values of ϕ and θ, as shown in Table 5.1. Such classification satisfies the conditions to apply the Bayesian model selection and averaging techniques, as $M_1, \ldots, M_5$ are mutually exclusive and add up to $\mathbf{M_1}$.

As the analysis is primarily designed to illustrate the methods rather than to obtain the best-possible migration predictions, which should possibly consider a wider class of models and include other explanatory variables, the forecasting models are limited to simple stochastic processes. Nevertheless, some of these models have already been used in forecasting international migration on the basis of time series within the frequentist approach. The examples, quoted after Keilman et al. (2001, p. 62), include a study of de Beer (1997), who modelled migration to and from the Netherlands by an AR(1) process (M_3), and net migration by an MA(1) process (M_4). In turn, Alho (1998) predicted migration for Finland using an ARIMA(0,1,1) model, combining a random walk with a moving average component. For Norway, Keilman et al. (2001) themselves modelled immigration by means of an ARMA(1,1) process (M_5), and emigration using a random walk (M_2).

As the models M_1 to M_4 can be obtained from M_5 by setting appropriate constraints on the parameters ϕ and θ, a possible simpler alternative to the Bayesian model selection and averaging procedures would be to make predictions on the basis

Table 5.1 Specification of forecasting models in the ARMA(1,1) class

	Parameter θ	
Parameter ϕ	$\theta = 0$	$\theta \neq 0$
$\varphi = 0$	M_1	M_4
$\varphi = 1$	M_2	M_5
$\varphi \neq 0 \wedge \varphi \neq 1$	M_3	M_5

Source: own elaboration

of the general ARMA(1,1) model. There are, however, two reasons for not limiting oneself to such an approach. Firstly, the character of various models is different: while M_1 and M_4 are always stationary, M_2 is non-stationary, and in M_3 or M_5 the stationarity depends on the value of ϕ. Furthermore, in various models the number of parameters subject to estimation differs: from two in M_1 and M_2, through three in M_3 and M_4, to four in M_5. Especially in the case of short time series, as in the current example, simpler explanations of the phenomena should be preferred according to Occam's principle, because the error of estimation of each additional parameter contributes to the ultimate error of forecast. The diversification proposed before, in (5.5a), (5.5b), (5.5c), (5.5d), (5.5e), and Table 5.1, allows for the examination of these two aspects together: the stationarity of the process, and the model complexity.

A similar rationale lies behind limiting the analysis to the proposed class of models **M**. As argued before, this class already covers a variety of potentially interesting combinations with respect to the model complexity and (non-)stationarity. Moreover, the data series available for the current study are too short to allow for meaningful inference in the case of models with more structural parameters (e.g., more complex ARIMA models), even within the Bayesian approach. In general, autoregressive models reflect some inertia of migratory phenomena (compare for example the notion of migrant networks and 'cumulative causation' theories in Chapter 3), as well as delays of reactions to changing circumstances and diffusion of information (a self-perpetuating mechanism). Additionally, ARMA processes can be seen as incomplete variants of a more general class of Koyck's 'distributed lag' models containing additional explanatory variables (x), the latter entering into the model with a special infinite distribution of lags (cf. Zellner, 1971, pp. 200–201).[7]

The empirical illustration of the presented discussion is given in Section 5.2, devoted to the estimation of posterior probabilities of both parameters and models, as well as to presenting forecasts from the individual models, as well as the averaged ones.

5.1.4 Computations: The Carlin–Chib Algorithm

In the model selection problems described before, Bayes' theorem is applied twice: to update the prior distributions of the parameters $\boldsymbol{\theta}_i$ for all models M_i given the

---

[7]Consider the following model: $y_t = c + \alpha(x_t + \varphi x_{t-1} + \varphi^2 x_{t-2} + \varphi^3 x_{t-3} + \ldots) + \varepsilon_t$, which can be alternatively transformed to: $y_t = c' + \alpha x_t + \varphi y_{t-1} + \varepsilon_t - \varphi \varepsilon_{t-1}$, thus, to a special case of ARMA(1,1) with an additional explanatory variable x. In such way, Koyck's transformation allows by-passing the problem of formal selection of the number of lags in the predictors, and thus to avoid a discussion on the time-span between the socio-economic impulse and migratory response, which remains beyond the scope of the current study. The difference of Koyck's model towards general ARMA proposed in (5.5) is that the latter do not consider an instantaneous influence of explanatory variables x, and allow for values of the moving average parameter θ that may differ from ϕ. The impact of such variables is analysed in an alternative fashion in the VAR(1) models presented in Chapter 6, where, however, such additional variables x are considered with a time-lag of one period.

data $\mathbf{x}$ according to (2.2), and at the same time to obtain the posterior probabilities of particular models following (5.1). In order to accommodate it within the framework of the Gibbs sampling procedure used for numerical computations, the *Model Choice via Markov Chain Monte Carlo* (MC^3) algorithm of Carlin and Chib (1995) has been applied, implemented within the WinBUGS 1.4 software (see Annex B for code). The method consists of an iterative sampling from full conditional distributions for model-specific parameters $\boldsymbol{\theta}_i$ and the model index μ, repeated sequentially until convergence to an ultimate solution is reached.

The full conditional distributions are given by Carlin and Chib (1995, pp. 475–477):

$$\begin{cases} p(\boldsymbol{\theta}_i \mid \boldsymbol{\theta}_{j\neq i}, \mu, \mathbf{x}) \propto \begin{cases} p(\mathbf{x} \mid \boldsymbol{\theta}_i, \mu = i) \cdot p(\boldsymbol{\theta}_i \mid \mu = i) & \text{for } \mu = i \\ p(\boldsymbol{\theta}_i \mid \mu \neq i) & \text{for } \mu \neq i \end{cases} \\ p(\mu = i \mid \boldsymbol{\theta}, \mathbf{x}) = \dfrac{p(\mathbf{x} \mid \boldsymbol{\theta}_i, \mu = i) \cdot p(M_i) \cdot \prod_{j \in \mathbf{M}} p(\boldsymbol{\theta}_j \mid \mu = i)}{\sum_{k \in \mathbf{M}} [p(\mathbf{x} \mid \boldsymbol{\theta}_k, \mu = k) \cdot p(M_k) \cdot \prod_{j \in \mathbf{M}} p(\boldsymbol{\theta}_j \mid \mu = k)]} \end{cases} \tag{5.6}$$

The model parameters $\boldsymbol{\theta}_i$ are therefore either sampled using a standard Gibbs procedure whenever $\mu = i$ or otherwise drawn from pre-defined linking densities (‘*pseudo-priors*’), $p(\boldsymbol{\theta}_i \mid \mu \neq i)$. The pseudo-priors can be, for example, replaced by the preliminary estimates of model-specific posteriors $p(\boldsymbol{\theta}_i \mid \mu = i, \mathbf{x})$ (*idem*), which suggestion has been followed in the examples presented in the current study. As in all Markov Chain Monte Carlo methods, the first S iterations of (5.6), until convergence, are discarded (the ‘burn-in’ phase), while further M are used to estimate the posterior distributions of the parameters and the model index.

In the current study, the presented MCMC algorithms have been applied to estimate the posterior distributions of parameters, perform formal selection and to perform HPD tests for the VAR models within the ‘from general to specific’ framework (Chapter 6). The listing of examples of the WinBUGS code used is provided in Annex B. On the basis of obtained posterior outcomes, the forecasts have been subsequently computed, and are presented in the next section.

5.2 Simple Time Series Forecasts: Individual and Averaged

5.2.1 Estimation of the Models and Calculation of Their Posterior Probabilities

In the case of simple stochastic processes belonging to the ARMA(1,1) model class ($\mathbf{M_1}$); that is, models $M_1, \ldots, M_5$ defined in (5.5a), (5.5b), (5.5c), (5.5d) and (5.5e), the following prior distributions, conditional on the model M_i, have been assumed. Let the constants c_i follow a rather non-informative,[8] diffuse Normal prior

[8] For a further discussion on informative versus hardly informative prior distributions, see Chapter 8.

Table 5.2 Categories of Gamma priors for the precision parameters in models $M_1 - M_5$

Migration flow	Sample N (period)	Prior precision class	Shape: r	Scale: μ	Mean: r/μ	Variance: r/μ^2
From Italy to Germany	20 (1985–2004)	Low	2.25	0.75	3.0	4.0
From Germany to Italy	14 (1991–2004)	Medium	4.00	0.40	10.0	25.0
From Poland to Germany	14 (1991–2004)	Very low	0.25	0.25	1.0	4.0
From Germany to Poland	14 (1991–2004)	Medium	4.00	0.40	10.0	25.0
From Switzerland to Germany	20 (1985–2004)	High	4.00	0.20	20.0	100.0
From Germany to Switzerland	14 (1991–2004)	High	4.00	0.20	20.0	100.0

Source: own elaboration

distribution N(0, 100^2). Further, it is assumed that the priors for the autoregression parameters ϕ_i are also Normal, yet more informative, following N(0.5, 1^2). This reflects a belief that the autoregressive parts of appropriate models are likely stationary. For θ_i, Normal priors N(0.5, 1^2) are taken as well, which indicates a similar belief with respect to invertibility[9] of the moving average components of the respective models.

Finally, precision parameters τ_i of the error terms $\varepsilon_i(t)$, being reciprocals of the variances, $\tau_i = 1/\sigma_i^2$, are assumed to follow Gamma distributions[10] $\Gamma(r, \mu)$ with the hyper-parameters[11] of shape (r) and scale (μ) varying so as to reflect beliefs in different precision of the logarithms of migration rates,[12] summarised in Table 5.2. In general, all such distributions reflect the prior assumptions on a relatively high uncertainty of the random term, which seem to be natural in the study of uncertain

[9] A moving average process MA(q) is called *invertible* if it can be represented by an infinite-order autoregressive process, AR(∞) (Granger & Morris, 1976, p. 247). Similarly, a stationary AR(p) process can be represented by an infinite-order moving average, MA(∞) (*idem*; compare also Koyck's transformation in Section 5.1).

[10] Parameterisation of the Gamma distribution follows the one used in the WinBUGS manual: for $x \sim \Gamma(r, \mu)$ the density is $p(x|r, \mu) = \mu^r \cdot x^{r-1} \cdot \exp(-\mu x)/\Gamma(r)$, where $x > 0$ and $\Gamma(\cdot)$ is Euler's Gamma function. The same notation is adopted e.g. in DeGroot (1970/1981) and Osiewalski (2001).

[11] *Hyper-parameters* are the parameters of prior distributions.

[12] Precisely, τ_i for log-transformed emigration rates concerns precision for the orders of magnitude for crude $m_{i\text{-}j}$, conditional on the systematic part of the model and the model itself, and exact to transforming natural logarithms into decimal ones. In such a way, unity on the scale of $\mu_{i\text{-}j}$ corresponds to $\log_{10}(e) = 0.43$ orders of magnitude of the raw emigration rates $m_{i\text{-}j}$, and thus one unit of τ_i to $0.43^{-0.5} = 1.52$ orders of magnitude of $m_{i\text{-}j}$.

migratory phenomena. A simple analysis of sensitivity of the forecasts to the selection of some alternative prior distributions for particular parameters is offered in Chapter 8.

Following the discussion in Section 5.1, two alternative types of prior distributions for the models $M_1, \ldots, M_5$ have been set: the uniform one, with $p(M_i) = 0.2$, and 'Occam's razor', with $p(M_i)$ proportional to $2^{-l_i}, l_i$ denoting the number of parameters of the i-th model. The estimation of model parameters and posterior probabilities has been performed using the Carlin and Chib (1995) model selection algorithm presented in Section 5.1, which was implemented within the MCMC Gibbs sampling framework in the WinBUGS 1.4 software (Spiegelhalter et al., 2003).[13] For computational simplicity, no constraints for ϕ_i and θ_i have been specifically set for models M_3, M_4, and M_5, as the values to be excluded (zero or one, respectively) have the probability of occurrence equal zero. The code used for the computations is listed in Annex B.[14]

Hence, for each of the series of log-transformed emigration rates under study, $\mu_{i-j}(t)$, five models $M_1, \ldots, M_5$ have been estimated. The posterior distributions of the parameters of each model have been calculated on the basis of 250,000 iterations of the MCMC algorithm, obtained after discarding the preceding iterations from the 'burn-in' phase of the procedure. After visual checks of convergence of the simulations, following the suggestions of Spiegelhalter et al. (2003), the length of the 'burn-in' phase has been universally established as 100,000. The checks were purely heuristic: stabilisation of the running quantiles from the posterior distributions was taken as a sign of convergence. The simulation has been repeated three times: firstly to obtain the posterior estimates of the parameters of particular models treated separately; secondly to estimate the posterior probabilities of the models using the Carlin and Chib (1995) algorithm, treating model-specific posteriors as pseudo-priors for other models; and finally to calculate an averaged forecast on the basis of the information obtained.

---

[13] As remarked by Dellaportas, Forster, and Ntzoufras (2002), the main drawback of the Carlin and Chib (1995) procedure is the necessity to specify many pseudo-priors during every iteration of the MCMC algorithm, making the method computationally demanding. Nevertheless, the Carlin-Chib approach has been followed in the current study, as all models under study are in fact particular examples of the ARMA(1,1) process, and due to the relatively simple form of the full conditional distributions of particular parameters. As an alternative for more complex tasks concerning Bayesian model selection and forecast averaging problems, the Reversible Jump MCMC algorithm (Green, 1995) can be used instead.

[14] In programming the Carlin-Chib method, the *Pines* example from an on-line WinBUGS manual has been followed, available at: www.mrc-bsu.cam.ac.uk/bugs/documentation/exampVol2/node20.html (accessed on 1 April 2007). The WinBUGS code for univariate AR(p), MA(q) and ARMA(p,q) models, which has been also helpful for the current study, is provided by Congdon (2003, pp. 172–189, Programmes 5.1–5.3). As to the moving average component, Congdon (2003, p. 187) argues that due to WinBUGS being based on the centred form of the Normal distribution, there is a need for an additional error term, $u(t) \sim$ iid N(0, σ_u^2), which for the MA(1) processes would yield the following model: $\mu_{i-j}(t) = c + \varepsilon(t) - \theta\varepsilon(t-1) + u(t)$. However, the current study applies a standard moving average model without the $u(t)$ term, which nevertheless uses the centred Normal form (see code in Annex B for details).

Table 5.3 Prior and posterior probabilities for models M_i: Uniform and 'Occam's razor' $p(M_i)$

Model (M_i)	M_1	M_2	M_3	M_4	M_5	Sum
Prior probabilities						
(A) Uniform prior, $p(M_i) \propto const.$	0.200	0.200	0.200	0.200	0.200	1
(B) 'Occam's razor' prior, $p(M_i) \propto 2 \wedge (-l_i)$	0.308	0.308	0.154	0.154	0.077	1
Migration from Italy to Germany						
$p(M_i \mid x)$, prior (A)	0.007	**0.680**	0.186	0.021	0.106	1
$p(M_i \mid x)$, prior (B)	0.008	**0.832**	0.115	0.013	0.032	1
Migration from Germany to Italy						
$p(M_i \mid x)$, prior (A)	0.271	**0.277**	0.219	0.128	0.105	1
$p(M_i \mid x)$, prior (B)	0.363	**0.372**	0.145	0.086	0.034	1
Migration from Poland to Germany						
$p(M_i \mid x)$, prior (A)	**0.316**	0.218	0.211	0.160	0.096	1
$p(M_i \mid x)$, prior (B)	**0.423**	0.293	0.143	0.108	0.032	1
Migration from Germany to Poland						
$p(M_i \mid x)$, prior (A)	0.171	**0.368**	0.240	0.119	0.103	1
$p(M_i \mid x)$, prior (B)	0.231	**0.494**	0.161	0.080	0.035	1
Migration from Switzerland to Germany						
$p(M_i \mid x)$, prior (A)	0.160	**0.414**	0.216	0.101	0.109	1
$p(M_i \mid x)$, prior (B)	0.210	**0.546**	0.141	0.067	0.036	1
Migration from Germany to Switzerland						
$p(M_i \mid x)$, prior (A)	0.000	**0.764**	0.162	0.001	0.074	1
$p(M_i \mid x)$, prior (B)	0.000	**0.884**	0.094	0.001	0.021	1

Notes: Boldface indicates highest posterior probabilities of models for particular migratory flows; underlined figures denote other cases with $p(M_i \mid x) > p(M_i)$. Some figures in the table may not exactly add up to 1 due to rounding.
Source: own elaboration in WinBUGS

Both prior distributions on $\mathbf{M_1}$ and respective posterior probabilities $p(M_i \mid x)$ calculated in WinBUGS using the Carlin and Chib (1995) MC^3 algorithm, are shown in Table 5.3. The posterior odds ratios R_{ij} defined in (5.2), which can be calculated on the basis of probabilities shown in Table 5.3, yield that for all flows both types of priors lead to the selection of the same model: for μ_{PL-DE} (and perhaps also μ_{DE-IT}, with almost as high posterior probabilities of M_1 as those of M_2) – random deviations from a constant (M_1), for other flows – random walks with drift (M_2). The simplicity of both processes is the reason why the posterior probabilities under the 'Occam's razor' prior are visibly higher than under the uniform one for all flows under study. In such cases, where the posterior probabilities are similar under equiprobable $p(M_i)$, Occam's prior strongly penalises models characterised by more parameters. The 'special case' of migration from Poland to Germany can either suggest a different character of this particular flow as compared with the other ones, or (more likely) reflect a peculiar characteristic of a small sample.

Applying formal criteria for R_{ij}; for example setting $O_L = 1/20$ and $O_R = 20$ after Raftery et al. (1996), can help identify models with the relatively biggest disagreement with data. For migration from Germany to Switzerland (μ_{DE-CH}), the ratios that do not fall into the $[O_L, O_R]$ window include all R_{ij} involving models M_1 and M_4. Almost the same applies for models of migration from Italy to Germany (μ_{IT-DE}), indicating the implausibility of using the constant model M_1, as well as the moving average M_4, for the three above-mentioned cases. The posterior probabilities of all ARMA(1,1) processes M_5 are also generally very low, especially under the 'Occam's razor' model prior, casting doubts as to the relevance of such models for forecasting migration flows under study. On the other hand, next to the selected highest-posterior probability models, in several cases the AR(1) processes also receive some data support indicated by $p(M_i|x) > p(M_i)$, as shown in Table 5.3.

Results of the estimation of particular parameters of models $M_1 - M_5$ are presented in Table 5.4, summarising the posterior distributions of parameters with their median values and the quantiles of rank 0.025 and 0.975, delimiting the 95% credible intervals. All estimates have been obtained from the first-round MCMC simulations. As the univariate posterior distributions of c_i and ϕ_i seem symmetric and unimodal, the intervals derived on the basis of respective quantiles should have the highest posterior density (HPD) property defined in (2.3), although the HPD regions as such are not available from the WinBUGS software.

As shown in Table 5.4, the credible intervals for c_i, ϕ_i, and θ_i that do not cover zero, which in the frequentist framework would indicate the significance of the estimates, consider a minority rather than a majority of models. This reflects high uncertainty of the processes under study, some of which do not perfectly match the proposed framework based only on simple models from the ARMA(1,1) class. It gives a rationale for the future research to further extend the model space $\mathbf{M_1}$, for example to ARIMA models, both with and without constant terms, as suggested by Alho and Spencer (2005, p. 217), although with a reservation, that such an analysis should be generally based on longer time series. In the small-sample cases presented here, the role of data in the final (posterior) knowledge on particular parameters is already relatively weak, as indicated in Table 5.4, and the role of prior information therefore quite high, which provides an additional argument for performing the analysis within the Bayesian framework.

In addition, for migration from Germany to Italy and Switzerland, the same calculations were performed on a longer data series, covering the period 1985–2004. The aim of this exercise was to identify which effect is stronger: whether the truncation of the time series increases the predictive uncertainty due to the short-sample effect, or decreases the uncertainty because of the enhancement of the statistical properties of the sample. The latter case would provide an argument for treating the intensity of migration from the united Germany after 1991 distinctly from the West German emigration beforehand.

With respect to particular parameters, the constants c_i proved to be much less diffuse than it has been assumed a priori, which is not surprising given the hardly informative character of the prior distributions. Next to Table 5.4, comparisons of prior and posterior distributions of the autoregressive coefficients ϕ_3 for the AR(1) models and the moving average parameters θ_4 for the MA(1) processes are presented in Fig. 5.1. The posterior estimates of the former (ϕ_3) appear to be to some degree consistent with the priors, at least with respect to their central location, in

Table 5.4 Summaries of posterior distributions of parameters of the forecasting models: median, 2.5 and 97.5% quantiles (estimated by MCMC)

Flow	Constant c_i			Autoregression parameter φ_i			Moving average parameter θ_i			Precision $\tau_i = 1/\sigma_i^2$		
Model M_i	2.5%	Median	97.5%	2.5%	Median	97.5%	2.5%	Median	97.5%	2.5%	Median	97.5%
Migration from Italy to Germany												
M_1	**–0.59**	**–0.42**	**–0.26**	-	-	-	-	-	-	4.35	8.38	14.36
M_2	–0.17	–0.04	0.10	-	-	-	-	-	-	6.56	12.65	21.67
M_3	–0.31	–0.03	0.25	**0.38**	**1.02**	**1.64**	-	-	-	6.21	12.14	20.95
M_4	**–0.67**	**–0.43**	**–0.20**	-	-	-	**–1.07**	**–0.66**	**–0.07**	5.41	10.66	18.58
M_5	–0.44	–0.05	0.28	**0.12**	**0.96**	**1.68**	–0.95	–0.38	0.45	6.33	12.35	21.44
Migration from Germany to Italy												
M_1	**–0.92**	**–0.80**	**–0.68**	-	-	-	-	-	-	11.37	22.97	40.45
M_2	–0.13	–0.01	0.11	-	-	-	-	-	-	11.43	23.03	40.69
M_3	–1.48	–0.40	0.69	–0.85	0.51	1.87	-	-	-	11.21	22.70	40.25
M_4	**–0.95**	**–0.80**	**–0.64**	-	-	-	–0.98	–0.24	0.68	11.42	23.02	40.76
M_5	–1.57	–0.41	0.71	–0.96	0.49	1.89	–0.97	–0.21	0.73	11.25	22.73	40.35
Migration from Poland to Germany												
M_1	**0.79**	**0.94**	**1.10**	-	-	-	-	-	-	5.43	13.69	27.71
M_2	–0.16	0.00	0.16	-	-	-	-	-	-	5.13	12.91	26.23
M_3	–0.28	0.53	1.32	–0.38	0.44	1.27	-	-	-	5.37	13.83	28.63
M_4	**0.76**	**0.95**	**1.19**	-	-	-	–1.08	–0.28	0.67	5.52	13.97	28.68
M_5	–0.26	0.58	1.51	–0.54	0.39	1.26	–1.05	–0.18	0.72	5.32	13.77	28.61

Table 5.4 (continued)

Flow	Constant c_i			Autoregression parameter φ_i			Moving average parameter θ_i			Precision $\tau_i = 1/\sigma_i^2$		
Model M_i	2.5%	Median	97.5%	2.5%	Median	97.5%	2.5%	Median	97.5%	2.5%	Median	97.5%
Migration from Germany to Poland												
M_1	–0.14	0.00	0.13	-	-	-	-	-	-	8.45	17.03	29.99
M_2	–0.14	–0.01	0.12	-	-	-	-	-	-	9.10	18.37	32.47
M_3	–0.14	–0.01	0.12	–0.09	0.59	1.26	-	-	-	9.17	18.78	33.55
M_4	–0.18	0.00	0.20	-	-	-	–1.09	–0.44	0.42	8.86	18.09	32.35
M_5	–0.17	–0.01	0.17	–0.36	0.50	1.21	–0.96	–0.16	0.76	9.12	18.62	33.26
Migration from Switzerland to Germany												
M_1	**0.05**	**0.12**	**0.18**	-	-	-	-	-	-	29.43	53.87	89.12
M_2	–0.05	0.01	0.08	-	-	-	-	-	-	31.66	58.00	95.91
M_3	–0.05	0.05	0.15	–0.16	0.66	1.47	-	-	-	31.00	57.22	95.17
M_4	**0.03**	**0.12**	**0.21**	-	-	-	–1.00	–0.43	0.34	30.43	56.19	93.46
M_5	–0.07	0.05	0.18	–0.36	0.62	1.53	–0.96	–0.30	0.59	31.12	57.55	95.86
Migration from Germany to Switzerland												
M_1	**–2.16**	**–2.03**	**–1.89**	-	-	-	-	-	-	8.83	17.85	31.44
M_2	–0.03	0.05	0.14	-	-	-	-	-	-	22.27	44.88	79.31
M_3	–0.62	0.27	1.14	**0.68**	**1.11**	**1.52**	-	-	-	21.16	43.53	77.86
M_4	**–2.20**	**–2.01**	**–1.83**	-	-	-	**–1.29**	**–0.78**	**–0.24**	13.48	28.22	51.70
M_5	–0.91	0.26	1.29	**0.53**	**1.10**	**1.59**	–0.99	–0.23	0.67	21.02	43.03	77.10

Notes: Boldface denotes 95% credible intervals for the estimates of c_i, φ_i, and θ_i, which do not cover zero.
Source: own elaboration in WinBUGS

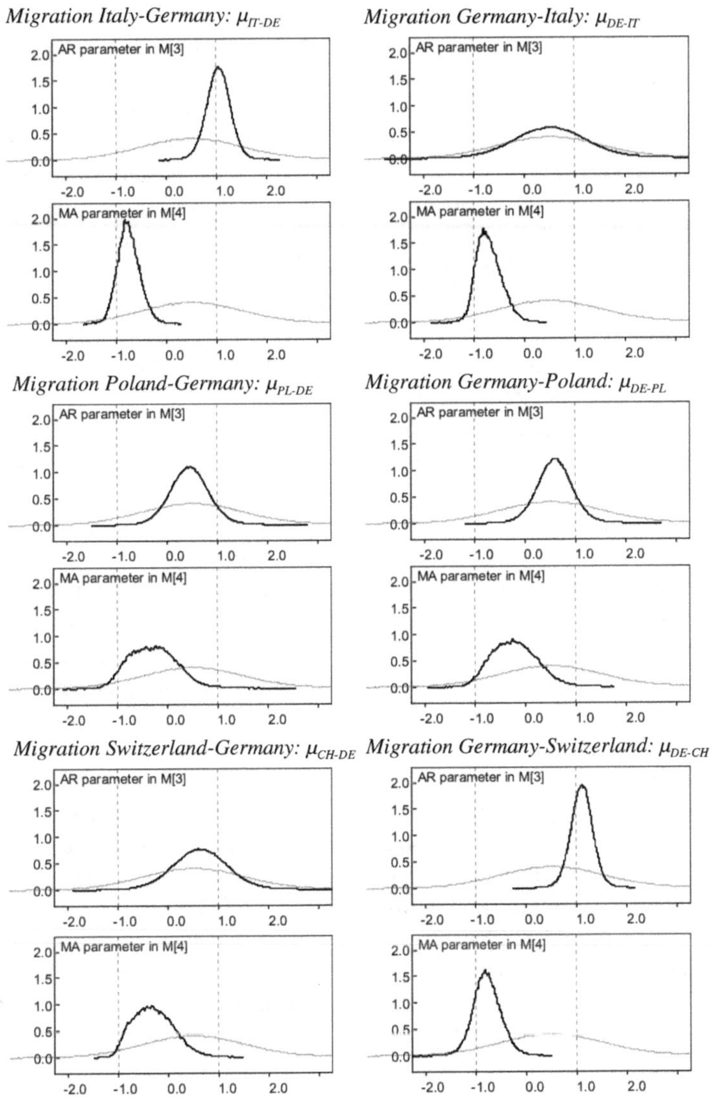

Fig. 5.1 Prior and posterior distributions of parameters: φ in M_3 (AR) and θ in M_4 (MA).
Notes: Grey lines depict prior, while black ones – posterior distributions. Dashed lines indicate $(-1, 1)$ interval limits related to stationarity of AR(1) or invertibility of MA(1) processes.
Source: own elaboration in WinBUGS

four AR(1) models: for μ_{DE-IT}, μ_{PL-DE}, μ_{DE-PL}, as well as μ_{CH-DE}. Nonetheless, even in these cases the posterior probability of non-stationarity; that is, of ϕ_3 being outside of the $(-1, 1)$ interval, remains relatively high. For μ_{DE-PL}, $p(|\varphi_3| \geq 1\,|\mathbf{x})$ slightly exceeds 0.1, while for μ_{CH-DE} and μ_{DE-IT} this probability is about 0.25.

For the remaining two models (μ_{IT-DE} and μ_{DE-CH}) the values of ϕ_3 are very likely greater than one, with the respective posterior probabilities higher than 0.5 (see Table 5.4), providing even stronger arguments for non-stationarity. Only for μ_{PL-DE}, $p(|\varphi_3| \geq 1\,|\mathbf{x})$ equals slightly more than 0.05, indicating a rather stationary character of the process, which somehow coincides with the results of the model selection procedure summarised in Table 5.3. In turn, the posterior estimates of the moving average parameters θ_4 are all likely negative and in disagreement with prior beliefs with respect to their central location (Fig. 5.1). However, at the same time they indicate a relatively likely invertibility of the MA(1) processes under study, with $p(\theta_4\,|\geq 1|\,\mathbf{x})$ ranging between about 0.025 for μ_{CH-DE} and μ_{DE-IT}, and less than 0.1 for μ_{DE-CH}. On the other hand, there seems to be little point in discussing the parameter inference in models M_4 in more detail, as in the presented examples such models are anyway hardly probable a posteriori.

In all cases, the posterior means of the precision parameters $\tau_i = 1/\sigma_i^2$ are substantially higher in comparison to the values assumed a priori, as shown in Fig. 5.2 for the most popular model according to the posterior odds criterion; that is, for the random walk with drift (M_2). This indicates that the prior assumptions on a relatively high uncertainty of the random term are more pessimistic than the data indicate. Even small samples of 14 or 20 observations were able to modify the prior beliefs to a certain extent, and give some weight to the data in the ultimate calculations of the posterior distributions of the parameters. These issues are also discussed in more detail in Chapter 8, where the estimation and forecasting is performed under diffuse (hardly informative) priors for precision.

The results for migration from Germany to Italy and Switzerland proved to be sensitive to change in the sample size. This is especially true with respect to model probabilities for μ_{DE-IT}, where for 20 observations (1985–2004) the random walk model M_2 was selected unambiguously, the posterior probability equalling 0.64 under a flat prior and 0.81 under 'Occam's razor'. For both flows, the estimated precision was somewhat higher for the longer sample, which also translated to lower predictive uncertainty.[15] Such an outcome may at first indicate the presence of a small-sample effect rather than improvement of statistical properties of the sample, but as either sample is in fact small, the dominant role of respective prior distributions can to some extent obscure the picture. This issue is further elaborated in Chapter 8.

5.2.2 Predictions Based on the Formally-Selected and Averaged Stochastic Processes

Forecasts for the migration flows under study for the period 2005–2015 based on the models presented above have been calculated using the WinBUGS code listed in Annex B. Summaries of the predictive distributions of respective emigration rates,

[15] Selected numerical results based on 20 observations (1985–2004) are available in Bijak (2008b).

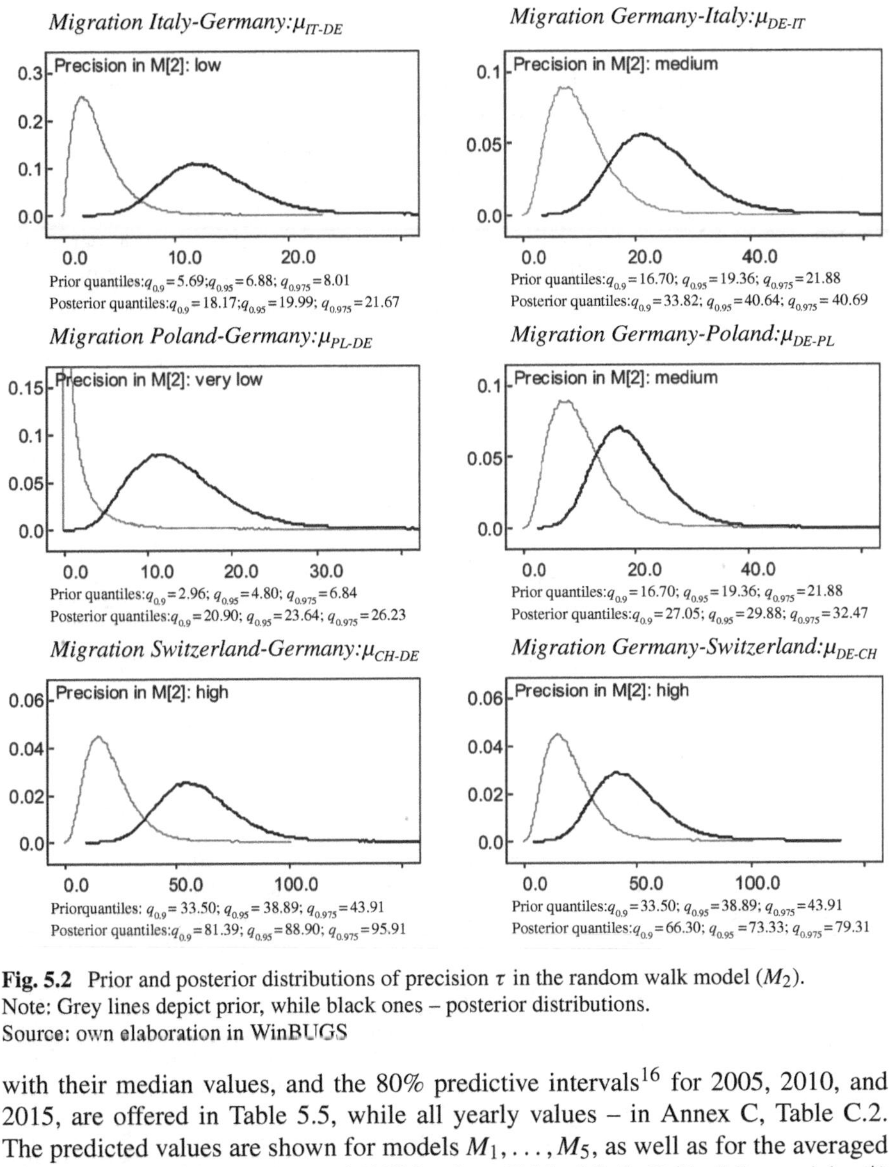

Fig. 5.2 Prior and posterior distributions of precision τ in the random walk model (M_2).
Note: Grey lines depict prior, while black ones – posterior distributions.
Source: own elaboration in WinBUGS

with their median values, and the 80% predictive intervals[16] for 2005, 2010, and 2015, are offered in Table 5.5, while all yearly values – in Annex C, Table C.2. The predicted values are shown for models $M_1, \ldots, M_5$, as well as for the averaged forecasts, applying posterior probabilities from Table 5.3. In Table 5.5, asterisks (*)

[16] In probabilistic population forecasting, 80 per cent predictive intervals are commonly applied, either instead of, or alongside the traditionally-used 95 per cent intervals (Alho, 1998; Keilman et al., 2001; Lutz et al., 2004). The rationale is to avoid unnecessary amplification of the uncertainty assessment, already very high in demographic predictions. Lutz et al. (2004, p. 37) argue that 'the forecast distributions are themselves uncertain at the extremities. The 80 per cent intervals are far more robust to the technicalities in the forecasting methodology than the 95 per cent intervals'. As argued in Chapter 11, such intervals can also provide additional warning to the forecast users, as the probability that the process will fall beyond their limits cannot be neglected.

indicate forecasts based on models with very low posterior probabilities, with $p(M_i|x) < 0.05$ under 'Occam's' prior. They are presented merely for the sake of comparison with the remaining ones, especially as their results in terms of predictions visibly differ from the outcome of the more likely models. Grey shading denotes implausible values, over five times exceeding the within-sample maxima [$\max(m_{IT-DE}) = 0.89$, $\max(m_{DE-IT}) = 0.50$, $\max(m_{PL-DE}) = 3.72$, $\max(m_{DE-PL}) = 1.48$, $\max(m_{CH-DE}) = 1.28$ and $\max(m_{DE-CH}) = 0.20$]. Furthermore, '∞' refers to nearly impossible values (rates of over 1,000 per 1,000 inhabitants). The predictive distributions of the logarithms of respective emigration rates forecasted for 2005, 2010 and 2015 are also depicted in Fig. 5.3.

Table 5.5 Summaries of predictive distributions of emigration rates forecasted for 2005, 2010 and 2015: median, 10 and 90% quantiles (estimated by MCMC)

	2005			2010			2015		
Model	10%	Median	90%	10%	Median	90%	10%	Median	90%
Rates of emigration from Italy to Germany, m_{IT-DE}									
M_1	*0.41	*0.65	*1.04	*0.41	*0.65	*1.04	*0.41	*0.65	*1.04
$\mathbf{M_2}$ **(A, B)**	**0.24**	**0.35**	**0.52**	**0.11**	**0.29**	**0.82**	**0.05**	**0.24**	**1.12**
M_3	0.22	0.35	0.56	0.00	0.28	0.84	0.00	0.23	0.96
M_4	*0.34	*0.51	*0.79	*0.39	*0.65	*1.07	*0.39	*0.65	*1.07
M_5	*0.23	*0.35	*0.55	*0.00	*0.33	*1.02	*0.00	*0.31	*1.19
M_{avg}(A)	0.24	0.36	0.54	0.07	0.30	0.87	0.02	0.26	1.09
M_{avg}(B)	0.24	0.36	0.53	0.09	0.30	0.84	0.04	0.25	1.09
Rates of emigration from Germany to Italy, m_{DE-IT}									
M_1	0.34	0.45	0.59	0.34	0.45	0.59	0.34	0.45	0.60
$\mathbf{M_2}$ **(A, B)**	**0.33**	**0.44**	**0.58**	**0.19**	**0.42**	**0.94**	**0.12**	**0.40**	**1.36**
M_3	0.33	0.44	0.59	0.21	0.44	0.77	0.08	0.44	0.97
M_4	0.34	0.46	0.61	0.33	0.45	0.61	0.33	0.45	0.62
M_5	*0.34	*0.45	*0.61	*0.21	*0.44	*0.87	*0.09	*0.44	*1.27
M_{avg}(A)	0.33	0.44	0.59	0.25	0.44	0.72	0.18	0.44	0.85
M_{avg}(B)	0.33	0.44	0.59	0.25	0.44	0.73	0.19	0.44	0.86
Rates of emigration from Poland to Germany, m_{PL-DE}									
$\mathbf{M_1}$ **(A, B)**	**1.77**	**2.57**	**3.72**	**1.78**	**2.57**	**3.71**	**1.78**	**2.57**	**3.72**
M_2	2.49	3.64	5.33	1.23	3.62	10.71	0.69	3.61	18.80
M_3	1.99	3.00	4.53	1.64	2.67	5.02	1.56	2.63	5.17
M_4	1.80	2.80	4.21	1.72	2.60	3.94	1.72	2.60	3.96
M_5	*1.97	*3.09	*4.81	*1.60	*2.68	*5.29	*1.53	*2.65	*5.46
M_{avg}(A)	1.92	2.95	4.55	1.64	2.70	5.45	1.50	2.66	6.13
M_{avg}(B)	1.92	2.94	4.57	1.63	2.71	5.80	1.45	2.67	7.00
Rates of emigration from Germany to Poland, m_{DE-PL}									
M_1	0.72	1.00	1.39	0.72	1.00	1.39	0.72	1.00	1.38
$\mathbf{M_2}$ **(A, B)**	**0.91**	**1.25**	**1.71**	**0.48**	**1.18**	**2.90**	**0.28**	**1.12**	**4.36**
M_3	0.82	1.14	1.58	0.64	1.02	1.79	0.59	1.00	1.89
M_4	0.77	1.09	1.52	0.70	1.01	1.45	0.70	1.01	1.45
M_5	*0.80	*1.14	*1.61	*0.63	*1.01	*1.72	*0.59	*0.99	*1.76
M_{avg}(A)	0.81	1.15	1.61	0.60	1.04	2.09	0.48	1.02	2.58
M_{avg}(B)	0.82	1.16	1.62	0.59	1.05	2.25	0.43	1.02	2.92

Table 5.5 (continued)

Model	2005 10%	2005 Median	2005 90%	2010 10%	2010 Median	2010 90%	2015 10%	2015 Median	2015 90%
Rates of emigration from Switzerland to Germany, m_{CH-DE}									
M_1	0.94	1.12	1.35	0.94	1.12	1.35	0.94	1.12	1.35
M_2 (A, B)	**1.05**	**1.25**	**1.49**	**0.83**	**1.34**	**2.17**	**0.71**	**1.44**	**2.93**
M_3	1.00	1.21	1.45	0.90	1.19	1.99	0.87	1.19	3.31
M_4	0.97	1.16	1.39	0.92	1.12	1.37	0.92	1.12	1.38
M_5	*1.01	*1.22	*1.48	*0.87	*1.19	*2.16	*0.84	*1.18	*3.85
M_{avg}(A)	1.00	1.21	1.46	0.88	1.20	1.96	0.83	1.20	2.56
M_{avg}(B)	1.00	1.21	1.46	0.88	1.21	1.98	0.81	1.22	2.58
Rates of emigration from Germany to Switzerland, m_{DE-CH}									
M_1	*0.08	*0.13	*0.22	*0.08	*0.13	*0.22	0.08	0.13	*0.22
M_2 (A, B)	**0.16**	**0.22**	**0.30**	**0.11**	**0.28**	**0.69**	**0.09**	**0.36**	**1.44**
M_3	0.15	0.23	0.33	0.11	0.45	265.60	0.10	1.30	∞
M_4	*0.11	*0.17	*0.26	*0.08	*0.13	*0.23	0.08	0.13	*0.23
M_5	*0.16	*0.23	*0.35	*0.10	*0.45	* ∞	0.08	1.24	* ∞
M_{avg}(A)	0.18	0.22	0.27	0.16	0.29	0.76	0.15	0.39	3.62
M_{avg}(B)	0.18	0.22	0.27	0.16	0.29	0.57	0.15	0.38	1.16

Notes: Averaged forecasts are indicated as M_{avg}(A) and (B) for uniform and 'Occam's razor' model priors. Boldface denotes values yielded by models selected on the basis of the respective posterior odds (Table 5.3). NB: Data for 2005 are: $m_{IT-DE} = 0.35$, $m_{DE-IT} = 0.35$, $m_{PL-DE} = 4.17$, $m_{DE-PL} = 1.28$, $m_{CH-DE} = 1.26$ and $m_{DE-CH} = 0.22$. Further explanations in text.
Source: own elaboration in WinBUGS, data for 2005: Eurostat/NewCronos

In general, the predictive distributions are not symmetric, and the quantile-based predictive intervals do not have the HPD property. This especially concerns predictive distributions yielded by the averaged forecasts, which in some cases may be multimodal and skewed, being in fact discrete mixtures of particular predictive distributions yielded by Gaussian likelihood and Gamma prior for precision,[17] which were obtained from all five models $M_1 - M_5$. The graphs for 2005 additionally include a visualisation of the actual observations for that year, with the values of

[17] In simple, linear models, predictive distributions under Normal likelihood and Gamma prior for precision are Student's t (for analytical derivation, see e.g. Bernardo and Smith, 2000, p. 440). Following the parameterisation used in WinBUGS, Student's t distribution with mean μ, scale parameter τ, and υ degrees of freedom has density function: $p(x\,|\,\mu, \tau, \upsilon) = \Gamma[0.5(\upsilon + 1)]/\Gamma(0.5\upsilon) \cdot (\tau/\upsilon\pi)^{0.5} \cdot [1 + \tau/\upsilon \cdot (x - \mu)^2]^{-0.5(\upsilon+1)}$, where $\upsilon \geq 1$ and $\Gamma(\cdot)$ is Euler's Gamma function (cf. Bernardo and Smith, 2000, p. 122). Normal distribution is obtained as a limit for $\upsilon \to \infty$, whereas the case with $\upsilon = 1$ is known as the Cauchy distribution. For the t distribution only the k first positive moments exist, where $k < \upsilon$, and thus the Cauchy distribution has none. This is related to the property of the t distribution, in which the less are degrees of freedom, the heavier are the tails. The log-t distributions for random variables $Y = \exp(t)$, as for migration rates predicted under simple models, are also heavy tailed and do not have positive moments, what can be shown using integration by substitution and expanding $\exp(x)$ into Taylor's series. This is an additional argument for discussing the predictive outcomes in terms of quantiles rather than moments (means, standard deviations, etc).

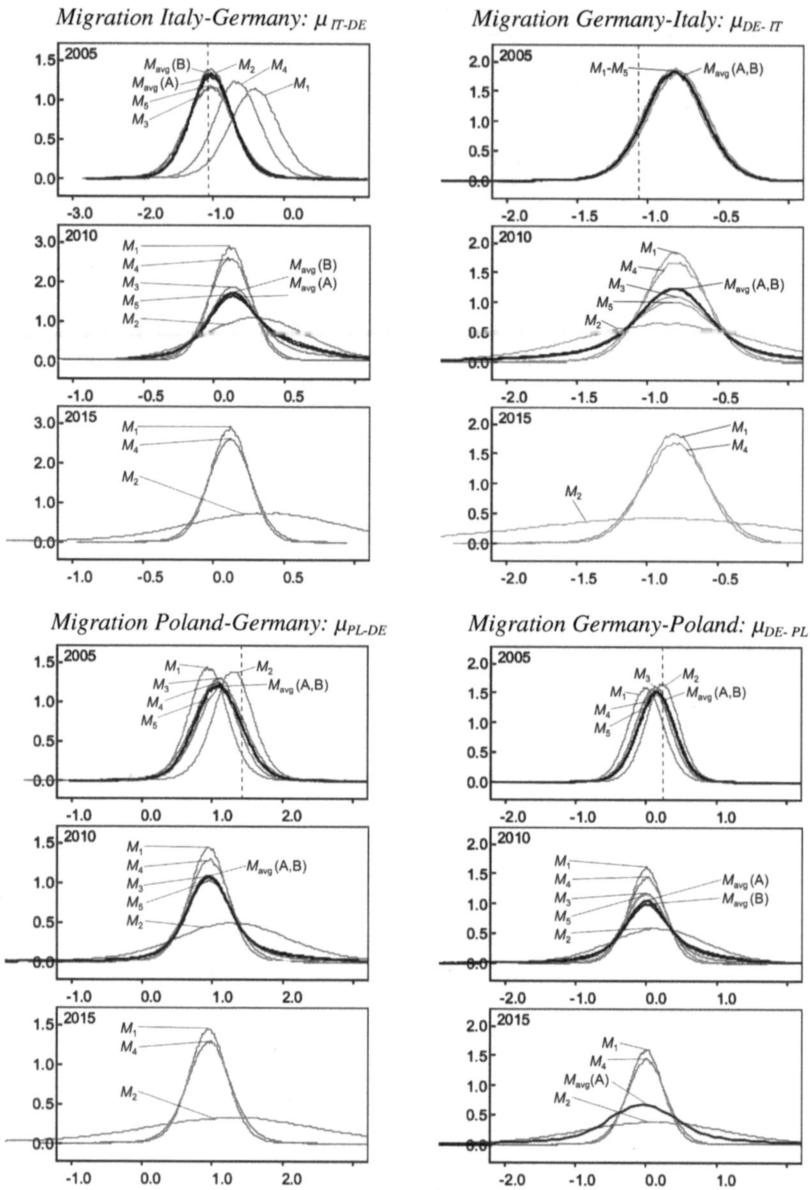

Fig. 5.3 Various predictive distributions of log-transformed rates: 2005, 2010 and 2015.
Notes: M_{avg} (black lines) depicts averaged forecasts, corresponding to (**A**) uniform, or (**B**) 'Occam's razor' prior. Dashed vertical lines indicate 2005 observations of log-transformed migration rates (Tables 5.5 and 5.6). Some distributions for M_3, M_5 and M_{avg} in 2015 not shown due to numeric problems: high variance and heavy tails.
Source: own elaboration in WinBUGS, data for 2005: Eurostat/NewCronos

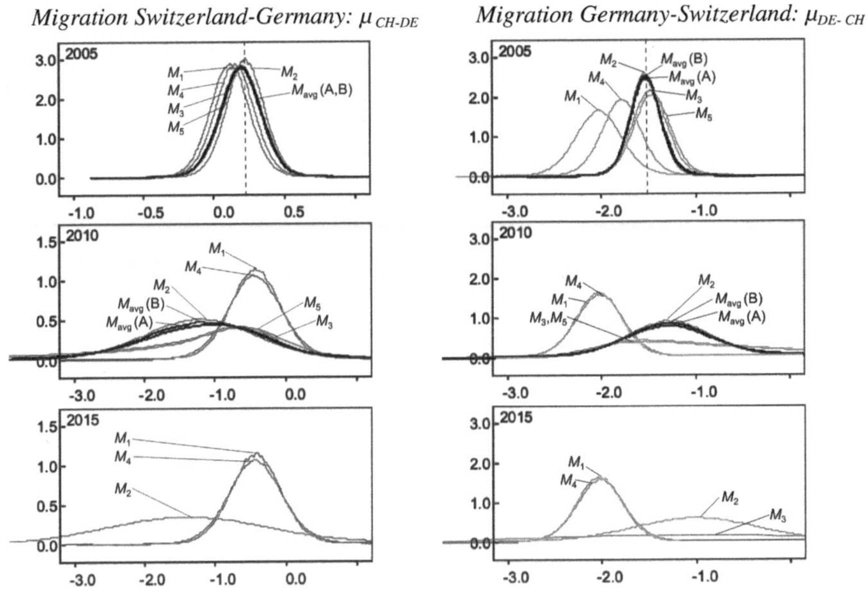

Fig. 5.3 (continued)

the respective migration rates listed in a note underneath Table 5.5. The issues of *ex-post* prediction errors for the first year of the forecast horizon are discussed in more detail in the next subsection.

On the example of forecasts of migration from Germany to Switzerland and from Italy to Germany, it is worth noting that models with very low posterior probabilities (M_1 and M_4 in Fig. 5.3) are almost entirely excluded from the calculations of the averaged forecasts. It was even hardly necessary to remove M_1 and M_4 from the modelling framework, as they have effectively been eliminated due to their strong disagreement with the data.

From the demographic point of view, the upper limits of the 80% credible intervals for m_{DE-CH}, especially those yielded by models M_3 and M_5, seem implausibly high. The likely non-stationary autoregressive components of M_3 and M_5, and non-negligible posterior probabilities of these models, especially under the flat prior (Table 5.3), render the respective averaged forecasts also hardly reasonable. In addition, too wide predictive intervals could be also yielded by random walks, non-stationary by definition. An example is migration from Poland to Germany, where M_2 gained high data support. Moreover, in the instance of non-stationary models with the AR(1) component, some MCMC problems towards the end of the forecast horizon appear, visualised in an imprecise mixing of particular predictive distributions in the averaged one. In particular, this concerns differences in means and second moments, most likely due to numerical problems concerning sampling from heavy-tailed predictive distributions for M_3 and M_5.

Figure 5.4 illustrates the predicted median values, as well as the limits of 80% predictive intervals for emigration rates to and from Germany for the period

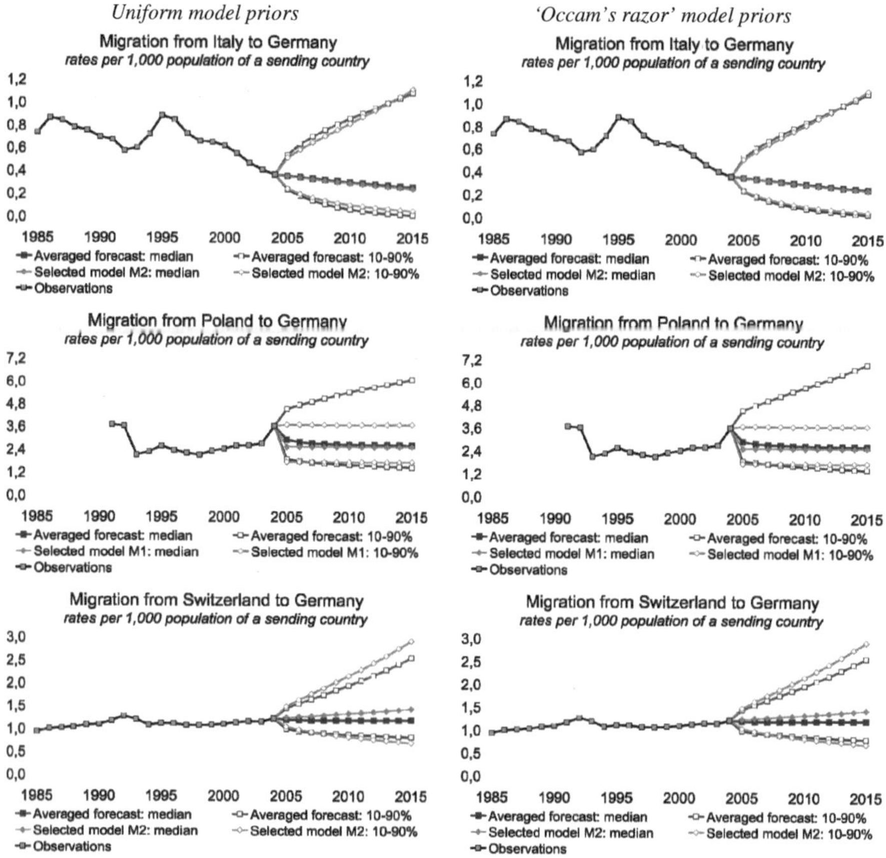

Fig. 5.4 Forecasted migration to and from Germany, 2005–2015: selected and averaged forecasts. Source: own elaboration in WinBUGS. Data series until 2004: Eurostat/NewCronos; DESTATIS (see Annex A)

2005–2015. The graphs show predictions from formally-selected models and averaged forecasts, both under the uniform and Occam's priors, next to the historical data series, on the basis of which the estimation was made. From the figures it seems that in the majority of cases the median forecast trajectories obtained from the models selected on the basis of the posterior odds criterion are almost the same as the averaged ones. This conclusion is valid for both model priors. The most visible exceptions concern migration from Switzerland to Germany under either $p(M_i)$, where the selected random walk trajectory obtained from M_2 has visibly higher values than the one based on the averaged forecast. Some differences, albeit to a lesser extent, can be also seen for flows from Germany to Italy, as well as from Germany to Poland.

There is no rule as to for which forecast the *ex-ante* uncertainty estimate is higher. In particular, the predictive intervals of averaged forecasts can be broader than the ones obtained from the formally-selected models, which would reflect taking

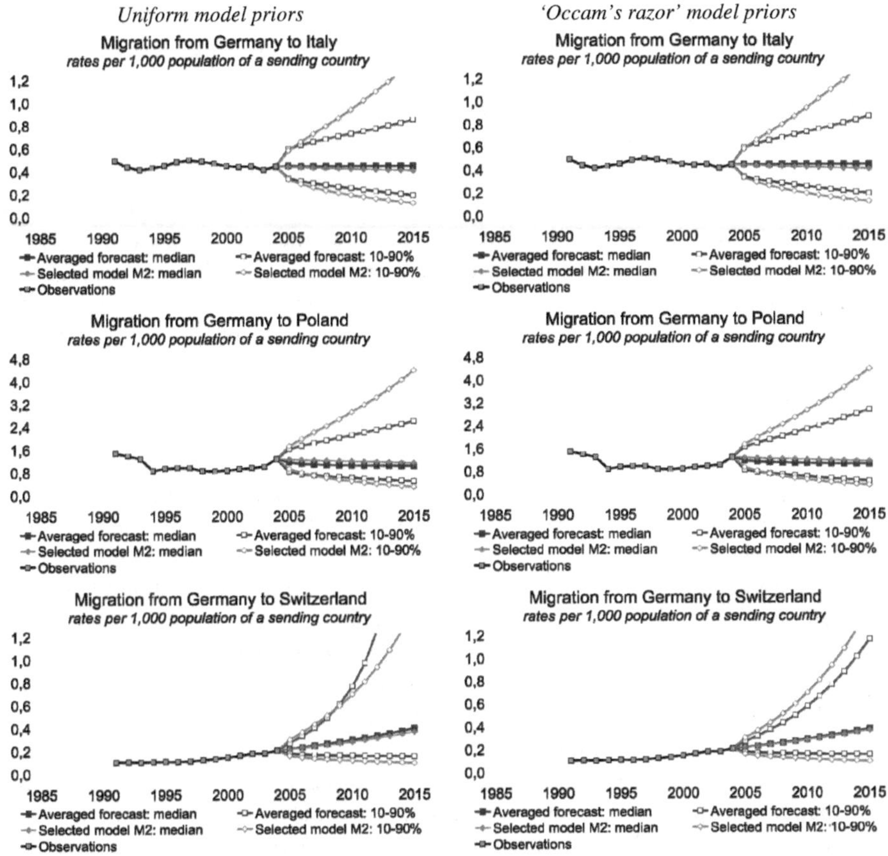

Fig. 5.4 (continued)

into account the uncertainty of the model choice process. Nonetheless, the opposite pattern seems to be prevalent. It concerns for example situations with relatively high posterior probabilities for models other than the selected one, where the selection criteria (posterior odds) indicate a high-variance model, as M_2 for migration from Switzerland to Germany, or from Germany to Poland or to Italy.

5.2.3 Interpretation of Forecasts and the Comparison of Ex-Post *Errors for 2005–2007*

On the basis of median trajectories derived from the respective predictive distributions (Fig. 5.4), the obtained migration forecasts can be summarised as follows. In the period 2005–2015 it is expected that the propensity to migrate from Italy to Germany and vice versa will visibly diminish. In 2005, the both-ways migration between Germany and Poland is predicted to return to the levels observed before the increase in 2004, which was likely related to the accession of Poland to the European Union. From 2005 onwards the intensity of migration between these countries is expected to remain almost constant until the end of the forecast horizon.

Such an outcome might have indeed occured due to external circumstances. There was a considerable migration from Poland to the United Kingdom and Ireland after the opening of the labour markets in 2004. Hence, the supply of migrants, who might potentially want to wait until Germany lifts the restrictions, what is expected in 2011, has likely reduced. For these reasons, no dramatic increase in the Polish-German migration can be currently envisaged. This interpretation, however, is valid merely by coincidence, as it is based on additional information, far exceeding the inference from the presented forecasting models given the observed data and assumed priors.

Finally, the relative magnitude of flows between Germany and Switzerland in both directions is envisaged to slightly increase throughout the whole period 2005–2015, although for flows from Switzerland to Germany this conclusion holds mainly for the forecast based on the selected model M_2. All in all, from the demographic point of view, the median forecasts of the intensities of migratory movements under study seem reasonable.

Most averaged forecasts also yield plausible uncertainty ranges, slightly increasing with time and spanning realistic magnitudes of possible future migration flows. The exceptions concern migration from Germany to Switzerland, where the upper limits of the 80% predictive intervals are very high and increase exponentially (see Fig. 5.4). This is due to the fact that the averaged predictions include the outcome of models M_3 and M_5, characterised by a visible non-stationary autoregressive component. Both of these models produce predictive distributions characterised by very large variances, which renders the forecasts made on their basis unreliable, as indicated in Table 5.5. Moreover, in such cases the logarithmic transformation of variables additionally contributes to an overestimation of the upper bounds of predictive intervals due to the way the models are formulated.

Interestingly, for most of the flows under study, except for migration from Poland to Germany (and possibly also from Germany to Italy, cf. Table 5.3), the formal model selection criteria yield the random walk M_2, which models rather unpredictable phenomena. This non-stationary process is characterised by a rapidly increasing uncertainty span which, in migration studies is consistent with the intuition of the researcher, at least in the short term. High posterior probabilities of this random model suggest that hardly any theory or hypothesis may be capable of explaining the respective observed flows in a reasonable manner, especially given the shortness of the time series of available migration data.

It is also worth noting that the plausible uncertainty ranges of migration predictions are relatively wide. This is not surprising, given the inevitable problems underlying high errors of international migration forecasts that have been mentioned in Chapter 1. Although median forecasts are merely the continuation of past trends, there is a high probability that the actual migration developments will substantially deviate from the central trajectories. Therefore, apart from the median

Table 5.6 Comparison of absolute and relative *ex-post* errors for 2005–2007: model class $\mathbf{M_1}$

	m_{IT-DE}	m_{PL-DE}	m_{CH-DE}	m_{DE-IT}	m_{DE-PL}	m_{DE-CH}	Average
Absolute errors							
Model M_1	*** 0.31	*** 1.59	** 0.26	** 0.12	* 0.37	*** 0.15	0.47
Model M_2	0.01	**0.52**	**0.11**	**0.11**	**0.13**	0.05	**0.16**
Model M_3	0.01	1.29	0.18	0.11	0.28	0.03	0.32
Model M_4	** 0.26	** 1.49	** 0.25	** 0.12	* 0.34	** 0.13	0.43
Model M_5	**0.01**	1.27	0.18	0.12	0.29	**0.02**	0.32
Averaged forecast (A)	0.01	1.33	0.18	0.12	0.26	0.05	0.32
Averaged forecast (B)	0.01	1.32	0.17	0.12	0.25	0.05	0.32
Relative errors							
Model M_1	*** 89.1%	*** 38.2%	** 18.4%	** 37.5%	* 27.1%	*** 50.9%	43.5%
Model M_2	2.9%	**12.5%**	**7.7%**	**32.4%**	**9.4%**	16.3%	**13.5%**
Model M_3	4.0%	31.1%	12.9%	35.2%	19.9%	8.3%	18.5%
Model M_4	** 74.5%	** 35.8%	** 17.3%	** 38.3%	* 24.2%	** 45.2%	39.2%
Model M_5	**2.1%**	30.5%	12.3%	37.3%	20.9%	**8.0%**	18.5%
Averaged forecast (A)	3.0%	31.8%	12.5%	35.9%	18.9%	14.7%	19.5%
Averaged forecast (B)	3.0%	31.6%	12.1%	35.7%	18.0%	15.6%	19.3%
Observed values:							
2005	0.35	4.17	1.26	0.35	1.28	0.22	-
2006	0.34	4.29	1.39	0.33	1.37	0.27	-
2007	0.35	4.03	1.49	0.31	1.47	0.34	-

Notes: (A) denotes uniform and (B) Occam's prior. Column-wise minima are in boldface and maxima underlined. Asterisks (*,**,***) indicate the 2005–2007 realisations falling outside the 80% predictive intervals respectively for one, two or three years of observation.
Source: own elaboration; data for 2005–2007: Eurostat/NewCronos

predictions, the presented forecasts provide the policy makers with an important piece of information: uncertainty related to migration predictions between the countries under study is high and this should be properly acknowledged in the policy decisions.

At the time of revising this chapter observed values of m_{i-j} became available for 2005–2007; they have been used for a simple comparison of *ex-post* forecast errors for those years yielded by various models. The results, presented in Table 5.6 both in terms of mean absolute and relative (percentage) errors of prediction, *MAE* and *MAPE* (see also Chapter 8), indicate that highest discrepancies between the observed and forecasted (median) values have been noted for the constant model (M_1) for all migration flows under study. On the other hand, the 2005–2007 observations are on average closest to the median forecast in four cases for the random walks (M_2), and in the further two – for the ARMA(1,1) models (M_5). In general, considering all six flows under study, random walk models yielded the most accurate forecasts for 2005–2007.

The forecasts yielded by the constant models did not fit into the 80% predictive intervals, at least for one observation period. It is worth noting that, as M_1 was formally selected to forecast migration from Poland to Germany, such a high mismatch between the observed value of m_{PL-DE} and the one forecasted on its basis already in the first years of the forecast period can indicate weak plausibility of forecasts generated by M_1. In practice this means that either the forecasted return to the pre-accession trajectory will happen later than in 2007, or the trend itself has changed. In the latter case, although judging only by the *ex-post* performance for 2005–2007, this can indicate that the second-best model in terms of posterior probability; that is, the random walk with drift, could produce much more reliable forecasts than M_1. A more detailed analysis of *ex-post* and *ex-ante* accuracy of various forecasts from $\mathbf{M_1}$-class models is offered in Chapter 8.

Chapter 6
Bayesian VAR Modelling 'from General to Specific'

The current chapter outlines the second of proposed perspectives on migration forecasting, which applies the 'from general to specific' modelling principle in the context of nested vector autoregression (VAR) models. In this way, the impact of various theory-based interdependent variables on migration can be tested. Similarly to the previous chapter, Section 6.1 outlines the theoretical foundations of VAR modelling, while Section 6.2 illustrates the approach with empirical forecasts for emigration rates among the countries under study.

6.1 VAR Processes and Lindley's Tests for Restrictions

6.1.1 Selection of Predictors in Econometric Models: Rationale for the VAR Modelling

According to Sanderson (1998), including the socio-economic context in population forecasting models (that is, additional knowledge on the processes under study) can potentially improve demographic predictions. Nonetheless, in the econometric models, there is a problem with the selection of potential explanatory variables. The empirical analysis of Jennissen (2004) shows that two macroeconomic determinants that had a significant impact on immigration (both net and gross) for a number of Western and Northern European countries were differences in the GDP per capita levels (neoclassical perspective), and in unemployment rates (Keynesian perspective). As both these variables describe pull factors of migration, in the current research the sending-country unemployment rates (a push factor) have been analysed next to the GDP gap, instead of the differences in unemployment rates. Such a variable can be a proxy of the uncertainty of the household income in the home country (Jennissen 2004, p. 57), which is an important migration factor according to the new economic theory of migration (see also Chapter 3). Unemployment in the receiving country has also not been considered due to the likely existence of the irregular features of the dual labour market, which can considerably obscure the picture.

J. Bijak, *Forecasting International Migration in Europe: A Bayesian View*,
The Springer Series on Demographic Methods and Population Analysis 24,
DOI 10.1007/978-90-481-8897-0_6,

As the current study makes an attempt to model migration intensity separately for each of the origin-and-destination-specific flows, it does not directly examine the effect of migrant networks (which could be modelled by a proxy – the population stock of given origin present at the destination). The main rationale for excluding this potential predictor is that changes in population stocks are a result of two-way flows expressed in terms of numbers of migrants rather than emigration rates, which is contrary to theoretical considerations of demographic forecasting followed in the current study. Coherent forecasts of the numbers of migrants would require additional demographic modelling with respect to two components of natural change (births and deaths), which remains far beyond the scope of the analysis.

Nevertheless, the migrant network hypothesis can be to some extent indirectly acknowledged in the autoregressive models, where the perpetuating effect of the autoregression coefficient ϕ can be taken as another proxy of the 'cumulating' of the number of migrants. This notion has been already explored in the study of Sinn et al. (2001). Despite the limitations of such an approach, its simplicity renders it potentially useful in the studies of migratory phenomena, for which the availability of consistent and reliable data is problematic. For this reason, in the current chapter various autoregressive models are explored, within different modelling frameworks.

Another problematic issue concerning econometric forecasts is that independent variables, like the GDP or unemployment, are potentially at least as difficult to predict as migration – the dependent variable. A way to overcome the 'how to predict the predictors' problem is to use the vector autoregressive (VAR) models (Litterman, 1979; Sims, 1980), which do not require a separate modelling and forecasting of all the variables, as they are examined jointly. The VAR models are presented and discussed in more detail in the following subsections.

6.1.2 VAR Models and the 'from General to Specific' Approach

Vector autoregressive (VAR) models are multi-dimensional generalisations of univariate autoregressive processes, examples of which have been presented in Chapter 5. In this way the structure of models is kept relatively simple, while, as noted by Fertig and Schmidt (2000, p. 2),

> imposing more and more structure on the estimation of the determinants of aggregate migration flows has important consequences for the forecasting of future migration flows; more structure typically reduces uncertainty within sample if the invoked assumptions are correct, but may not necessarily lead to better forecasts.

In many macroeconomic applications, the VAR models have performed better in terms of forecast errors than large structural models (Greene, 2000, p. 741).

On the other hand, the joint analysis of all variables without going deeper into causal relationships makes the VAR models often considered as 'atheoretical', as they do not reflect any particular economic theory (Cooley & LeRoy, 1985). The same reservations also hold for other time series models, which relate future values of the variable under study only to its past developments, irrespective of the broader theoretical context. Another problem concerns Bayesian VAR analyses based on *Minnesota priors* of Litterman (1979), reflecting a belief that the series under study are generated by independent random walk processes.[1] Numerous studies have shown that such prior assumptions generate well-performing forecasts, in terms of low forecast errors. However, Ramos (1996) observed that the 'success' of Litterman's priors is not due to the fact that they reflect realistic beliefs, but rather because the construction of the prior 'covers' misspecification problems, reducing the variance of the forecast error.

Nevertheless, given the problems with the lack of a comprehensive and easy-to-operationalise migration theory outlined in Chapter 3, the feature of 'atheoreticity' is not a strong argument against using the VAR models in migration forecasting. Moreover, the aim of the current study is to forecast migration rather than to verify any particular theory. The available data series are also not sufficiently long to allow for examining various causal relationships of migration. For these reasons, next to the univariate time series models presented in the previous section, the current study examines vector autoregression models based on two main economic determinants of migration (income differentials and unemployment rates), and tests several possible restrictions on their parameters within the simplified 'from general to specific' approach.

In general terms, the vector autoregression model of order p, VAR(p), is defined as (e.g., Greene, 2000, p. 740, see also model (4.13) in Chapter 4):

$$\mathbf{x_t} = \mathbf{c} + \left(\sum_{i=1}^{p} \mathbf{A_i} L^i \right) \mathbf{x_t} + \boldsymbol{\varepsilon}_\mathbf{t}. \qquad (6.1)$$

In the above equation, $\mathbf{x_t}$ depicts the n-dimensional variable under study, $\mathbf{c}$ is a vector of constants, $\mathbf{A_i} = [\alpha_{ij}]_{n\times n}$ are matrices of cross-variable coefficients, and L is the lag operator (see also explanations to (4.12) in Chapter 4); whereas ε_t denotes the error term, usually assumed to be an iid n-dimensional Gaussian white noise, with mean $\mathbf{0}$ and covariance matrix $\boldsymbol{\Sigma}$, $\mathbf{N}_n(\mathbf{0}, \boldsymbol{\Sigma})$.

The selection of variables that are examined jointly in the vector $\mathbf{x_t}$ within a VAR, as well as the determination of the model order p, may follow the 'from general to specific' methodology outlined for example by Hendry (1995, p. 270; 344*nn*), or from the Bayesian perspective, by Bauwens, Lubrano, and Richard (1999, pp. 52–56). In this approach, the modelling process starts from the relatively broadest plausible model, involving all potential predictors, and the maximum number of lags that is found sensible, given the length of the time series. The last factor is important, as the number of parameters that are to be estimated is large: the coefficients matrices $\mathbf{A_i}$ alone account for n^2p parameters; there are also n constants and $1/2 \cdot n \cdot (n+1)$ elements of the covariance matrix of $\boldsymbol{\varepsilon}_\mathbf{t}$. For this reason, due to short time series of origin-destination migration flow data available for most of European countries, the analysis is limited to VAR(1) models.

[1] See also Section 5.3 for more details on the construction of Minnesota priors.

The 'broadest plausible' model constitutes a departure point for further analysis, where various constraints on parameters are tested in successive steps, in order to check whether the original model can be reduced to a simpler one. In this way, the procedure follows a nested structure of models: from the most general, to more detailed ones – hence the name.

In testing the admissibility of the reductions of the initial model, various methods can be applied. One option is the Granger-causality test, which indicates whether 'lagged values of a variable, say x_t, have explanatory power in a regression of a variable y_t on lagged values of y_t and x_t' (Greene, 2000, p. 742). More generally, for a VAR(1) model, in order to examine the impact of a subgroup of variables, $\mathbf{x_2}$, on the key ones of interest, $\mathbf{x_1}$, the model (6.1) can be partitioned as follows, with $\mathbf{x_t}' = [\mathbf{x_{1,t}}' \; \mathbf{x_{2,t}}']$ (*idem*, p. 743):

$$\begin{bmatrix} \mathbf{x_{1,t}} \\ \mathbf{x_{2,t}} \end{bmatrix} = \begin{bmatrix} \mathbf{c_1} \\ \mathbf{c_2} \end{bmatrix} + \begin{bmatrix} \mathbf{A_{11}} & \mathbf{A_{12}} \\ \mathbf{A_{21}} & \mathbf{A_{22}} \end{bmatrix} \cdot \begin{bmatrix} \mathbf{x_{1,t-1}} \\ \mathbf{x_{2,t-1}} \end{bmatrix} + \begin{bmatrix} \boldsymbol{\varepsilon}_{1,t} \\ \boldsymbol{\varepsilon}_{2,t} \end{bmatrix}, \\ \text{with Var} \begin{bmatrix} \boldsymbol{\varepsilon}_{1,t} \\ \boldsymbol{\varepsilon}_{2,t} \end{bmatrix} = \begin{bmatrix} \boldsymbol{\Sigma}_{11} & \boldsymbol{\Sigma}_{12} \\ \boldsymbol{\Sigma}_{21} & \boldsymbol{\Sigma}_{22} \end{bmatrix}. \tag{6.2}$$

In particular, we can set $\mathbf{x_{1,t}} = \mu_{i-j}(t)$, where $\mu_{i-j}(t)$ denotes log-transformed rates of migration from country i to j in year t, calculated per 1,000 of population at risk, defined in Chapter 1. The Granger-test hypothesis regarding the impact of lagged variables $\mathbf{x_2}$ on $\mathbf{x_1}$ is $H_0: \; \mathbf{A_{12}} = \mathbf{0}'$. In the traditional, sampling-theory framework, this hypothesis can be verified for example using a likelihood ratio test, or other tests for restrictions (e.g., the Wald test, or the Lagrange multiplier test, *idem*, pp. 152–160), although in the current study a Bayesian approach of Lindley's HPD testing is followed, which is presented in the following subsection.

In addition to the impact of lagged variables $\mathbf{x_2}$ on $\mathbf{x_1}$, the direct (non-lagged) impact of particular variables on each other can be assessed using the properties of conditional distributions, under the Normality assumption for $\boldsymbol{\varepsilon}_t$. With the partition of the original model $\mathbf{x_t}$ given by (6.2), the conditional distribution of $\mathbf{x_1}$ given $\mathbf{x_2}$ is Normal (e.g., Greene, *idem*, p. 87):

$$\mathbf{x_{1,t}} \,|\, \mathbf{x_{2,t}} \sim \mathrm{N}(\boldsymbol{\mu}_{1,t} + \boldsymbol{\Sigma}_{12}\boldsymbol{\Sigma}_{22}^{-1}[\mathbf{x_{2,t}} - \boldsymbol{\mu}_{2,t}],\; \boldsymbol{\Sigma}_{12}\boldsymbol{\Sigma}_{22}^{-1}\boldsymbol{\Sigma}_{21}), \tag{6.3}$$

where, by (6.2), $\boldsymbol{\mu}_{1,t} = \mathbf{c_1} + \mathbf{A_{11}x_{1,t-1}} + \mathbf{A_{12}x_{2,t-1}}$, and $\boldsymbol{\mu}_{2,t} = \mathbf{c_2} + \mathbf{A_{21}x_{1,t-1}} + \mathbf{A_{22}x_{2,t-1}}$. The elements of vector $\boldsymbol{\beta} = \boldsymbol{\Sigma}_{12}\boldsymbol{\Sigma}_{22}{}^{-1}$ can be interpreted as coefficients of regression of $\mathbf{x_1}$ on $\mathbf{x_2}$, and tested in an analogous way to the parameters by lagged variables, $\mathbf{A_{12}}$; the respective hypothesis on the direct impact of $\mathbf{x_2}$ on $\mathbf{x_1}$ being H_0: $\boldsymbol{\beta} = \mathbf{0}'$. Such a test is thus equivalent to testing $\mathbf{A_{12}} = \mathbf{0}'$ and $\boldsymbol{\Sigma}_{12} = \mathbf{0}'$.

The results of both presented tests can be ultimately applied in the 'from general to specific' framework, allowing for the removal from the initial model of such variables $\mathbf{x_2}$, which are neither causal for $\mathbf{x_1}$ in the sense of Granger, nor have a significant direct impact on $\mathbf{x_1}$, in terms of the regression coefficients $\boldsymbol{b}$. In principle, a comprehensive econometric study of the VAR models involving multi-dimensional time series should additionally contain several other elements of the analysis. Such issues primarily include the tests of cointegration between the pairs of variables, which allow identifying long-term relationships in the case of non-stationary or trended series (Greene, *idem*, pp. 789–796), as well as the tests of exogeneity of some of the variables to the system (the problem is discussed in a Bayesian framework in Osiewalski & Steel, 1996, as well as in Osiewalski, 2001, pp. 25–30).

Nevertheless, in the current study, the time series are too short to allow for the establishing of fundamental, long-term causal relationships between migration flows and their determinants[2]. Any attempt to address such issues in the VAR modelling based on small samples would produce ambiguous results. Therefore, most of these theoretical possibilities are not explored here, being not very useful given the available data. The interpretation of the results of the tests of Granger causality, as well as of the direct impact of migration determinants on migration rates, also needs to be very cautious. Therefore, the (lack of) significance of respective model parameters is treated in the current study as a justification for the selection of a particular structural form of a model, rather than as an indication of long-term relationships, that would potentially provide evidence for any of migration theories.

6.1.3 Inference on the Impact of Additional Variables on Migration

In the Bayesian framework, one option to test the significance of restrictions imposed on particular parameters of the model is to compare various competing (in this case, nested) models using the formal selection techniques, as outlined in Chapter 5. An alternative option is D. V. Lindley's approach to the testing of significance of particular sets of parameters on the basis of the HPD credible regions (for definition, see Chapter 2). The latter method, which is employed in the current study, has been described for example by Zellner (1971, pp. 298–302).

Lindley's methodology is akin to the construction of the sampling-theory tests of significance for model parameters. Given the unimodal posterior distribution of the parameter θ, $p(\theta\,|\mathbf{x})$, the null hypothesis H_0: $\theta = \theta_0$ is tested at the significance level γ, by constructing a HPD interval (a, b) such that $p(a<\theta<b\,|\mathbf{x}) = 1-\gamma$. If the value of θ_0 lies within such an interval, H_0 is accepted at the γ-level of significance (*idem*). For a joint examination of a significance of a vector of several regression parameters $\boldsymbol{\beta}$, as indicated in the previous subsection, the test statistics can be obtained in a similar manner to the traditional, frequentist variant of a Wald test for restrictions. A null hypothesis H_0: $\boldsymbol{\beta} = \mathbf{0}$ implies then a test statistic (cf. Greene, 2000, pp. 153–156):

[2] In migration studies interesting properties of the estimated VAR models have been tested by Gorbey et al. (1999, pp. 78–84), who, however, had longer time series at their disposal – about 80 quarterly observations.

$$\psi(\boldsymbol{\beta}) = [\boldsymbol{\beta} - E(\boldsymbol{\beta} \mid \mathbf{x})' \cdot [Var(\boldsymbol{\beta} \mid \mathbf{x})]^{-1} \cdot [\boldsymbol{\beta} - E(\boldsymbol{\beta} \mid \mathbf{x})], \tag{6.4}$$

and examination, whether $\psi(\mathbf{0})$ is located within the HPD region for $\psi(\boldsymbol{\beta})$. In (6.4), $E(\boldsymbol{\beta} \mid \mathbf{x})$ and $Var(\boldsymbol{\beta} \mid \mathbf{x})$ respectively denote the posterior mean and covariance matrix of $\boldsymbol{\beta}$, given the data $\mathbf{x}$. A sampling-theory equivalent of $\psi(\boldsymbol{\beta})$ would follow an asymptotic chi-squared distribution with k degrees of freedom, with k being the dimension of $\boldsymbol{\beta}$, but in the Bayesian framework with informative priors the posterior distribution of $\psi(\boldsymbol{\beta})$ has to be obtained using numerical methods, like for example the MCMC, or other.

It has to be noted that for all $\boldsymbol{\beta} \in \mathbf{R}^k$, $\psi(\boldsymbol{\beta}) \geq 0$, as $\psi(\boldsymbol{\beta})$ is a positive-definite quadratic form, with $\psi(\boldsymbol{\beta}) = 0$ if and only if $\boldsymbol{\beta} = E(\boldsymbol{\beta} \mid \mathbf{x})$. This implies that in the testing of H_0: $\boldsymbol{\beta} = \mathbf{0}$, the realisation of the test statistic $\psi(\mathbf{0}) = 0$ would indicate a 'perfect insignificance' of $\boldsymbol{\beta}$, whereas only large values of $\psi(\mathbf{0})$, located in the right tails of the $\psi(\boldsymbol{\beta})$ distributions, would lead to the rejection of H_0. Therefore, the performed HPD test should be one-sided, with respective credible intervals having the form $[0, a)$, where a is determined by the fixed posterior probability $(1 - \gamma)$.

Lindley's HPD tests are usually used under non-informative prior distributions for the parameter(s) in question, when the posterior odds of particular hypotheses may be difficult to calculate. In such cases, an additional advantage of the HPD approach is that 'for many, but not all, problems, Lindley's procedure leads to tests that are computationally equivalent to sampling-theory tests' (Zellner, 1971, p. 296). Otherwise, the prior information has a significant impact on the test results. Nevertheless, Lindley's tests can be also performed for informative prior distributions, and are sometimes used (e.g., by Bauwens et al., 1999, pp. 62–64) as an alternative to the approach based on the posterior odds, which remains a central inferential tool in the orthodox Bayesian statistics. Under informative priors, the computational equivalence with frequentist tests (if there was any under diffuse priors) is usually lost.

In the application of the presented methodology to forecasting European migration flows, let the general VAR(1) model consist of three variables: logarithms of rates of emigration from country i to j, income differentials measured by PPP-adjusted GDP per capita, as well as unemployment rates in the sending country, $\mathbf{x_t} = [\mu_{i-j}(t), \ \ln(\mathrm{Y_j(t)/Y_i(t)}), \ln(\mathrm{U_i(t)})]'$. The analysis is limited to three variables for the reasons listed in the first part of the current section, as well as due to the problem of a small sample versus high number of VAR parameters – even in such a simple model, there are 18 parameters to be estimated. Thus, the general model is:

$$\mathbf{x_t} = \mathbf{c} + \mathbf{A} \cdot \mathbf{x_{t-1}} + \boldsymbol{\varepsilon}_\mathbf{t}, \tag{6.5}$$

with the notation as in (6.1), and iid $\boldsymbol{\varepsilon}_\mathbf{t} \sim \mathrm{N}_3(\mathbf{0}, \ \boldsymbol{\Sigma})$. Following (6.2), the modelled vector can be partitioned into $x_{1,t} = \mu_{i-j}(t)$, and $\mathbf{x_{2,t}} = [\ln(Y_j(t)/Y_i(t)), \ln(U_i(t))]'$, so that $\mathbf{x_t}' = [x_{1,t}\, \mathbf{x_{2,t}}']$. The analysis of the impact of the two key determinants (GDP differentials and unemployment) on the intensity of migration flows is offered in Section 6.2, together with the examples of forecasts based both on general and reduced models.

Additionally, conditional forecasts can be calculated from (6.5), by assuming that given trajectories of income differentials and sending-country unemployment rates, $\mathbf{x}^*_{2,t}$, come true with probability one, and thus can be treated as given (non-stochastic). The conditional forecasts, obtained from $x_{1,t}|\mathbf{x}_{2,t} = \mathbf{x}^*_{2,t}$, can serve as a formal representation of a 'what-if' variant analysis, thus forming a statistical alternative to purely judgemental scenarios produced in a deterministic fashion (see Chapter 4). The trajectories $\mathbf{x}^*_{2,t}$ can be determined using various *ad hoc* solutions, such as: expert judgement, simple methods of trend extrapolation, additional VAR models for $\mathbf{x}_{2,t}$ alone, or auxiliary forecasts from unconditional VARs (6.5). As in all cases single trajectories are needed rather than whole predictive distributions, the former can be obtained as the respective predictive means or medians. This approach is followed in the empirical example in Section 6.2.

6.2 Example: Migration Forecasts from General and Reduced VARs

6.2.1 Applying the Reduction Approach: Model Specification, Estimation, and Testing

With respect to the modelling framework outlined in Section 6.1, the analysis focuses on VAR(1) models (6.5), with log-transformed income differentials and unemployment rates in the sending countries as additional variables, potentially helpful in explaining migratory developments. In such models, the prior distributions of the constants c_k, where $k \in \{1, 2, 3\}$ denotes the variable index, respectively for $\mu_{i-j}(t)$, $\ln(Y_j(t)/Y_i(t))$ and $\ln(U_i(t))$, are assumed to be diffuse with $c_k \sim$ N(0, 100^2), and reflect very vague knowledge as to their values.

What concerns the elements of the matrix **A** is that the parameters by own-variable lags (a_{ii}) are a priori supposed to follow N(1, 1^2), depicting a belief in a likely random walk character of each of the variables separately. In turn, the parameters by cross-variable lags, α_{12} and α_{13}, follow N(0.5, 1^2), reflecting an initial hypothesis of a positive impact of the lagged values of the income differentials and unemployment in the sending country on migration flows. For inverse relationships, it is simplistically assumed that $\alpha_{21} \sim$ N(0, 0.1^2) and $\alpha_{31} \sim$ N(0, $1/3^2$), reflecting a rather firm belief in the non-existence of a lagged impact of migration on income differentials, and to a lesser extent, on unemployment. The prior convictions on relationships between both determinants of migration are rather vague, with α_{23}, $\alpha_{23} \sim$ N(0.1, 2^2).

Such prior distributions as proposed above tacitly assume independence of various parameters of the matrix **A**. A more general alternative could for example consist in putting a matrix variate Normal distribution on **A**, which can be operationalised via a multivariate Normal for the vectorisation of **A**. Besides, the proposed prior distributions for particular elements of **A** differ from the widely-used *Minnesota priors* of Litterman (1979), applied for example by Gorbey et al. (1999). The main reason for using a purely subjective prior instead of a partially

Table 6.1 Categories of Wishart priors for precision matrices in VAR(1) models: **P** and $E(\mathbf{T})$

Migration flow	Prior precision class	p_{ii}	p_{ij} for $i \neq j$	Prior mean τ_{ii}	Prior mean τ_{ij} for $i \neq j$
From Italy to Germany	Low	1.00	−0.020	3.00	0.06
From Germany to Italy	Medium	0.30	−0.010	10.02	0.35
From Poland to Germany	Very low	1.50	−0.030	2.00	0.04
From Germany to Poland	Medium	0.30	−0.010	10.02	0.35
From Switzerland to Germany	High	0.15	−0.005	20.05	0.69
From Germany to Switzerland	High	0.15	−0.005	20.05	0.69

Source: own elaboration

data-based one is an attempt to retain coherence in the Bayesian approach, following the argumentation presented in Chapter 4.

The precision matrix, conditional on the systematic part of the model, $\mathbf{T} = \mathbf{\Sigma}^{-1}$ is assumed a priori to follow a Wishart distribution[3] with $k = 3$ degrees of freedom and the matrix parameter $\mathbf{P} = [p_{ij}]_{3\times 3}$ varying according to the flow-specific uncertainty class, as defined in Table 5.2. The matrices **P** and the corresponding prior expected values of the respective parameter matrices **T** ($E(\mathbf{T}) = k\ \mathbf{P}^{-1}$) are shown in Table 6.1. All these assumptions also reflect a general belief in a relatively low precision of the random term, albeit higher for the three variables independently than for their interrelations, and have been designed to be as consistent as possible with the ones presented in the previous section with respect to the combination of central location and dispersion[4].

[3]The Wishart distribution is a multivariate generalisation of the Gamma distribution (in particular, of the chi-squared distribution). Again, following the parameterisation used in the WinBUGS manual, for $x \sim \chi^2(k)$, the density is $p(x\,|k) = 2^{-k/2} x^{k/2-1} \exp(-x/2)/\,\Gamma(k/2)$, where $x > 0$ is Euler's Gamma function. In general, for a random vector $\mathbf{x} \sim \text{Wishart}(\mathbf{P}, k)$, $p(\mathbf{x}\,|\mathbf{P}, k) \propto (\det\ \mathbf{P})^{k/2}\,|\mathbf{x}|^{(k-n-1)/2} \exp[-1/2 \cdot \text{tr}(\mathbf{Px})]$, where $\text{tr}(\cdot)$ is a matrix trace (a sum of the diagonal elements) and **x** is an n-dimensional symmetric and positive definite matrix. As in the case of τ in Chapter 5, elements of **T** refer to the precision of log-transformed variables, and thus to the precision with respect to the orders of magnitude for crude emigration rates.

[4]The prior for the Wishart distribution parameter **P** can be for example elicited as follows: the symmetric matrix $\mathbf{P}/k$ can be decomposed into a matrix product $\mathbf{P}/k = \mathbf{D}\ \mathbf{R}\ \mathbf{D}$, where $\mathbf{D} = [d_{ij}]_{k\,x\,k}$ denotes a very highly probable a priori diagonal matrix of standard deviations and $\mathbf{R} = [r_{ij}]_{k\,x\,k}$ a very highly probable a priori symmetric matrix of Pearsonian correlations between particular variables (credit to Jacek Osiewalski). Under such transformation, in the precision classes listed in Table 7.1, the prior assumptions are thus: for very low precision: $d_{ii} = 0.71$ and $r_{ij} = -0.02$ (for $i \neq j$); for low precision: $d_{ii} = 0.58$ and $r_{ij} = -0.02$; for medium precision: $d_{ii} = 0.32$ and $r_{ij} = -0.03$; and for high precision: $d_{ii} = 0.22$ and $r_{ij} = -0.03$. Naturally, in all cases by definition holds: $r_{ii} = 1$ and $d_{ij} = 0$ for $i \neq j$.

From the point of view of the current analysis, of interest is the impact of $\mathbf{x_2}$ on x_1, both lagged and instantaneous. Therefore, Lindley's tests based on (6.4) have been performed for $\alpha_{12}, \alpha_{13}, b_2, b_3$, as well as jointly for $\boldsymbol{\alpha} = [\alpha_{12}\ \alpha_{13}]'$, and $\mathbf{b} = [b_2\ b_3]'$; the latter concerning regression coefficients derived from the conditional distribution (6.3), intended to measure the direct impact of both potential migration factors on log-transformed emigration rates.

The estimation has been executed with the Gibbs algorithm of WinBUGS software, using the code listed in Annex B. The simulation has been performed in three stages: firstly to obtain the posterior estimates of the parameters under study $\boldsymbol{\beta} \in \{\alpha_{12}, \alpha_{13}, \boldsymbol{\alpha}, b_2, b_3, \mathbf{b}\}$ as their expected values $E(\boldsymbol{\beta} \mid \mathbf{x})$; then to calculate the variance-covariance matrices $Var(\boldsymbol{\beta} \mid \mathbf{x})$; and finally to estimate the distributions of $\psi(\boldsymbol{\beta})$ and calculate forecasts from the VAR model. On the basis of visual convergence checks solely for the elements of $\boldsymbol{\beta}$ in the first two stages, the burn-in phase of the Gibbs procedure has been established as 150,000 and the sample size as 100,000 iterations. In the third stage, the length of the burn-in phase had to be raised to 200,000, and the sample size to 250,000 iterations, in order to ensure convergence for the forecasted values of particular variables under study. The results of Lindley's tests for VAR(1) models of particular migration flows are provided in Table 6.2 and the posterior summaries for all parameters of VAR(1) models are presented in Annex C, Table C.1.

As the empirical distributions of $\psi(\boldsymbol{\beta})$ are positively skewed, with their monotonously declining density functions close to the respective $\chi^2_{(1)}$ or $\chi^2_{(2)}$ distributions (see Section 6.1), and the test is one-sided, the quantile-based credible regions correspond to the respective HPD intervals. As shown in Table 6.2, for all flows under study the values of $\psi(\mathbf{0})$ fall into the respective one-sided 95%, or even 90% HPD regions, being located either in the left or central parts of the intervals, for most cases well below the respective medians, and definitely not in the right-tail H_0 rejection areas. This allows for reckoning that no significant impact of $\ln(Y_j(t)/Y_i(t))$ and $\ln(U_i(t))$ on $\mu_{i-j}(t)$ has been found, neither lagged, nor conditional. In the case of models for μ_{DE-IT} and μ_{DE-CH}, the test results point to the same conclusions for both short and long samples (14 and 20 observations, respectively).

As the impact of the additional variables on migration was found insignificant, in the light of the simplified 'from general to specific' approach the model for $\mu_{i-j}(t)$ can be reduced from an initial VAR(1) to an univariate AR(1), formulated as in (5.5c). The prior distributions for the constant c and autoregression parameter ϕ in the reduced models have been kept the same as in the general VARs, thus with $c \sim \mathrm{N}(0,\ 100^2)$ and $\varphi \sim \mathrm{N}(1,1^2)$. Note that the latter is only slightly different from the AR(1) model examined in Chapter 5, where $\varphi \sim \mathrm{N}(0.5,\ 1^2)$ was assumed. In order for the precision parameter τ to be coherent with the general VAR(1) models, it has been assumed that $\tau \sim v_{11} \cdot \chi^2_{(3)}$. Here, $\chi^2_{(3)}$ denotes a chi-squared distribution with three degrees of freedom and v_{11} is the left-uppermost element of a matrix $\mathbf{V} = \mathbf{P}^{-1}$, where $\mathbf{P}$ is the Wishart parameter for the precision matrix $\mathbf{T}$ in the VAR, defined in Table 6.1. In this way, the prior for τ is a marginal distributions of an appropriate $\mathbf{T}$ with respect to the first variable, its expected value $E(\tau)$ equalling the respective τ_{11} mean from Table 6.1, and variance $Var(\tau) = 2\ v_{11} \cdot E(\tau)$.

Table 6.2 Lindley's tests for parameters: results and distribution summaries of test statistics

Migration flow / Tested set of parameters $\boldsymbol{\beta}$	Calculated $\psi(\mathbf{0})$	Summary of the $\psi(\boldsymbol{\beta})$ distribution: Mean	Std. dev.	Median	90.0%	97.5%
Migration from Italy to Germany						
$\boldsymbol{\beta}=\alpha_{12}$: income differentials, lagged	**0.271**	1.004	1.417	0.459	2.711	5.055
$\boldsymbol{\beta}=\alpha_{13}$: unemployment, lagged	**0.065**	1.015	1.408	0.487	2.647	5.098
$\boldsymbol{\beta}=\boldsymbol{\alpha}$: both determinants, lagged	**0.370**	2.019	2.000	1.424	4.618	7.463
$\boldsymbol{\beta}=b_2$: income diff., instantaneous	**0.007**	0.996	1.534	0.424	2.668	5.277
$\boldsymbol{\beta}=b_3$: unemployment, instantaneous	**0.031**	1.000	1.538	0.426	2.677	5.262
$\boldsymbol{\beta}=\mathbf{b}$: both determinants, instantaneous	**0.038**	1.996	2.233	1.284	4.649	8.031
Migration from Germany to Italy						
$\boldsymbol{\beta}=\alpha_{12}$: income differentials, lagged	**0.239**	1.005	1.418	0.456	2.715	5.038
$\boldsymbol{\beta}=\alpha_{13}$: unemployment, lagged	**0.322**	1.006	1.603	0.413	2.662	5.510
$\boldsymbol{\beta}=\boldsymbol{\alpha}$: both determinants, lagged	**0.581**	2.012	2.139	1.345	4.657	7.797
$\boldsymbol{\beta}=b_2$: income diff., instantaneous	**0.012**	1.005	1.626	0.412	2.675	5.426
$\boldsymbol{\beta}=b_3$: unemployment, instantaneous	**0.004**	0.997	1.606	0.411	2.651	5.366
$\boldsymbol{\beta}=\mathbf{b}$: both determinants, instantaneous	**0.016**	2.002	2.369	1.249	4.674	8.325
Migration from Poland to Germany						
$\boldsymbol{\beta}=\alpha_{12}$: income differentials, lagged	**0.006**	0.991	1.401	0.449	2.683	5.013
$\boldsymbol{\beta}=\alpha_{13}$: unemployment, lagged	**0.490**	1.109	1.576	0.507	3.007	5.542
$\boldsymbol{\beta}=\boldsymbol{\alpha}$: both determinants, lagged	**0.527**	2.073	2.133	1.411	4.803	7.790
$\boldsymbol{\beta}=b_2$: income diff., instantaneous	**0.001**	0.996	1.599	0.412	2.649	5.391
$\boldsymbol{\beta}=b_3$: unemployment, instantaneous	**0.000**	1.005	1.613	0.411	2.685	5.417
$\boldsymbol{\beta}=\mathbf{b}$: both determinants, instantaneous	**0.002**	2.001	2.356	1.252	4.688	8.318
Migration from Germany to Poland						
$\boldsymbol{\beta}=\alpha_{12}$: income differentials, lagged	**0.830**	1.036	1.527	0.465	2.767	5.261
$\boldsymbol{\beta}=\alpha_{13}$: unemployment, lagged	**0.113**	1.015	1.494	0.451	2.693	5.208
$\boldsymbol{\beta}=\boldsymbol{\alpha}$: both determinants, lagged	**0.866**	2.051	2.187	1.372	4.711	7.921
$\boldsymbol{\beta}=b_2$: income diff., instantaneous	**0.002**	0.999	1.620	0.414	2.647	5.337
$\boldsymbol{\beta}=b_3$: unemployment, instantaneous	**0.014**	1.005	1.632	0.410	2.675	5.450
$\boldsymbol{\beta}=\mathbf{b}$: both determinants, instantaneous	**0.016**	2.004	2.391	1.248	4.686	8.359
Migration from Switzerland to Germany						
$\boldsymbol{\beta}=\alpha_{12}$: income differentials, lagged	**0.560**	1.010	1.466	0.451	2.721	5.115
$\boldsymbol{\beta}=\alpha_{13}$: unemployment, lagged	**0.526**	1.016	1.522	0.441	2.735	5.274
$\boldsymbol{\beta}=\boldsymbol{\alpha}$: both determinants, lagged	**0.600**	1.995	2.168	1.311	4.643	7.793
$\boldsymbol{\beta}=b_2$: income diff., instantaneous	**0.050**	0.999	1.556	0.421	2.667	5.316
$\boldsymbol{\beta}=b_3$: unemployment, instantaneous	**0.203**	1.004	1.538	0.427	2.696	5.306
$\boldsymbol{\beta}=\mathbf{b}$: both determinants, instantaneous	**0.232**	2.004	2.249	1.292	4.676	8.109
Migration from Germany to Switzerland						
$\boldsymbol{\beta}=\alpha_{12}$: income differentials, lagged	**0.019**	1.020	1.458	0.457	2.761	5.191
$\boldsymbol{\beta}=\alpha_{13}$: unemployment, lagged	**0.083**	1.027	1.788	0.424	2.673	5.562
$\boldsymbol{\beta}=\boldsymbol{\alpha}$: both determinants, lagged	**0.084**	2.022	2.373	1.309	4.641	7.936
$\boldsymbol{\beta}=b_2$: income diff., instantaneous	**0.004**	1.004	1.678	0.410	2.668	5.458
$\boldsymbol{\beta}=b_3$: unemployment, instantaneous	**0.070**	0.994	1.631	0.405	2.629	5.357
$\boldsymbol{\beta}=\mathbf{b}$: both determinants, instantaneous	**0.074**	1.998	2.439	1.239	4.675	8.352

Note: None of the computed $\psi(\mathbf{0})$ values is falling outside the respective 95% credible interval.
Source: own elaboration in WinBUGS

As for the univariate AR(1) models convergence of the MCMC algorithm in WinBUGS was visibly faster than in the case of VAR processes, the estimates and forecasts have been obtained on the basis of 250,000 iterations following the 100,000-long burn-in phase. The posterior summaries of the parameters of the reduced AR(1) models are offered in Table 6.3. For all flows the estimated models have relatively high posterior probabilities of non-stationarity. As compared with the AR(1) models estimated in Chapter 5, the change of the prior distribution for φ from $N(0.5, 1^2)$ to $N(1, 1^2)$ resulted in shifting the posterior more upwards, towards non-stationary regimes, at the same time increasing posterior probability of φ being greater than zero.

Posterior medians for precision τ, presented in Table 6.3, are also visibly higher than the ones based on respective Gamma priors assumed in the previous chapter, except for flows from Poland to Germany. In the latter case, this may be the effect of the priors being located around slightly smaller values, at least with respect to the means a priori; or, more importantly, it can be attributed to the fact that the $\Gamma(0.25, 0.25)$ priors from Chapter 5 have heavier tails than the $v_{11} \cdot \chi^2_{(3)} = \Gamma(1.5, 0.5/v_{11})$ priors used here[5]. In all other instances, the differences are likely due to heavier tails of the respective $v_{11} \cdot \chi^2_{(3)}$-based priors. Both

Table 6.3 Summaries of posterior distributions of the parameters of reduced AR(1) models: median, 2.5 and 97.5% quantiles (estimated by MCMC)

	Constant c			Autoregression parameter ϕ			Precision $\tau = 1/\sigma^2$		
Migration flows	2.5%	Median	97.5%	2.5%	Median	97.5%	2.5%	Median	97.5%
From Italy to Germany	−0.259	−0.010	0.236	**0.506**	**1.068**	**1.629**	7.94	15.88	27.99
From Germany to Italy	−1.169	−0.263	0.673	**−0.458**	**0.679**	**1.852**	17.86	41.57	80.63
From Poland to Germany	−0.624	0.395	1.371	−0.427	0.581	1.634	3.23	7.55	14.68
From Germany to Poland	−0.119	−0.009	0.101	**0.063**	**0.630**	**1.213**	11.79	27.84	54.39
From Switzerland to Germany	−0.036	0.043	0.120	**0.101**	**0.721**	**1.351**	54.54	109.00	192.20
From Germany to Switzerland	−0.360	0.320	0.995	**0.803**	**1.129**	**1.452**	32.89	77.92	152.50

Note: Boldface denotes 95% credible intervals for the estimates of c and ϕ, which do not cover zero.
Source: own elaboration in WinBUGS

[5]Note the relationship between the chi-squared and Gamma distributions: $\chi^2_{(k)} \equiv \Gamma(1/2 \cdot k,\ 1/2)$, as well as the scaling property of the Gamma distribution: $X \sim \Gamma(r,\ \mu) \Rightarrow k \cdot X \sim \Gamma(r, \mu/k)$.

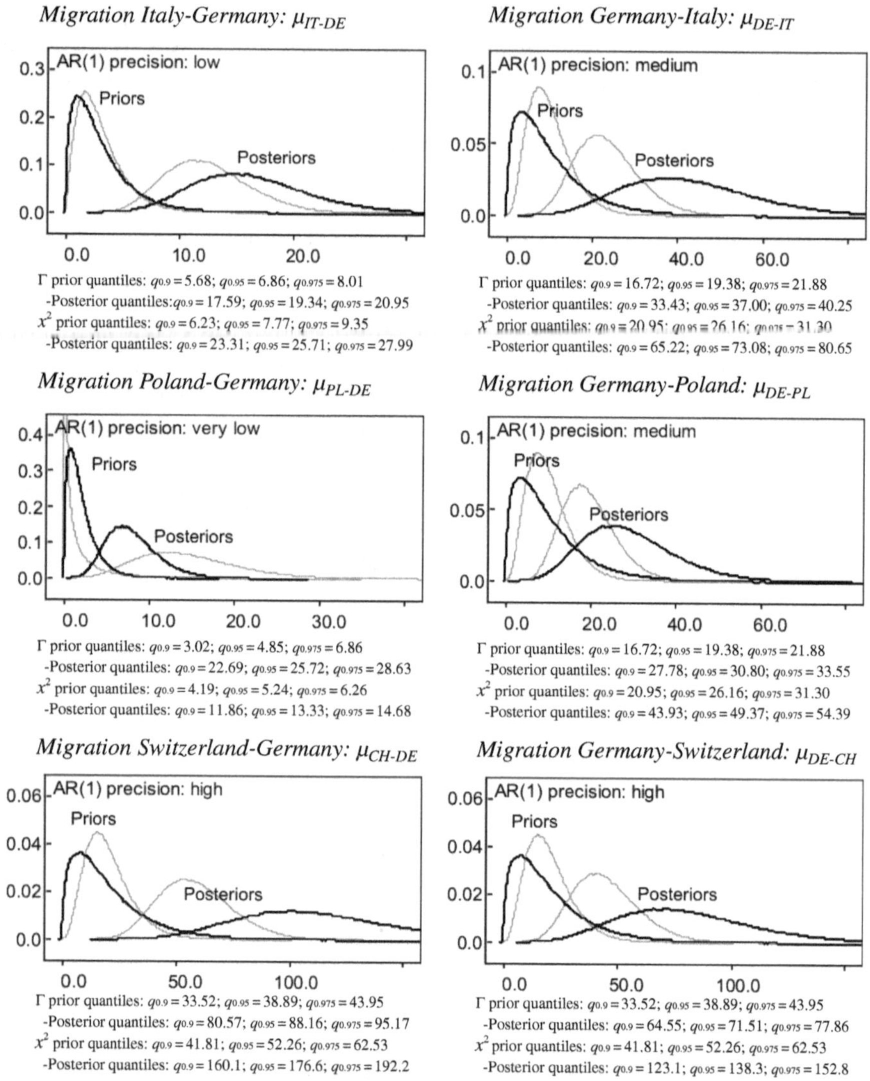

Fig. 6.1 Different prior and posterior distributions for precision in AR(1) models.
Note: *Grey lines* depict distributions related to Gamma priors from Chapter 5, while the *black* ones to the chi-squared priors from the current chapter.
Source: own elaboration in WinBUGS

types of prior distributions for τ (more concentrated Gamma and more diffuse chi-squared), as well as the respective posteriors are compared in Fig. 6.1. The results again indicate high sensitivity of the results to the selection of prior distributions, and especially to their tails, which is not surprising given the shortness of the time series.

6.2.2 Results of Forecasts from the General VAR and Marginal AR Models

Forecasts of the emigration rates under study for the period 2005–2015 yielded by general VAR(1) models with GDP per capita differentials and unemployment rates in the sending country have been computed alongside the ones obtained from the reduced univariate AR(1) models. The trajectories of the predicted median values, as well as the limits of the 80% predictive intervals for the forecasted emigration rates are also illustrated in Fig. 6.2, together with the historical data series. The graphs shown therein compare forecasts obtained from general VAR(1) and marginal AR(1) models. It is worth underlining that Fig. 6.2 presents the unconditional forecasts of a variable $x_{1,t}$ constituting the migration part of a VAR model according to partition introduced in (6.2). The summaries of predictive distributions of emigration rates, showing median values, and the 80% predictive intervals for 2005, 2010, and 2015, are offered in Table 6.4, while all yearly values are presented in Annex C, Table C.3a.

In addition, predictions from conditional distributions of log-transformed migration rates given the assumed trajectories of income differentials and sending-country unemployment rates, $x_{1,t}|\mathbf{x_{2,t}} = \mathbf{x^*_{2,t}}$, are shown in Fig. 6.3 and also summarised in Table 6.4. In this exercise, the pre-defined trajectories of migration determinants $\mathbf{x^*_{2,t}}$ are assumed arbitrarily as the medians of the respective predictive distributions of $\mathbf{x_{2,t}}$ in unconditional VAR models, calculated in an auxiliary run of the MCMC algorithm. As in conditional forecasts these trajectories are assumed to occur with probability one, both migration factors are treated de facto as non-random variables, illustrating only one of possible 'what-if' scenarios, which in the unconditional 'reality' bears a zero probability of realisation. Detailed summaries of such conditional predictive distributions are shown in Table C.3b, Annex C with the assumed input: median trajectories of GDP differentials and unemployment rates.

The forecasted trajectories presented in Fig. 6.2 indicate that in half of the cases, for m_{DE-IT}, m_{IT-DE}, and m_{CH-DE}, the median forecasts based on the general VAR(1) and the reduced AR(1) models are relatively similar. This outcome is also not surprising in the light of the results of Lindley's tests presented before, where the impact of both migration factors on the logarithms of emigration rates has been found insignificant.

For the remaining flows, m_{DE-PL}, m_{PL-DE}, and m_{DE-CH} however, the results visibly differ, with the VAR-based forecasts being higher in all three cases. With regard to the future values, a majority of the median forecasts yield plausible values, apart from the exploding ones for migration from Germany to Switzerland obtained from both the general VAR(1) and the reduced AR(1) models, although in the latter instance both median predictions yield values that are visibly higher than the ones observed in the data series.

The interpretation of median prediction trajectories derived from the reduced AR(1) models, indicated by solid grey lines in Fig. 6.2, can be recapitulated in the following way. Between 2005 and 2015, migration from Italy to Germany is

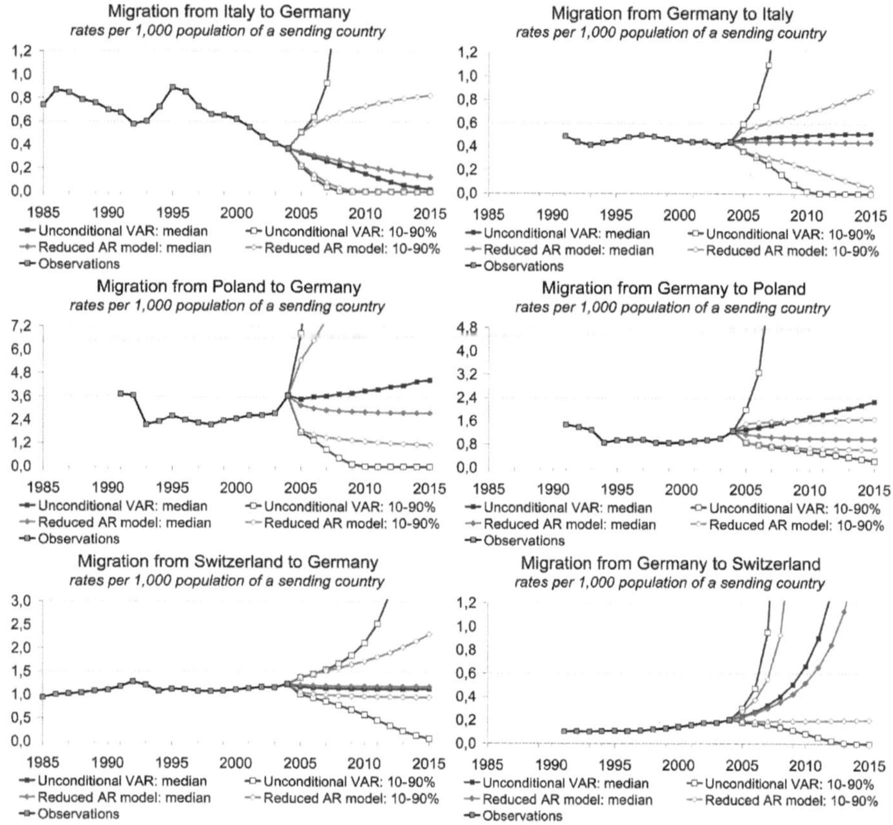

Fig. 6.2 Forecasted migration rates, 2005–2015: general VAR(1) and reduced AR(1) models. Source: own elaboration in WinBUGS. Data series until 2004: Eurostat/NewCronos; DESTATIS (see Annex A)

expected to decrease even more substantially than in the formally-selected and averaged forecasts presented in Chapter 5. On the contrary, population flows in the opposite direction are going to remain relatively stable throughout the forecast horizon. The same applies to both-ways migration between Germany and Poland, which are expected to remain almost constant until 2015, following the initial, post-EU-accession increase observed in 2004, and the forecasted return to the previously-observed or slightly higher levels starting from 2005. Steadiness of migration intensity applies also to the forecast of flows from Switzerland to Germany. However, the relative magnitude of population flows in the opposite direction is forecasted to increase very rapidly, due to a very high posterior probability of non-stationarity, for this particular model exceeding 0.75. Apart from the last case, all other median forecasts of the intensities of migratory movements under study seem reasonable from a demographic point of view. This conclusion holds even despite the fact that the posterior probabilities of non-stationarity are not negligible.

Table 6.4 Summaries of predictive distributions of emigration rates from VAR(1) and AR(1) models, for 2005, 2010 and 2015: median, 10 and 90% quantiles (MCMC estimates)

	2005			2010			2015		
Migration flow	10%	Median	90%	10%	Median	90%	10%	Median	90%
General VAR(1) model – unconditional forecast									
Italy–Germany, m_{IT-DE}	0.22	0.33	0.51	0.00	0.15	152.63	0.00	0.02	∞
Germany–Italy, m_{DE-IT}	0.36	0.46	0.59	0.02	0.50	142.31	0.00	0.51	∞
Poland–Germany, m_{PL-DE}	1.76	3.47	6.85	0.01	3.90	∞	0.00	4.47	∞
Germany–Poland, m_{DE-PL}	0.87	1.31	2.00	0.56	1.74	∞	0.25	2.30	∞
Switzerland–Germany, m_{CH-DE}	1.01	1.18	1.37	0.58	1.14	2.12	0.07	1.14	11.35
Germany–Switzerland, m_{DE-CH}	0.18	0.23	0.30	0.09	0.67	290.32	0.00	5.19	∞
General VAR(1) model – conditional forecast									
Italy–Germany, m_{IT-DE}	0.37	0.37	0.37	0.01	0.20	0.68	0.00	0.05	1.02
Germany–Italy, m_{DE-IT}	0.36	0.46	0.59	0.25	0.52	2.00	0.15	0.56	632.70
Poland–Germany, m_{PL-DE}	1.75	3.47	6.83	1.03	3.87	252.65	0.46	4.74	∞
Germany–Poland, m_{DE-PL}	0.86	1.30	1.98	0.71	1.75	140.33	0.65	2.55	∞
Switzerland–Germany, m_{CH-DE}	1.01	1.18	1.38	0.82	1.13	1.63	0.65	1.12	1.98
Germany–Switzerland, m_{DE-CH}	0.18	0.23	0.30	0.16	0.62	34.33	0.12	3.94	∞
Reduced AR(1) model									
Italy–Germany, m_{IT-DE}	0.23	0.34	0.51	0.00	0.22	0.74	0.00	0.13	0.83
Germany–Italy, m_{DE-IT}	0.36	0.44	0.54	0.22	0.44	0.69	0.05	0.44	0.88
Poland–Germany, m_{PL-DE}	1.83	3.15	5.46	1.30	2.82	14.01	1.11	2.77	82.27
Germany–Poland, m_{DE-PL}	0.88	1.15	1.51	0.68	1.02	1.63	0.63	1.00	1.68
Switzerland–Germany, m_{CH-DE}	1.06	1.21	1.39	0.97	1.19	1.73	0.95	1.19	2.31
Germany–Switzerland, m_{DE-CH}	0.19	0.23	0.28	0.19	0.52	4.61	0.20	2.28	∞

Note: Grey shading denotes implausible values, exceeding over 5 times the maximum levels observed in the sample (see notes before Table 5.5). '∞' refers to almost impossible values (rates over 1,000 per 1,000 inhabitants).
Source: own elaboration in WinBUGS

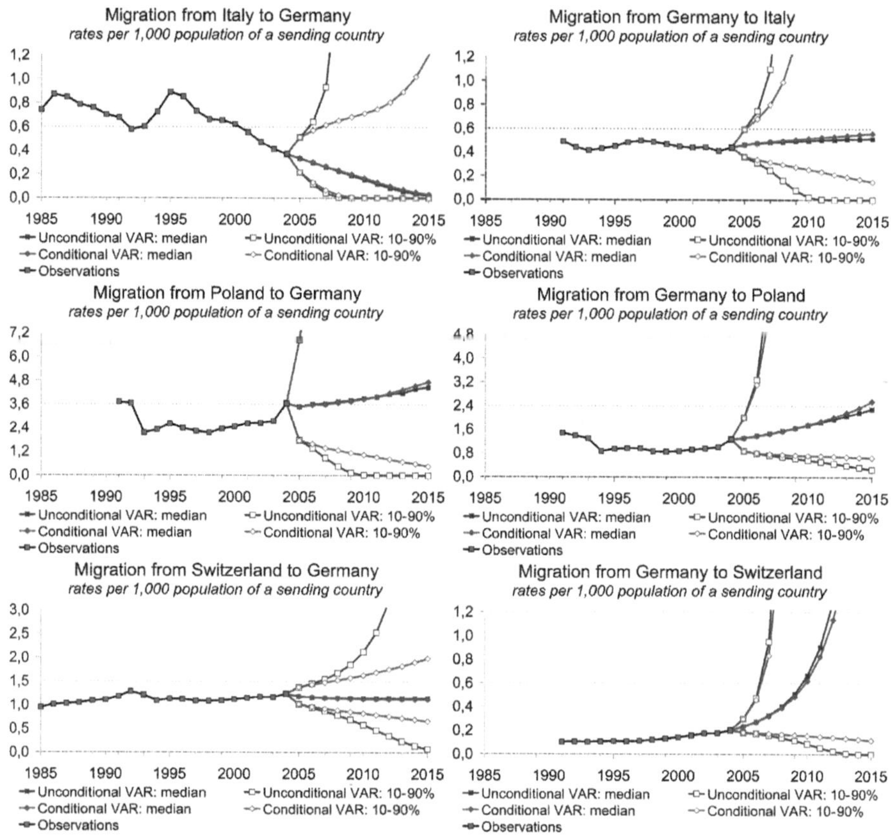

Fig. 6.3 Unconditional and conditional VAR(1) forecasts of migration rates, 2005–2015. Source: own elaboration in WinBUGS. Data series until 2004: Eurostat/NewCronos; DESTATIS (see Annex A)

As it can be also seen in Fig. 6.2, in all cases the reduction of the general three-dimensional VAR model to an univariate AR(1) resulted in a visible narrowing of the uncertainty span. The reduced models are also in some cases closer to stationarity than the original, multivariate ones. Apart from the German-Swiss and Polish-German migrations, the 80% credible predictive intervals for emigration rates, yielded by the reduced AR(1) models, seem plausible, which is definitely not the case of the unrestricted VARs. In a general case, this might be to some extent attributed to the fact that unconditional forecasts of emigration rates incorporate the uncertainty of predicting the explanatory variables: income differentials and unemployment, as well as of their interactions. The latter can be observed both with a time lag, through the parameter matrix **A** as well as instantaneously through the precision matrix **T**, in either case having an associated parameter uncertainty.

With respect to the outcome of conditional forecasts, from Fig. 6.3 it can be judged that, as it might be expected, the medians from both conditional and unconditional predictive distributions almost overlap. The slight exceptions concern strongly non-stationary processes towards the end of the forecast horizon, and are likely attributable to numerical inaccuracies. In all cases uncertainty of conditional forecasts is lower than in the case of the unconditional ones, which is not surprising, given that the former do not take into account the uncertainty of additional variables, treated here as given. Nevertheless, as in all instances the impact of both migration determinants on actual flows is insignificant, the differences in width between conditional and unconditional predictive intervals, although visible, are not substantial.

Another important problem concerns the forecasts of both explanatory variables used in the current analysis for conditioning: GDP differentials and sending-country unemployment rates. Judging solely by the medians from the predictive distributions of both predictors[6], these forecasts also need not be plausible. While the forecasts obtained from models for μ_{DE-IT} and μ_{CH-DE} indicate stable economic developments, the other ones implicitly assume changes in the macroeconomic setting of the countries under study. Thus, the models for μ_{IT-DE} predict an extremely good economic situation of Italy, with GDP growth coupled with a rapid decline in unemployment. The model for μ_{DE-PL} envisages economic convergence between Poland and Germany and some labour market difficulties in the latter country, while the forecast for μ_{DE-CH} implies a dramatic German economic crisis. Needless to say, the forecasts for μ_{DE-CH}, μ_{DE-IT} and μ_{DE-PL} strongly differ with respect to the foreseen sending-country economic developments.

Nevertheless, despite some inconsistencies, the above-mentioned forecasts can be seen as at least internally plausible (that is, within a particular model). However, this is definitely not the case with the forecasts for μ_{PL-DE}, which are neither plausible from an economic point of view, nor internally consistent. In particular, the forecasted extreme increase of median unemployment in Poland (from nearly 19% in 2004 to over 36% in 2015) is coupled with equally extreme economic growth (GDP per capita increase from about a half to 125% of the German one within just ten years). All outcomes such as the ones listed above additionally call for extraordinary attention in forecasting migration using additional socio-economic variables on the basis of very short time series, which are generally hardly suitable for such analyses. In practical applications, the macroeconomic variables could be modelled not only on the basis of longer series of higher-frequency data, but also using more complex models, such as for example structural VAR (SVAR) or vector error correction models (VECM).

[6]For yearly values of forecasted median trajectories of migration determinants, see Table C.3b in Annex C.

6.2.3 'From General to Specific' Modelling: Discussion of the Outcomes

The results of the presented VAR modelling and reduction according to the 'from general to specific' approach suggest that the impact of two main macroeconomic determinants: per capita GDP differentials and unemployment, proved to be insignificant in the cases under study. However, this conclusion, which at first sight seems contrary to migration theory, as well as empirical observations (especially for Polish-German migration) needs to be treated with caution.

To start with, it has to be noted that a general thesis about a lack of impact of important macroeconomic determinants on migration would be too far-going, given the limited information provided by the data samples. A more appropriate supposition would be that available migration statistics do not allow for empirical testing of particular theories, but nevertheless additional variables can contribute to prediction-making, at least with respect to the construction of formal and probabilistically-coherent 'what-if' scenarios. Nevertheless, even if (hypothetically) the reductions of general VAR models were illegitimate, which would suggest that forecasting can include variables recommended by the theories, the price to pay would be an extremely high predictive uncertainty, far exceeding any reasonable quantities. In contrast, unjustified reductions could potentially lead to errors and discrepancies both in the central tendencies, as well as in the uncertainty spans. Summing up, the outcomes obtained in the empirical example seem to be primarily attributable to the weak statistical properties of samples rather than flaws in theories or model misspecification. Both issues are elaborated in more detail below.

Firstly, proper attention should be paid to the forecasts of migration determinants in a dynamic setting, such as VAR models. In order to overcome some of the problems mentioned in the previous subsection, analyses aimed at producing policy-relevant empirical outcomes should involve a careful selection of appropriate priors, accompanied by a sensitivity analysis. Another possibility is to add more structure into VAR modelling in order to ensure more consistency of the outcomes. In particular, models for various flows can be run within a system of VAR equations rather than separately, applying proper deterministic constrains (e.g., unemployment in a given country has to be the same for all models involving this variable, a fraction of GDP in country A to the one for country B should be a reciprocal of the respective fraction for the reversed order of countries, etc.). For conditional forecasts, as mentioned before, predictions of macroeconomic variables alone could be based on longer data series (should these become available), including data of higher frequency than annual, or involve more complex models, such as for example VECM.

However, another argument for the limited usefulness of applying the VAR(1) models in forecasting migration rates, especially with respect to the uncertainty span, may be a too complex model structure given the small number of observations. Estimation of 18 parameters based on a time series comprised of 20, or even 14 observations, has to bear very high errors, and produce implausible results, even within the Bayesian framework and with the help of a priori assumptions. In the presented instances, the estimation of the VAR parameters yielded non-stationary models, at best being close to three-dimensional random walk processes[7]. As it can be seen in Fig. 6.2, in most of the examples shown above, apart from m_{DE-CH}, the reduction of an initial complex VAR reduced the *ex-ante* forecasts errors. This result provides an additional argument for using relatively simple models in the context of European international migration. The differences in predictive distributions obtained from general multivariate VAR(1) and marginal AR(1) processes are also an indication that some residual impact of both macroeconomic predictors on migration remains, despite negative results of Lindley's tests. This finding additionally supports all caveats concerning small-sample inference on the impact of additional variables on population flows.

Nevertheless, again given very small sample sizes, a verification of the above-mentioned conclusions from the point of view of the cointegration analysis, for example via the Augmented Dickey-Fuller test or the Johansen test (cf. Charemza & Deadman, 1992/1997) seems pointless. Inference on long-term relationships, such as cointegration between particular variables, on the basis of time series of 14 or 20 observations would not produce any meaningful results. In such cases neither does the 'lack of impact' preclude the existence of long-term relationships, nor the 'significant impact' a lack thereof, as either result can equally well reflect the deficiencies of data. Nevertheless, even if longer series of observations, or data of higher frequency than annual, were available, a full Bayesian analysis of cointegration would be very complex, both theoretically and computationally (cf. Wróblewska, 2009), and as such has not been considered in the current study.

The results delivered by the reduced AR(1) models compared with their counterparts produced by autoregressive processes from Chapter 5 again indicate the sensitivity of the results to the selection of priors for precision, already suggested by the graphs presented in Fig. 6.1. As regards the median trajectories, they are similar for both types of models in four cases: for m_{DE-IT}, m_{PL-DE}, m_{DE-PL} and m_{CH-DE}. For migration from Italy to Germany (m_{IT-DE}), the reduced AR(1) model presented herein yielded slightly lower median values than the one from Chapter 5, while for flows from Germany to Switzerland (m_{DE-CH}), the opposite holds. The uncertainty intervals in almost all reduced AR(1) models, except for the ones concerning migration from Poland to Germany and from Germany to Italy, are visibly narrower than in the case of models from the previous section. This is a consequence of generally higher precision obtained under chi-squared priors with somewhat heavier tails. Again, as shown in Fig. 6.1, m_{PL-DE} is an exception here.

Interestingly, in the case of migration rates observed for 2005–2007, as listed in Table 5.6, for two flows (m_{DE-CH} and m_{DE-PL}), the general VAR models yielded smaller *ex-post* errors than the $\mathbf{M_1}$-class models, discussed in Chapter 5. In terms of average accuracy, however, the VAR(1) were very close to the random walks, which were leading the competition amongst $\mathbf{M_1}$. Such a relatively good result can be

[7]For detailed posterior summaries of parameters of the VAR(1) models under study, see Table C.1 in Annex C.

due to including some theory-based information in the VAR processes, albeit at the expense of extremely high predictive uncertainty *ex ante*. In only one case, for flows from Germany to Italy, the 80% predictive intervals obtained from VAR models are too narrow to include the actual 2005 realisation of the forecasted variable, m_{DE-IT}, and the same applies to the respective forecasts from a reduced AR model for 2005 and 2006.

Chapter 7
Selected Approaches to Discontinuities in Trends

This chapter presents some basic possibilities of acknowledging unpredictable events and structural changes, which are important factors causing discontinuities in observed migration trends. Also here, the relevant models are proposed within the Bayesian framework. In particular, two classes of models are considered: simple historical analogies and extensions of autoregressive models to cases with non-constant conditional variance. After a theoretical introduction of both types of models in Section 7.1, empirical applications to the forecasts of migration flows under study are offered in Section 7.2.

7.1 From Deterministic Analogies to Stochastic Volatility

7.1.1 The Simplest Options: Dummy Variables and Forecasting by Analogy

As noted in Chapter 4, an important flaw of the existing migration theories, which can be also easily attributed to a vast majority of migration models in general, is the fact of ignoring forced migration and migration policy factors, as well as economic system shocks, which all are crucial for the actual magnitudes of observed population flows (Öberg & Wils, 1992; Kupiszewski, 2002a). As observed by Keilman (1990, p. 20, see also Chapter 2), the discontinuity in trends is a major source of prediction errors in forecasting, and this conclusion is by no means limited to modelling demographic phenomena. In the current section an attempt is made to touch upon this problem, presenting a handful of simple examples, which may be potentially useful for practical applications in migration studies.

One of the most straightforward options to deal with the sudden changes in the observed or forecasted series of the phenomenon under study is to include dummy variables in the models, equal zero for 'regular' circumstances, and one for periods with visibly higher or lower observations, in order to modify either the intercept, or other structural parameters. In the context of migration, the dummies can be, first of all, used only to accommodate atypical observations from the data sample, as is done for example in the studies of Keilman and Pham (2004b), Jennissen (2004), or Bijak (2008a). In this way, the estimated model is not biased by the data points,

J. Bijak, *Forecasting International Migration in Europe: A Bayesian View*, The Springer Series on Demographic Methods and Population Analysis 24, DOI 10.1007/978-90-481-8897-0_7, © Springer Science+Business Media B.V. 2011

which mark only temporary deviations from the long-term tendency, and such a tendency in the observed patterns is therefore more easily detectable. However, such an approach can be also dangerous in that removing atypical observations (outliers) from the sample can artificially reduce the predictive uncertainty assessment.

The presented option can be extended to the predicted future values of the variables under study. However, the limitation of such an approach is that the forecaster needs to make very strong assumptions concerning the future, notably with respect to the time, when the deviations from the predicted trend are expected to occur, as well as to their magnitude. This would require a constant monitoring of the whole political, economic and social settings in a given country and, as such, is deemed hardly feasible in practice. Although the strength of these propositions renders this approach of little use in 'pure' migration forecasting problems, it may be beneficial for example for dealing in a formal way with the 'statistical artefacts' in migration data. The potential applications may concern 'administrative adjustments' of migration figures on the basis of population censuses (see also Section 2.1 and Annex A). In such cases, the dummy variables may be used for separating migration based on population registers from the after-census corrections also in the forecast period.

Determining the value of a parameter by the dummy variable can also be performed without a reference to the past data on the phenomenon under study, but using an analogy instead. Such an approach has been used for example by Sinn et al. (2001). In their model of foreign population stocks in Germany, the authors calibrated the parameters by the dummies regarding the EU membership and the freedom of movement of the labour force of the then-accession countries, using the historical data on population from Greece, Italy, Portugal, Spain, and Turkey.[1] A probabilistic alternative to parameter calibration would be to subjectively assume certain probability distributions for the parameters in question, akin to 'pure' priors, but not subject to the a posteriori updating based on the data. However, in the fully Bayesian context, such an approach is inevitably deficient, as it does not explore the available sample information.

In order to combine the advantages of the mentioned possibilities, and remain within the fully Bayesian framework, in the current section a model based on historical analogies from other countries is proposed. The model assumes that the impact of certain, for instance political, events which are expected to change the trends of the variable under study, can be analogous to the effect of similar factors observed in the past in the other countries. For example, the recent enlargement of the European Union in 2004 to include ten new Member States can be considered as analogous to the preceding ones concerning Ireland in 1973, and Greece in 1981, as well as Portugal and Spain in 1986 (Kupiszewski, 1998; see also Section 5.1). Letting $\mu_{i-j}(t)$ represent the log-transformed migration rates under study, and $\mu^a_{i-j}(t-k)$ concern the countries, which were in an analogous situation in the past, k years before, the proposed model can have a form:

$$\mu_{i-j}(t) = c + \alpha\, \mu^a_{i-j}(t-k) + \beta\, z(t) + \varepsilon_t. \qquad (7.1)$$

[1] See explanations to equation (4.8) in Section 4.2.

The random component ε_t can be assumed for example as a Gaussian white noise, or a Gaussian autoregressive process AR(1): $\varepsilon_t = \varphi\ \varepsilon_{t-1} + v_t$, where $v_t \sim$ iid N(0, σ^2). The variable $z(t)$ is an optional dummy used to accommodate the atypical observations from the data sample for the explanatory 'analogy' flows. In general, the removal of outliers from the sample is rather discouraged, as it would lead to an unjustified and artificial reduction of the predictive uncertainty. However, in this particular case it can be assumed that conditions that caused nuisance, embodied in the analogy variable, had a one-off character, were limited to some country-specific events or circumstances, and thus are unlikely to replicate in the future for the migration flows under study.

Given the availability of appropriate data series, the current study analyses a model class $\mathbf{M_2}$ containing three models with linear analogies of the form (7.1), respectively related to Portugal (model M_7), Spain (M_8) and both Iberian countries together (M_9). These models are subsequently compared with the reference one, assuming no analogy (model M_6) using the formal selection criteria introduced in Chapter 5. The results of estimation and prediction are offered in Section 7.2.

7.1.2 Models with Changing Conditional Variance ARCH, GARCH, Stochastic Volatility

Apart from shifts in intercepts and trend parameters, another issue of concern in migration studies are changes in variance, which potentially may to some extent accommodate for unexpected events generating unusual waves of migration (see Chapter 1 for discussion). In the econometric literature, there are two most common approaches to this problem. The first one, assuming deterministic patterns of variance change, stems from the autoregressive conditional heteroskedasticity (ARCH) models of Engle (1982; after: Greene, 2000, pp. 800–802), and the generalised versions thereof – the GARCH models of Bollerslev (1986; *idem*). Given the basic underlying regression model $y_t = \boldsymbol{\beta}'\mathbf{x_t} + \varepsilon_t$, with ε_t usually assumed to follow N(0, σ_t^2) conditionally on its past history (usually denoted by $\boldsymbol{\psi}_{t-1}$), the ARCH model of order q is defined for the conditional variance σ_t^2 as akin to the moving average process MA(q) of the squared random component, only without the term ${\varepsilon_t}^2$ (e.g., Greene, *idem*, p. 800):

$$\sigma_t^2 = k + \sum_{i=1}^{q} \alpha_i \cdot \varepsilon_{t-i}^2. \tag{7.2}$$

The GARCH(p,q) generalisation additionally involves the dependence of the conditional variance on its own values from the past, given all information available at time $t-1$, $\boldsymbol{\psi}_{t-1}$. In such a way, the conditional error term $\varepsilon_t \mid \boldsymbol{\psi}_{t-1}$ is typically presumed to follow N(0, σ_t^2), where (*idem*, p. 801):

$$\sigma_t^2 = k + \sum_{i=1}^{q} \alpha_i \cdot \varepsilon_{t-i}^2 + \sum_{j=1}^{p} \beta_j \,.\, \sigma_{t-j}^2\,. \tag{7.3}$$

The second approach to acknowledging the changes in conditional variance is by nature stochastic, and such models are referred to as having the *stochastic volatility*, or *stochastic variance* (SV) property. The SV concept originates at least from the work of Clark (1973; after: Jacquier, Polson, & Rossi, 1994, p. 371). There exists a variety of the SV examples. Among them, the simplest one, based on the underlying model $y_t = \boldsymbol{\beta}'\mathbf{x}_t + \varepsilon_t$, again usually with $\varepsilon_t \sim N(0, \sigma_t^2)$, can be obtained by defining the log-transformed conditional variance in terms of a Gaussian autoregressive process AR(1):

$$\ln(\sigma_t^2) = k + \gamma \cdot \ln(\sigma_{t-1}^2) + \xi_t, \text{ with } \xi_t \sim \text{iid } N(0, \rho^2). \tag{7.4}$$

As noted for example by Congdon (*idem*, p. 210), the character of the (G)ARCH and SV processes makes them very useful in modelling financial variables, such as stock prices or exchange rates, where the variance changes over time, and is characterised by visible clusters of observations with high, respectively low volatility. The possible relevance of (G)ARCH and SV models for demography has been discussed by Alho and Spencer (2005, pp. 220–223), albeit only from a theoretical perspective, without empirical applications. Nevertheless, the mentioned features of these multi-level (hierarchical) models make them potentially useful for migration forecasts, especially when longer time series of data can become available. In particular the SV property in general can be interpreted as the unpredictable character of changes of the phenomena under study, which in migration research is a very natural initial premise.

Despite the current study being based on relatively small samples of observations, an empirical illustration of assuming simple (G)ARCH or SV patterns for the variance of migration between Poland and Germany is presented in Section 7.2, with a comparison of various models according to the posterior odds criterion. For the analysis of potential changes in variance, a model class $\mathbf{M_3}$ consisting of four basic models has been defined. The underlying constant-variance model M_{10} is assumed to follow a Gaussian autoregressive process of order 1, AR(1): $\mu_{i-j}(t) = c + \varphi\, \mu_{i-j}(t-1) + \varepsilon(t)$. In its three extensions, M_{11}, M_{12}, and M_{13}, the disturbances $\varepsilon(t)$ have been respectively assumed to follow ARCH(1), GARCH(1,1), and SV processes, according to the definitions in (7.2), (7.3) and (7.4). Processes of higher orders are not considered due to the small sizes of the samples.

7.2 Example: Forecasts from Models with Discontinuities

7.2.1 Application to Polish-German Flows: Models with Analogy to Iberian Migration

In order to illustrate various models designed to deal with discontinuous trends, the forecasts of both-ways migration between Poland and Germany are examined. These flows have been selected both due to the visible fluctuations of their values in the past, and also those expected for the future. In particular, this is related to the fall of the Iron Curtain and the economic transformation in Poland, which in the period 1988–1990 generated a high wave of migration to and from Germany. Moreover, the migratory processes prior to and after 1989 inevitably differed with respect to their character, the former being to an extent political (or merely perceived as such, which might have masked the economic character), while the latter – mostly economic. In addition, there is the recent issue of European integration and the accession of Poland to the EU in 2004. What follows is the introduction of the full freedom of the movement of the labour force between Poland and Germany, likely expected for 2011; and also very likely to cause the deviation of migratory processes from their long-term trends.

First of all, consider a migration forecasting model (7.1), designed to include an analogy to the past experience of other countries. In such a model, let all variables be expressed in terms of log-transformed emigration rates. In the current section, three such analogies are examined, concerning the both-ways migration between Germany and Portugal (model M_7), Spain (M_8) or both Iberian countries jointly (M_9), in either case taken with a time lag of 18 years.[2] These models are then compared with a reference constant model with autoregressive random term (M_6) using the formal selection criteria introduced in Chapter 5, all four adding up to a model space $\mathbf{M_2}$ for the purpose of performing a Bayesian model selection. The series of Polish-German migration concern the period 1992–2004, and the forecast is made for 2005–2015, while the corresponding periods for the Portuguese or Spanish migration to and from Germany are respectively 1974–1986 and 1987–1997. An atypical observation of migration from Germany to Portugal for 1984 was removed from the sample using a dummy variable, as shown in (7.1). For the remaining flows, no dummy was used.

In such a way, the assumed analogy is as follows: Portugal and Spain joined the then-EEC in 1986, and the labour markets of the other then-EEC countries, including Germany, have fully opened to the Portuguese and Spanish citizens in 1992 (Kupiszewski, 1998). Poland joined the EU in 2004, and judging by the political declarations made at the time of preparing the study, the German labour market is expected to open for the Poles by 2011 (for discussion, see for example Bijak et al., 2007). Nevertheless, in the current model the full liberalisation of the flow of workers is envisaged to happen in 2010, in order to preserve the analogy in terms of the timing. Despite visible similarities, such as the presence of political and economic system transformations in all countries concerned (in Spain and Portugal since the mid-1970s and in Poland since late 1980s), the analogy is not perfect. Most importantly, Poland is a neighbour of Germany, and long before the EU enlargement many Poles had settled in the latter country forming important migrant networks, which was not the case with the citizens of Portugal or Spain.

[2] Hence, the data for the countries of analogy concern the period 1974–1997. The figures come from the German statistical registration, as reported by the Federal Office for Statistics, Wiesbaden, and have been obtained via: www.destatis.de/genesis (accessed on 21 September 2006).

For each of the flows and possible analogies, the prior distributions for the parameters of the model (7.1), assuming autoregressive errors $\varepsilon_t = \varphi\ \varepsilon_{t-1} + \nu_t$, and $\nu_t \sim$ iid N(0, σ^2), are as follows: $c \sim$ N(0, 100^2), $\varphi \sim$ N(0.5, 1^2) and $1/\sigma^2 = \tau \sim \Gamma(0.25,\ 0.25)$ for migration from Poland to Germany and $\tau \sim \Gamma(4,\ 0.4)$ for the counter-flow in the opposite direction. Such priors favour the stationarity of the random component, albeit with a large degree of uncertainty, and assume its relatively low precision, especially in the case of migration from Poland to Germany. The 'analogy parameter' is believed to be Normal, $\alpha \sim$ N(0.5, 1^2), which indicates a presumption of a likely positive relationship between the modelled migration flow and its past counterpart, although again with relatively high uncertainty. Similarly, the dummy parameter β is a priori set to follow N(0, 1^2), reflecting a belief in no impact of the atypical observation, albeit allowing for a possibility of error in this reasoning.

The models have been estimated using the same MC^3 Carlin and Chib (1995) algorithm as presented and applied in Chapter 5, with 200,000 iterations in the burn-in phase, and a further 250,000 in the final sample. The WinBUGS programme code is listed in Annex B, and the posterior estimates of parameters are summarised in Table 7.1, while the update of prior distributions of the analogy parameters α into posteriors are illustrated in Fig. 7.1. Both Table 7.1 and Fig. 7.1 show that in the models of log-transformed Polish-German migration rates assuming analogy to the similar Portuguese and Spanish experience lagged 18 years, hardly any structural parameters can be seen as a posteriori significant. This indicates that the presence of a linear analogy between the flows did not earn support from the observed data, which pulled the prior beliefs in positive values of the parameter α towards zero.

This conclusion is additionally supported by the results of the Bayesian model selection procedure over the space $\mathbf{M_2}$, which could be conducted, as the lagged analogous migration rates are treated here as exogenous variables and do not affect conditioning on the data $\mathbf{x}$. Due to the shortness of time series, the model choice should a priori penalise more complex models by using the 'Occam's razor' priors (see Chapter 5), whereby $p(M_i)$ equals 0.2 for each of models with analogy (not counting parameters by dummies), and 0.4 for the reference one.

By a formal comparison, the calculated posterior probabilities for models with analogies (M_7–M_9) proved to be much smaller than for the reference ones (M_6), and also smaller than the respective $p(M_i)$, indicating hardly any data support for such analogies as defined in $\mathbf{M_2}$. More precisely, $p(M_6 \mid \mathbf{x})$ totalled 0.69 for migration from Poland to Germany and 0.88 for flows from Germany to Poland, $p(M_7 \mid \mathbf{x})$ being equal respectively 0.07 and 0.01, $p(M_8 \mid \mathbf{x}) = 0.14$ and 0.09, while $p(M_9 \mid \mathbf{x}) = 0.10$ and 0.02. All these results do not support the existence of a simple, linear Polish-Iberian analogy of emigration flows. If anything, the posterior probabilities, as well as the results shown in Fig. 7.1 indicate a very slight analogy to both-ways Spanish migration rather than the Portuguese one, though still associated with substantial uncertainty. Such results indicate that either the analogy does not exist, or is not as direct and simple (linear), as stipulated by the definition of the model class $\mathbf{M_2}$.

Table 7.1 Summaries of posterior distributions of the parameters of models with analogy: median, 2.5 and 97.5% quantiles (estimated by MCMC)

	Constant c			Autoregression parameter ϕ			Precision $\tau = 1/\sigma^2$		
Model (analogy)	2.5%	Median	97.5%	2.5%	Median	97.5%	2.5%	Median	97.5%
Migration from Poland to Germany									
M_6: No analogy (reference)	**0.74**	**0.96**	**1.21**	−0.43	0.51	1.20	5.63	14.32	29.51
M_7: Analogy to Portugal	**0.69**	**1.06**	**1.49**	−0.27	0.64	1.24	5.25	13.82	29.25
M_8: Analogy to Spain	**0.42**	**1.42**	**2.52**	−0.26	0.64	1.22	5.43	14.38	30.46
M_9: Analogy to both countries	**0.52**	**1.23**	**2.05**	−0.25	0.65	1.24	5.31	14.06	29.82
Migration from Germany to Poland									
M_6: No analogy (reference)	−0.20	0.01	0.22	−0.22	0.58	1.22	9.09	18.45	32.86
M_7: Analogy to Portugal [a]	−0.80	0.00	0.77	−0.22	0.56	1.28	7.94	16.71	30.61
M_8: Analogy to Spain	−0.33	0.16	0.64	−0.26	0.56	1.21	8.77	18.02	32.37
M_9: Analogy to both countries [a]	−0.32	0.09	0.48	−0.31	0.52	1.28	7.99	16.86	30.77

	Analogy parameter α			Dummy coefficient β		
Model (analogy)	2.5%	Median	97.5%	2.5%	Median	97.5%
Migration from Poland to Germany						
M_6: No analogy (reference)	-	-	-	-	-	-
M_7: Analogy to Portugal	−0.24	0.11	0.52	-	-	-
M_8: Analogy to Spain	−0.33	0.28	0.96	-	-	-
M_9: Analogy to both countries	−0.30	0.19	0.76	-	-	-
Migration from Germany to Poland						
M_6: No analogy (reference)	-	-	-	-	-	-
M_7: Analogy to Portugal [a]	−0.38	−0.01	0.36	−0.58	0.01	0.58
M_8: Analogy to Spain	−0.24	0.13	0.47	-	-	-
M_9: Analogy to both countries [a]	−0.31	0.10	0.47	−0.51	−0.04	0.43

Notes: [a]With dummy. Boldface denotes 95% credible intervals for the estimates of c, ϕ, α and β, which do not cover zero.
Source: own elaboration in WinBUGS

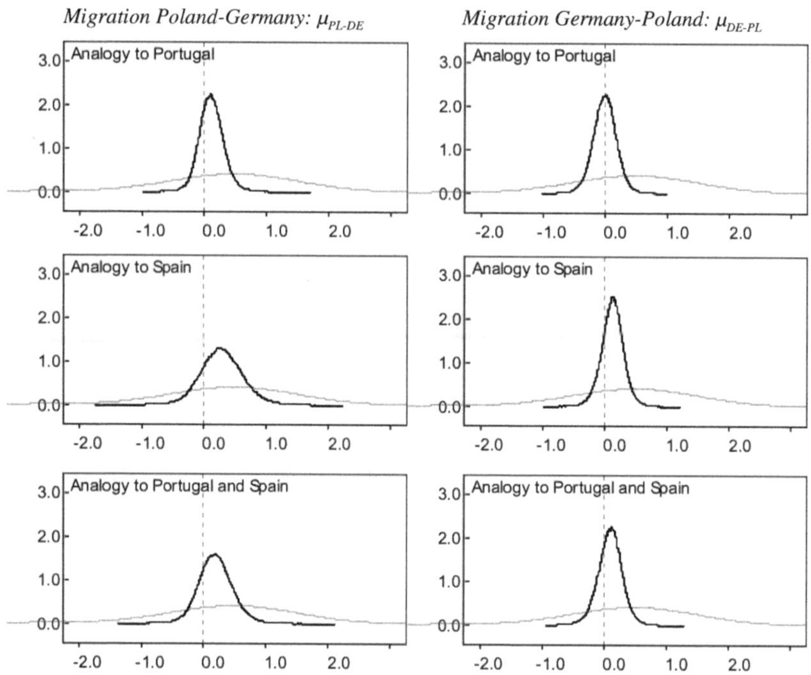

Fig. 7.1 Prior and posterior distributions of the analogy parameters α. Notes: Grey lines depict prior, while black ones – posterior distributions. Dashed vertical lines indicate zero. Source: own elaboration in WinBUGS

7.2.2 Models with Changing Conditional Variance: Model Selection for AR(1) Extensions

In an attempt to capture in the model three periods of different institutional background and potentially also migration volatility: before, during and after the system transformation, and three extensions of a Gaussian AR(1) model with non-constant conditional variance are considered, as defined in Section 7.1, formulae (7.2) – (7.4). These include the AR(1)–ARCH(1) model M_{11}, the AR(1)–GARCH(1,1) model M_{12} and the simplest AR(1)–SV (stochastic volatility) process, M_{13}. Thus, the proposed class of models, $\mathbf{M_3}$, encompasses the most straightforward processes characterised by either deterministic (ARCH/GARCH) or stochastic (SV) changes in conditional variance, which are compared with the constant-volatility reference AR(1) model, labelled as M_{10}.

In order to capture relatively longer periods with changes in volatility, in both cases the time series of 20 observations (1985–2004) has been used both for Polish-German and German-Polish flows. The German data prior to 1991 concern only West Germany (the then Federal Republic). Also here, a comparison of competing models has been performed in the Bayesian model selection framework on the basis of the posterior odds criterion, as outlined in Chapter 5. The prior probabilities for particular models $M_i, i \in \{10, 11, 12, 13\}$, follow Occam's principle. Hence, given the numbers of parameters l_i for particular models: $l_{10} = 3$, $l_{11} = 4$ and $l_{12} = l_{13} = 5$, the prior probabilities, proportional to 2^{-l_i} and respectively equal $p(M_{10}) = 0.5$, $p(M_{11}) = 0.25$, $p(M_{11}) = 0.25$ and $p(M_{12}) = p(M_{13}) = 0.125$. In this way, the models with more parameters are a priori penalised as less appropriate for a study based on a relatively small number of observations.

For all parameters, the following prior distributions have been set: in the basic level of the hierarchy, thus in the underlying AR(1) models, the constants c are assumed to be diffuse, following N(0, 100^2), and the autoregression parameters φ to favour stationary processes, by presuming $\varphi \sim$ N(0.5, 1^2). The precision $\tau_1 = 1/\sigma_1^2$ of the random component in M_{10}, follows a priori the same Gamma distributions as in Table 5.1: $\Gamma(0.25, 0.25)$ for μ_{PL-DE} and $\Gamma(4, 0.4)$ for μ_{DE-PL}. Again, this reflects beliefs in a relatively low precision of the statistical noise, albeit somewhat higher for migration from Germany to Poland. It is worth noting that although the proposed model M_{10} is the same as its counterpart defined in Chapter 5, the results of its estimation may differ, as the current one is based on different (longer) series of data.

In the extended models M_{11}, M_{12}, and M_{13}, the respective second-level constants, k_1, k_2 and k_3, defined as in (7.2), (7.3) and (7.4), are a priori assumed to follow a Gamma distributions $\Gamma(1, 0.1)$, with mean equal ten and variance 100. The second-level parameters α, β and γ are presumed to follow $\Gamma(10, 20)$, thus to be strongly concentrated around the mean 0.5 with variance 0.025. Such somewhat artificial assumptions are necessary in order to enable numerical calculations, practically enforcing the stationarity of the modelled volatility and not allowing it to explode over time. The reciprocal ψ of the second-level variance in M_{13}, ρ^2, has been assumed a prior distribution $\Gamma(10, 1)$, with mean and variance equal ten. In general, in hierarchical models, all above-mentioned priors are conditional on the respective main-level specifications, but this can be overcome by assuming independence of the parameters under study. As compared to M_{10}, they reflect a belief in a higher, albeit changing, precision of the error term. Additionally, artificial assumptions for $t = 1$ have been made: $\tau_{2,t=1}$, $\tau_{3,t=1} \sim \Gamma(1, 0.1)$ and $\ln(_{4,t=1}) \sim$ N(0, 1^2).

The estimation is MCMC-based, with 100,000 burn in and 250,000 sample iterations, and the WinBUGS code is listed in Annex B. The remarks on the use of the Carlin and Chib (1995) algorithm for model selection, made in Chapter 5 remain in force. The posterior estimates of parameters calculated under the above assumptions are listed in Table 7.2. Here, one remarkable result is the a posteriori significance of the main-level autoregression parameter, φ. For many other parameters, significance is in a way imposed on them by using relatively tight Gamma priors (as on k, α, β, and γ). Such 'enforcement' has been done in order to avoid numerical problems with exploding values of the main-level variance, which otherwise have been encountered while running the MCMC procedure. The posterior probabilities of non-stationarity of the main-level AR(1) processes are not negligible, although relatively small, with the exception of M_{11} (and also M_{12} for μ_{DE-PL}), where stationarity is a posteriori very likely. As compared with the empirical results presented

Table 7.2 Summaries of posterior distributions of the parameters of AR(1) model extensions: median, 2.5 and 97.5% quantiles (estimated by MCMC)

	Main-level constant c			Main-level AR(1) parameter ϕ			Main-level precision $\tau = 1/\sigma^2$		
Migration flow and model M_i	2.5%	Median	97.5%	2.5%	Median	97.5%	2.5%	Median	97.5%
M_{10} for μ_{PL-DE}: reference AR(1)	−0.15	0.29	0.73	**0.43**	**0.77**	**1.11**	3.61	7.70	14.08
M_{11}: AR(1)–ARCH(1)	−0.04	0.41	0.78	**0.22**	**0.57**	**1.03**		see Table C.5[a]	
M_{12}: AR(1)–GARCH(1,1)	−0.11	0.25	0.65	**0.35**	**0.75**	**1.12**		see Table C.5[a]	
M_{13}: AR(1)–SV	−0.09	0.28	0.64	**0.44**	**0.76**	**1.09**		see Table C.5[a]	
M_{10} for μ_{DE-PL}: reference AR(1)	−0.08	0.05	0.17	**0.42**	**0.77**	**1.13**	8.59	15.86	26.46
M_{11}: AR(1)–ARCH(1)	−0.09	0.00	0.12	**0.28**	**0.64**	**0.94**		see Table C.5[a]	
M_{12}: AR(1)–GARCH(1,1)	−0.10	−0.00	0.11	**0.34**	**0.71**	**1.03**		see Table C.5[a]	
M_{13}: AR(1)–SV	−0.07	0.04	0.15	**0.43**	**0.78**	**1.13**		see Table C.5[a]	
	Second-level constant k			ARCH(1) parameter α			GARCH(1,1) parameter β		
Migration flow and model M_i	2.5%	Median	97.5%	2.5%	Median	97.5%	2.5%	Median	97.5%
M_{11} for μ_{PL-DE}: AR(1)–ARCH(1)	0.01	0.05	0.16	0.27	0.53	0.91	–	–	–
M_{12}: AR(1)–GARCH(1,1)	0.00	0.01	0.07	0.25	0.48	0.82	0.22	0.40	0.60
M_{11} for μ_{DE-PL}: AR(1)–ARCH(1)	0.01	0.03	0.08	0.25	0.50	0.87	–	–	–
M_{12}: AR(1)–GARCH(1,1)	0.00	0.01	0.05	0.23	0.45	0.79	0.21	0.41	0.63
	Second-level constant k			SV autoregressive parameter γ			SV 2nd-level precision $\psi = 1/\rho^2$		
Migration flow and model M_i	2.5%	Median	97.5%	2.5%	Median	97.5%	2.5%	Median	97.5%
M_{13} for μ_{PL-DE}: AR(1)–SV	4.51	11.99	27.88	0.27	0.57	0.98	4.15	8.91	16.30
M_{13} for μ_{DE-PL}: AR(1)–SV	8.60	20.83	41.57	0.25	0.51	0.93	4.51	9.34	16.77

Notes: [a]For models M_{11}–M_{13} the estimated and forecasted precision varies, for details see Table C.5 in Annex C. Boldface denotes 95% credible intervals for the estimates of c and ϕ, which do not cover zero.
Source: own elaboration in WinBUGS

in Chapters 5 and 6, non-constant conditional variance in the AR(1) models helped the data to transform the prior beliefs on ϕ towards a more likely stationarity, than in the cases presented before.

For both flows under study, the results of the Bayesian model selection procedure were rather unambiguous. In the case of μ_{PL-DE}, the posterior probabilities (5.1) obtained for particular models from $\mathbf{M_3}$ equalled: $p(M_{10} \mid \mathbf{x}) = 0.116$, $p(M_{11} \mid \mathbf{x}) = 0.104$, $p(M_{12} \mid \mathbf{x}) = 0.045$ and $p(M_{13} \mid \mathbf{x}) = \mathbf{0.735}$, while for the opposite flow, μ_{DE-PL}, respectively $p(M_{10} \mid \mathbf{x}) = \mathbf{0.711}$, $p(M_{11} \mid \mathbf{x}) = 0.019$, $p(M_{12} \mid \mathbf{x}) = 0.003$ and $p(M_{13} \mid \mathbf{x}) = \mathbf{0.268}$. With respect to Polish-German migration, the supremacy of the AR(1)$-$SV model is clear, despite using Occam's prior. As concerns flows in the opposite direction, although the model M_{10} is a posteriori most probable, it has to be noted that the data also support the AR(1)$-$SV solution, magnifying the prior probability over twice. Hence, the formal selection criteria indicate that, among the four considered alternatives, simple AR(1)$-$SV models (7.4) can be appropriate for the processes under study, alongside a constant-variance model for μ_{DE-PL}. These results indicate that changes in the patterns of migration between Poland and Germany are indeed unpredictable due to the heavy tails of the error distributions.

7.2.3 Predictions Prepared with Models Acknowledging Discontinuity in Trends

Forecasts attempting to accommodate the discontinuity of trends in Polish-German migration flows have been prepared for 2005–2015 following the methodology presented in Section 7.1, hence using all models encompassed by the classes $\mathbf{M_2}$ and $\mathbf{M_3}$. The summaries of predictive distributions in terms of the predicted median values and quantile-based 80% intervals of emigration rates for 2005, 2010 and 2015, yielded by all eight models ($M_6$$-$$M_9$ and $M_{10}$$-$$M_{13}$) are listed in Table 7.3. The summaries for all years are shown in Table C.4, Annex C. The sample size in the 'analogy' models covers 13 yearly values (1992–2004), while for the processes with changes in conditional variance – 20 observations (1985–2004).

The trajectories of forecasted emigration rates for the period 2005–2015 yielded by models with analogy, as well as by AR(1)$-$SV and their AR(1) reference are shown in Fig. 7.2 together with the historical data series. For the 'analogy' models, the data concerning migration between Portugal or Spain and Germany are shown on the same vertical scale, but moved on the horizontal axis forward by 18 years (hence, the data for 1974–1997 are shown under 1992–2015).

The first six graphs in Fig. 7.2 indicate that, despite low posterior probabilities of the respective models, the Portuguese and the averaged Iberian migratory experience from 18 years before may nonetheless provide some interesting input for the forecasts of Polish-German and, albeit to a lesser extent, German-Polish migration. Specifically, these models predict a slight increase of the forecasted median emigration rates after the full opening of the German labour market, assumed around 2010–2011. At the same time, higher uncertainty is allowed especially for the

Table 7.3 Predictive summaries for emigration rates from the analogy models and AR(1)–SV, for 2005, 2010 and 2015, median, 10 and 90% quantiles (MCMC estimates)

	2005			2010			2015		
Migration flow	10%	Median	90%	10%	Median	90%	10%	Median	90%
Poland to Germany, m_{PL-DE}									
M_6: no analogy (reference) [a]	1.96	2.99	4.41	1.71	2.62	4.04	1.71	2.63	4.05
M_7: Analogy to Portugal [a]	1.99	2.99	4.42	1.79	2.93	4.88	1.77	3.24	6.22
M_8: Analogy to Spain [a]	1.97	2.94	4.31	1.77	2.76	4.34	1.85	2.96	4.80
M_9: Analogy to PT+ES [†] [a]	1.98	2.97	4.37	1.81	2.90	4.72	1.85	3.28	6.03
M_{10}: Reference AR(1) [b]	2.23	3.62	5.88	1.53	3.60	9.01	1.37	3.60	10.73
M_{11}: AR(1)−ARCH(1) [b]	2.04	3.18	4.91	1.57	2.64	4.96	1.49	2.59	5.03
M_{12}: AR(1)−GARCH(1,1) [b]	2.36	3.46	4.86	1.67	3.05	6.94	1.49	2.97	9.37
M_{13}: AR(1)−SV [b]	2.44	3.54	5.14	1.74	3.31	7.18	1.60	3.28	8.40
Germany to Poland, m_{DE-PL}									
M_6: no analogy (reference) [a]	0.80	1.12	1.55	0.69	1.01	1.48	0.69	1.01	1.48
M_7: Analogy to Portugal [a] [¶]	0.73	1.12	1.69	0.66	1.01	1.56	0.63	1.00	1.58
M_8: Analogy to Spain [a]	0.74	1.06	1.50	0.60	0.91	1.40	0.65	0.96	1.42
M_9: Analogy to PT+ES [†] [a] [¶]	0.71	1.05	1.54	0.61	0.94	1.45	0.69	1.03	1.53
M_{10}: Reference AR(1) [b]	0.90	1.26	1.76	0.69	1.24	2.36	0.63	1.24	2.71
M_{11}: AR(1)−ARCH(1) [b]	0.84	1.17	1.60	0.68	1.02	1.61	0.65	1.00	1.61
M_{12}: AR(1)−GARCH(1,1) [b]	0.87	1.19	1.59	0.64	1.04	1.79	0.58	1.02	1.89
M_{13}: AR(1)−SV [b]	0.94	1.25	1.67	0.72	1.21	2.24	0.68	1.21	2.62

Notes: [†] PT-Portugal, ES-Spain; [¶] Dummy for 1984; [a] Sample = 13 (1992–2004); [b] Sample = 20 (1985–2004).
No implausible values, exceeding over 5 times the maximum levels observed in the sample, have been forecasted.
Source: own elaboration in WinBUGS

above-median values, as indicated by the higher bounds of the 80% credible intervals, resulting from heavier upper tails of the predictive distributions. In turn, the AR(1)−SV models, shown in the two last graphs of Fig. 7.2, do not predict any visible shifts in future emigration rates. Nevertheless, in contrast to the 'pure' AR(1) processes, the stochastic volatility property accommodates the high-variance past behaviour of the series under study, and yields forecasts, that are slightly lower in terms of the median values, but more accurate ex ante than the ones produced by the reference model. The last conclusion is especially valid for m_{PL-DE}.

From the demographic point of view, the estimated median values and 80% credible intervals seem plausible, and generally comparable with the outcome of similar AR(1) models introduced before (see Table 5.5). The differences, especially in the estimated width of the uncertainty span, can be to a large extent explained by discrepancies in the size and composition of the samples. Notably, this concerns the Polish high-migration transition period of 1988–1990, included in the estimation of the AR(1)–SV and AR(1) in the current section.

As a note aside, the models based on Iberian analogies can be also interpreted using the category of spatial barriers or borders, widely used in human geography

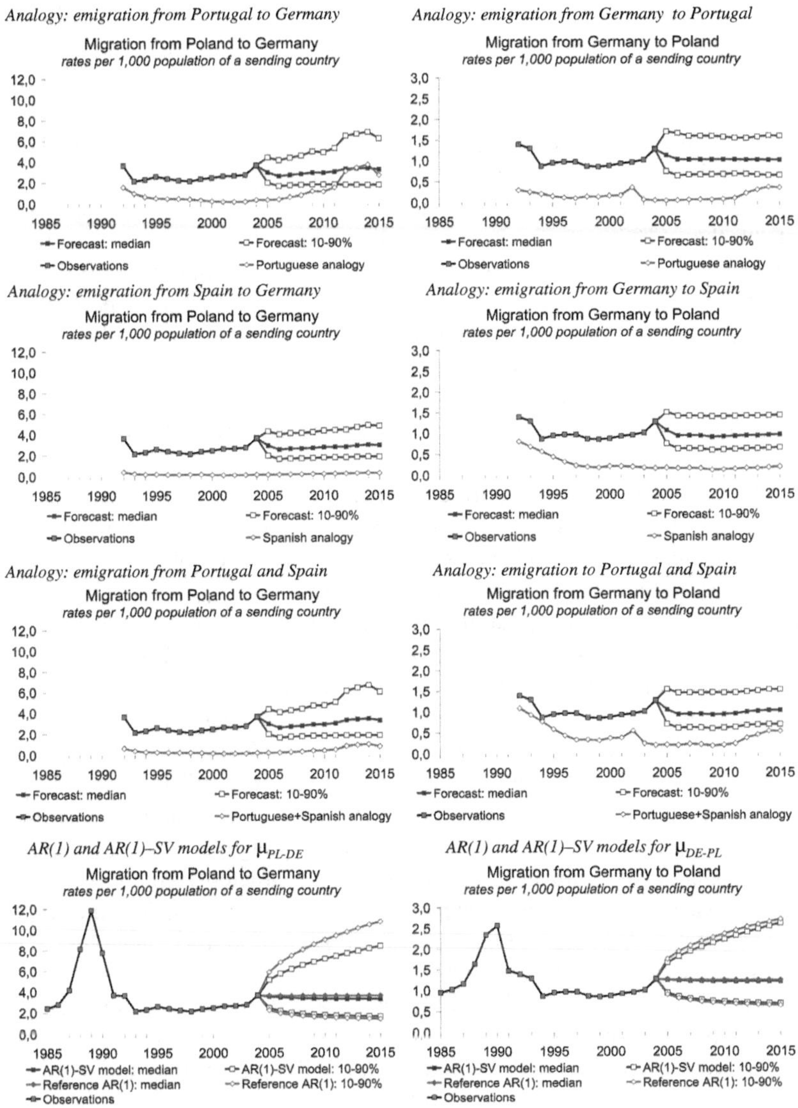

Fig. 7.2 Forecasted migration rates, 2005–2015: analogy and AR(1)–SV models.
Note: The analogous Portuguese and Spanish trajectories for 1974–1997 are shown moved forward by 18 years.
Source: own elaboration in WinBUGS. Data series until 2004: Eurostat/NewCronos; DESTATIS (see Annex A)

(e.g. Abler, Adams, & Gould, 1972, pp. 397–408). In international migration studies such barriers, formed along national boundaries, can have various features: administrative, political, economic, cultural, linguistic, and other. Rykiel (1986, p. 397) noted that the barriers may be subject to erosion due to the integration processes that take place between the countries or regions in question. In the presented example concerning the EU enlargement, these are mainly the administrative barriers that are gradually being dissolved.[3]

The main difficulty in interpreting the results of Polish-German migration forecasts in a full analogy to the Portuguese and Spanish migratory experience from 18 years before lies in the differences in the very nature and dynamics of the integration processes. In the two Iberian countries, the opening of labour markets in the rest of the then-EEC countries occurred all at once. The migration that followed could be thus interpreted as an instantaneous realisation of the 'migratory demand' that had been previously blocked by the existence of administrative barriers. In the case of migration from Poland and other 'new' EU Member States, the opening of Western European labour markets is a step-wise process that started from Ireland, Sweden and the United Kingdom in 2004, and is likely to finish in Austria and Germany by 2011 or thereabouts. In addition, some countries liberalise their labour market regulations gradually, making the barriers more and more permeable and the restrictions less and less acute. Furthermore, Poland initially introduced reciprocal political restrictions against the 'old' EU nationals with respect to the freedom of employment. Hence, in the Polish case, the effect of substitution between competing EU origins or destinations of migration is likely to occur.[4]

With respect to forecasts assuming non-constant conditional variance, data support for the AR(1)–SV models rather than for their ARCH(1) and GARCH(1,1) alternatives, as indicated by the posterior probabilities, suggests that not only migration as such, but also its volatility may have a random character. This is in a way similar to the prevalence of random walk models in Chapter 5, formally selected for all but one population flows, under either model prior. The outcome of both Bayesian model selection procedures puts a strong emphasis on the randomness of the migratory processes under study. This is an interesting result in the context of the current research, as it confirms the very uncertain character of international migration in Europe. Nevertheless, as shown in the current section, random models can in many instances produce reasonable forecasts of such migratory phenomena, which otherwise might be difficult to predict.

The *ex-post* forecast errors for 2005–2007 yielded by models from classes $\mathbf{M_2}$ and $\mathbf{M_3}$ analysed in the current section are listed in Table 7.4. In the case of both migratory flows, errors for models from $\mathbf{M_3}$ are visibly lower than the ones for $\mathbf{M_2}$.

---

[3] I am very grateful to Marek Kupiszewski for drawing my attention to this interpretation.

[4] Notably, upon the accession to the EEC, migration from Portugal shifted its primary destination from Switzerland to Germany (Kupiszewski, 1998). Hence, a substitution effect could be observed, akin to what can be expected for Poland with respect to post-accession migration to the British Isles, which at least to some extent might have substituted other directions. However, in the light of recent data on post-accession migration, it seems that the substitution hypothesis is far too simplistic. Migration between Poland and Germany also increased after 2004 as a result of an overall boost of mobility after the EU enlargement, yet to a lesser extent than migration between Poland and the UK. This effect can be explained by different selectivity patterns of migrations to both destinations, whereby Polish migrants to the UK seem to be better educated than migrants to Germany (Kaczmarczyk and Okólski, 2008).

Table 7.4 Absolute and relative ex-post errors for 2005–2007: model classes $\mathbf{M_2}$ and $\mathbf{M_3}$

Model class	$\mathbf{M_2}$: models with analogies to Iberian migration				$\mathbf{M_2}$: models with non-constant conditional variance			
Model	M_6: ref.	M_7: PT	M_8: ES	M_9: PT+ES	M_{10}: ref.	M_{11}: ARCH	M_{12}: GARCH	M_{13}: SV
Migration from Poland to Germany, $m_{PL-DE}(2005) = 4.17$; $m_{PL-DE}(2006) = 4.29$; $m_{PL-DE}(2007) = 4.03$								
- Absolute errors	1.42	1.38	<u>1.45</u>	1.41	**0.55**	1.20	0.84	0.69
- Relative errors	34.0%	33.1%	<u>34.8%</u>	33.8%	**13.3%**	28.8%	20.2%	16.6%
Migration from Germany to Poland, $m_{DE-PL}(2005) = 1.28$; $m_{DE-PL}(2006) = 1.37$; $m_{DE-PL}(2007) = 1.47$								
- Absolute errors	0.33	0.32	0.39	<u>0.40</u>	**0.12**	0.26	0.23	0.13
- Relative errors	23.4%	23.0%	28.2%	<u>28.4%</u>	**8.5%**	18.3%	16.4%	9.5%

Notes: Row-wise minima are in boldface and maxima underlined.
Source: own elaboration; data for 2005–2007: Eurostat/NewCronos

The lowest values have been obtained for the constant-variance AR(1) models M_{10}, followed by the AR(1)–SV processes M_{13}. Notwithstanding, in most instances the 2005–2007 realisations fall inside the respective 80% predictive intervals, except for all analogy models ($\mathbf{M_2}$) for m_{PL-DE}, all of which fell outside the intervals in 2006, as well as for models M_8 and M_9 for m_{DE-PL} in 2007.

Chapter 8
Evaluation of Presented Forecasts of European Migration

The current chapter focuses on two aspects of evaluating the forecasts of migration flows, presented earlier in Chapters 5, 6 and 7. Firstly, Section 8.1 deals with the sensitivity of the results to changes in the assumed prior distributions and in the information they carry. The focus of the discussion is on the precision parameters of the assumed random processes. Secondly, in Section 8.2, selected Bayesian forecasts are compared with their frequentist counterparts in terms of errors *ex ante* and *ex post*, in the latter case computed for forecasts for 2000–2007 calculated on the basis of series truncated in 1999.

8.1 Robustness of Forecasts Against Certain Changes In Priors

8.1.1 Role of Sensitivity Analysis in the Bayesian Approach: Basic Remarks

Due to the immanent presence of subjective elements in the Bayesian inference in a form of a priori distributions, an important element of the analysis is the examination of sensitivity, or robustness, of the results derived from the changes in the assumptions. The term 'robustness' usually relates to the study of the impact of priors on the final outcome of the inference (posterior or predictive distributions), and as such is also understood in the current study. However, it should be noted that also, from a purely statistical point of view, the likelihood of the observations, $p(x \mid \theta)$ in equation (2.2), can be subject to a sensitivity analysis.[1] Examples of such studies for elliptical families of distributions, among others encompassing Gaussian, Student's *t* and Cauchy cases, can be found in the research of Osiewalski and Steel (1993b, c) or, more recently, in applications to spatial predictions, of Kim and Mallick (2003).

[1] Poirier and Tobias (2004, p. 12) observed that any statistical model consists of two inherently subjective components: '(i) a window (likelihood) for viewing observables in the world, and (ii) a prior reflecting a professional position of interest'. They also noted that, even if there is an agreement among the researchers with respect to the likelihood function, there are high chances of 'agreement to disagree' about prior distributions, especially in complex models with multi-dimensional parameter space (*idem*, credits to Jacek Osiewalski for drawing my attention to this).

J. Bijak, *Forecasting International Migration in Europe: A Bayesian View*, The Springer Series on Demographic Methods and Population Analysis 24,
DOI 10.1007/978-90-481-8897-0_8,

Nevertheless, even when limiting the discussion to assumptions a priori, the robustness analysis within the Bayesian framework can still be very complex and multi-dimensional. Męczarski (1998) provides a comprehensive survey of methods of robustness analysis, including properties of particular classes of prior distributions, notably, the conjugate families,[2] methods based on nuisance parameters, or constraints set on the priors, methods optimal with respect to the robustness of posterior distributions, empirical Bayes techniques,[3] and discrete or continuous mixtures of distributions, as well as non-parametric methods.

Despite the multitude of available methods of robustness analysis, Męczarski (*idem*, p. 179) noted that many of them may be difficult to implement in practical applications using a popular software. A simpler alternative is the 'informal' sensitivity analysis (e.g., Congdon, 2003, p. 21), in which a handful of competing informative or non-informative prior distributions are examined, in order to assess the robustness of the posterior outcome. For computational reasons, the latter approach is applied in the current study, with focus on the sensitivity of predictive distributions obtained from two basic models yielded by the selection procedures outlined in Chapters 5 and 6, namely, the random walks and the AR(1) processes.

An important element of the robustness assessment is the application of such prior distributions that carry as little information as possible, so as to 'let the data speak for themselves'. Such priors are relatively 'objective', as opposed to the informative ones, and hence sometimes referred to as the 'Bayesian Holy Grail' (Bernardo & Smith, 2000, pp. 357–359). Theoretical foundations of non-informative priors have been laid by Jeffreys (1939/1961, 1946; *idem*). The departure point was the observation that, although the smallest amount of information on the parameters θ is usually carried by uniform distributions, there is also a need for invariance of such $p(\theta)$ to bijective transformations of θ. The Jeffreys rule states that given the sample space $\mathbf{X}$, parameter space $\mathbf{\Theta}$, and the family of probability distributions $p(x \mid \theta)$, depending on θ, such an invariant non-informative prior density $\pi(\theta)$ can be derived as (*idem*):

$$\pi(\theta) \propto h(\theta)^{0.5}, \text{ where } h(\theta) = -\int_{\mathbf{X}} p(x \mid \theta) \frac{\partial^2}{\partial \theta^2} \ln p(x \mid \theta) dx. \tag{8.1}$$

The Jeffreys rule[4] (8.1) can be generalised into the case of multivariate $\boldsymbol{\theta}$; however, such generalisation does not have certain intuitively-desired properties (details

[2]The parametric family of distributions $\mathbf{F}$ is conjugate with respect to the sample distribution $p(x \mid \theta)$, if and only if both the prior $p(\theta)$ and posterior $p(\theta \mid x)$ belong to $\mathbf{F}$ (e.g., Męczarski, 1998, pp. 21–22). Examples of such families are: Normal distributions for Normal samples (for with regard to the mean μ), Beta distributions for binomial samples (for the probability of success p), Gamma distributions for Poisson samples (with respect to λ), etc. (*idem*).

[3]The non-orthodox Bayesian character of empirical methods should be borne in mind, cf. Chapter 2.

[4]As indicated by Bernardo and Smith (2000, pp. 314, 361), under asymptotic Normality conditions, $h(\theta)$ is Fisher's information about θ (or Fisher's information matrix $\mathbf{H}(\boldsymbol{\theta})$ for multivariate $\boldsymbol{\theta}$, when $\pi(\boldsymbol{\theta}) \propto \det[\mathbf{H}(\boldsymbol{\theta})]^{0.5}$). For its formal derivation, see a part devoted to likelihood-based estimation in Section 8.2.

in: Bernardo & Smith, 2000, p. 361). Furthermore, in some cases, the Jeffreys priors are improper – they do not integrate to unity over the whole parameter space $\mathbf{\Theta}$. In order to overcome the latter problem in practical applications involving the available software, several solutions can be proposed. For example, Congdon (2003, pp. 2–3, 21) advocates the use of "just proper" diffuse priors with large variances, controlling for possible problems with the convergence of the MCMC algorithm. Such priors for structural variables may be for example Normal N(0, M^2), while for the inverse variance (precision) parameters, Gamma priors $\Gamma(M^{-1}, M^{-1})$ with mean 1 and variance M can be assumed. In both cases, M denotes a large real number, such as for example $M = 1{,}000$ or 10,000 (*idem*).

Such a simplified approach is followed in the empirical analysis presented in the current section. Following the argumentation presented above, in order to assess the robustness of the results against the changes in prior distributions, an assumption is made that the structural parameters of the models follow N(0, 100^2), while the precision parameters – $\Gamma(0.001, 0.001)$. Such distributions are sufficiently dispersed to serve as reference for forecasts prepared using informative priors, such as the ones presented in the previous chapters.

As stated before, for the clarity of presentation, as well as due to numerical reasons, the analysis is primarily limited to two classes of models: random walks and Gaussian AR(1) models. It is worth recalling that random walks have been formally selected for forecasting in Chapter 5 for almost all flows under study, under both prior distributions over the model space. Autoregressive models AR(1), in turn, are a universal outcome of the 'from general to specific' VAR reduction procedure shown in Chapter 6. The approach followed here is very simplistic, as it does not address the issues of the robustness of the model selection or reduction procedures themselves on the changes in priors. These issues, although potentially very interesting and important, are problematic in terms of the intensity of computations required for the MCMC algorithms to converge, and thus have been left out from the current study.

8.1.2 Robustness of Forecasts Yielded by Selected Models

The forecasts yielded by random walks and Gaussian autoregressive models AR(1) were re-calculated, assuming a priori $c \sim$ N(0, 100^2), $\varphi \sim$ N(0, 100^2) and $\tau \sim \Gamma(0.001,\ 0.001)$. Additionally, for m_{PL-DE} and m_{DE-IT}, the constant models (5.5a) have been estimated, being the only non-random walk models bearing high posterior probabilities under the 'Occam's razor' prior in Chapter 5. Due to the diffuse character of the priors, the calculations have been performed with 500,000 burn-in iterations of the MCMC algorithm, and estimates computed on the basis of the following sample of 250,000. Results of the estimation of the parameters of both types of models are summarised in Table 8.1. A comparison between the posterior distributions for τ obtained for the random walk models both under diffuse priors (in the current chapter), as well as under informative ones (Chapter 5), is presented in Fig. 8.1.

From a comparison of these estimates with the ones presented in Tables 5.4 and 6.3, it can be seen that the posterior distributions of c and ϕ based on diffuse

Table 8.1 Summaries of posterior distributions of the parameters of random walk and AR(1) models, obtained under diffuse priors: median, 2.5 and 97.5% quantiles (MCMC)[5]

	Constant c			Autoregression parameter φ			Precision $\tau = 1/\sigma^2$		
Migration flows	2.5%	Median	97.5%	2.5%	Median	97.5%	2.5%	Median	97.5%
Random walk with drift									
From Italy to Germany	−0.09	−0.04	0.02	–	–	–	35.90	75.45	137.20
From Germany to Italy	−0.04	−0.01	0.03	–	–	–	105.50	271.40	557.30
From Poland to Germany	−0.11	−0.00	0.11	–	–	–	10.54	27.10	55.65
From Germany to Poland	−0.10	−0.01	0.08	–	–	–	17.37	44.67	91.72
From Switzerland to Germany	−0.01	0.01	0.04	–	–	–	212.20	446.00	811.00
From Germany to Switzerland	**0.02**	**0.05**	**0.08**	–	–	–	134.10	344.80	707.90
Autoregressive model AR(1)									
From Italy to Germany	−0.13	−0.01	0.11	**0.80**	**1.07**	**1.35**	33.64	72.53	133.90
From Germany to Italy	−0.81	−0.39	0.03	**0.00**	**0.52**	**1.05**	124.20	336.80	712.90
From Poland to Germany	−0.03	0.53	1.10	−0.15	0.43	1.02	12.85	34.85	73.77
From Germany to Poland	−0.09	−0.01	0.07	**0.17**	**0.60**	**1.03**	20.69	56.08	118.70
From Switzerland to Germany	**0.01**	**0.05**	**0.08**	**0.39**	**0.69**	**0.99**	248.80	563.40	990.20
From Germany to Switzerland	**0.03**	**0.33**	**0.63**	**0.99**	**1.13**	**1.28**	159.70	432.80	916.30

Note: **Boldface** denotes 95% credible intervals for the estimates of c and φ, which do not cover zero.
Source: own elaboration in WinBUGS

[5] For the constant model M_1 for μ_{PL-DE}, the median c was estimated as 0.94 (quantile-based 95% interval ranging from 0.84 to 1.05) and precision τ as 30.89 (95% interval from 12.00 to 63.55). For μ_{DE-IT}, the median c was −0.80 (95% interval from −0.84 to −0.77) and τ as 258.7 (95% interval: 100.5 to 532.3).

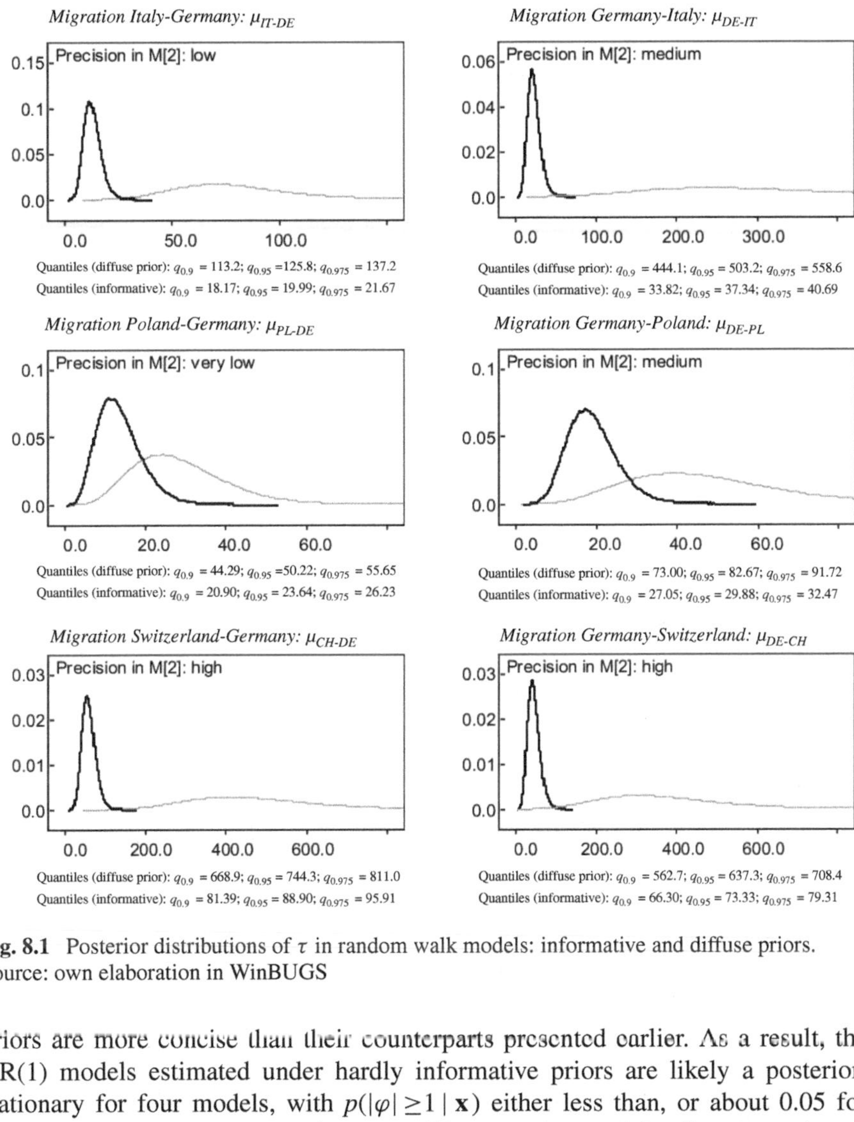

Fig. 8.1 Posterior distributions of τ in random walk models: informative and diffuse priors. Source: own elaboration in WinBUGS

priors are more concise than their counterparts presented earlier. As a result, the AR(1) models estimated under hardly informative priors are likely a posteriori stationary for four models, with $p(|\varphi| \geq 1 \mid \mathbf{x})$ either less than, or about 0.05 for m_{DE-IT}, m_{DE-PL}, m_{CH-DE} and m_{PL-DE}. The posterior precision (inverse variance) τ in the case of a dispersed Gamma prior also proves to be much higher and much more uncertain than in the models estimated in Chapters 5 and 6 (see also Fig. 8.1), indicating a substantial role of data in modifying prior beliefs on the uncertainty of the random term.

The median forecasts yielded by particular models, as well as the limits of the 80% predictive intervals for migration to and from Germany in the period 2005–2015, are illustrated respectively in Figs. 8.2 and 8.3, where they are compared with the results produced under the informative priors defined in Chapter 5. Figure 8.4 illustrates a similar comparison for cases, where the constant model (M_1) proved

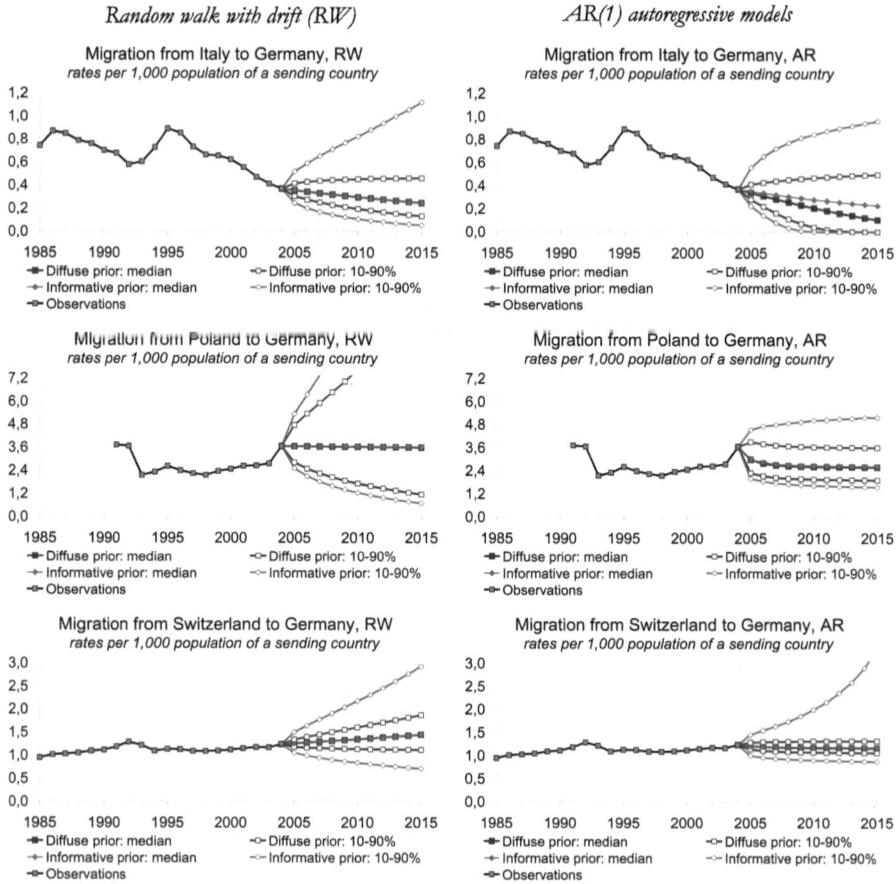

Fig. 8.2 Comparison of forecasted migration rates for 2005–2015 under informative and diffuse priors: random walks and Gaussian AR(1) models, migration to Germany.
Source: own elaboration in WinBUGS. Data series until 2004: Eurostat/NewCronos; DESTATIS (see Annex A)

highly-probable: for migration from Poland to Germany and from Germany to Italy. Forecasts yielded by the constant models for m_{PL-DE} and m_{DE-IT} oscillate around the medians of 2.57 and 0.45, respectively, with the 80% credible intervals 2.01–3.29 in the former case, and 0.41–0.49 in the latter. The rates forecasted using random walks and AR(1) models under diffuse priors are listed in detail in Table C.6 (Annex C).

In many cases, the median forecast trajectories produced by the models based on informative and diffuse priors are similar, the only visible exceptions being the AR(1) models for m_{IT-DE}, with higher values yielded under the informative a priori distributions, as well as for m_{DE-CH}, under the diffuse ones. Nonetheless, in all presented cases, the empirical outcome of comparison with respect to the 80%

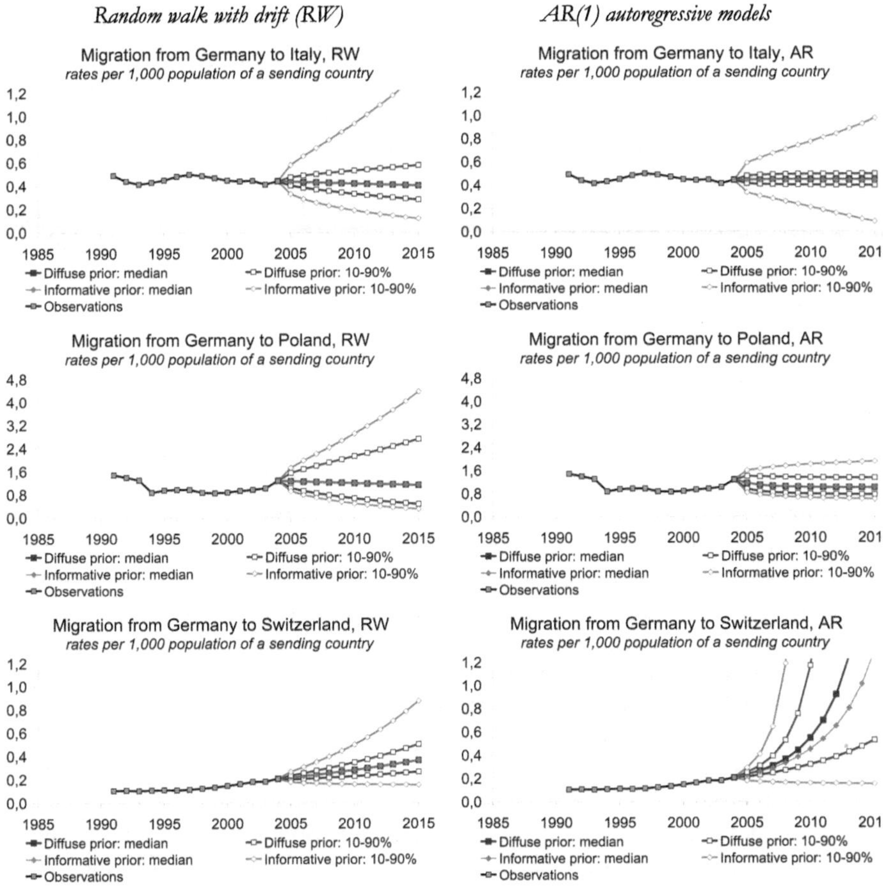

Fig. 8.3 Comparison of forecasted migration rates for 2005–2015 under informative and diffuse priors: random walks and Gaussian AR(1) models, migration from Germany

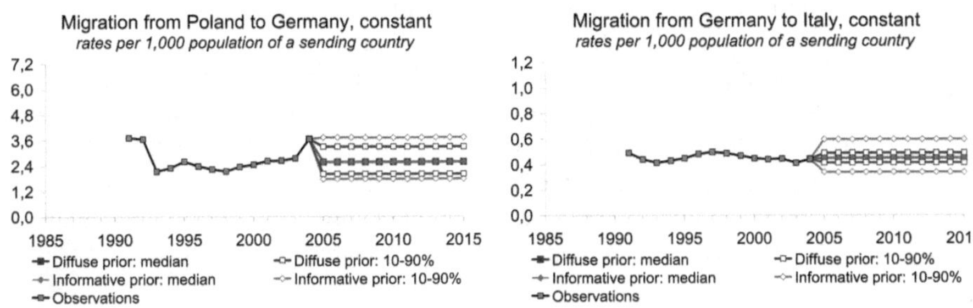

Fig. 8.4 Comparison of forecasted migration rates, constant models (m_{PL-DE} and m_{DE-IT}). Source: own elaboration in WinBUGS. Data series until 2004: Eurostat/NewCronos; DESTATIS (see Annex A)

predictive intervals is unambiguous: the uncertainty spans obtained under hardly informative (diffuse) priors are in all cases much narrower.

Similarly, as in Chapter 5, the results of the analysis for μ_{DE-IT} and μ_{DE-CH} were compared with the ones obtained for a longer sample of 20 observations (1985–2004). Unlike before, in the current exercise lengthening the series reduced the precision τ on average by a factor of two (between 1.4 and 2.5 times) for the random walk and AR(1) models. This indicates that for both flows truncation of the series improves statistical properties of the sample, something that could not be observed under informative priors due to their strong impact on the posterior distributions.

8.1.3 Discussion and Tentative Conclusions

The presented examples show that the forecasts constructed in the current study are not robust to changes in prior distributions, especially with respect to their uncertainty characteristics. Such an outcome is not surprising, given the small size of samples which are available for the purpose of forecasting origin-and-destination-specific European migration flows. As stated in Chapter 2, in the Bayesian inferential paradigm the smaller the sample, the greater is the weight of the subjective prior information in the ultimate posterior outcome. The shortness of available time series also explains the observations made in Chapter 6 with respect to forecasts yielded by AR(1) models based on different a priori distributions for the precision parameter τ.

Moreover, sensitivity of the outcome on the impact of the sample size, with respect to emigration from Germany, seems to confirm the rationale for limiting the analysis in these cases only to the post-unification period. Due to the improvement of the statistical properties of the 1991–2004 sample in comparison to the 1985–2004 series, as indicated by higher precision of the random term, it seems that the intensity of emigration from Germany to Italy and Switzerland visibly differed for the former FRG until 1990 and for the united country thereafter.[6] This outcome can be to some extent attributed to the fact that after the unification citizens of the former GDR were much more likely to migrate to West Germany rather than abroad.

The outcome of the analysis shown in the current section, especially concerning the *ex-ante* assessments of the uncertainty of forecasts, indicates a disagreement between the informative priors introduced in Chapters 5 and 6 and the data. The forecasts made under diffuse a priori distributions, in particular concerning precision τ, yield much more accurate forecasts in terms of narrower 80% credible intervals from the respective predictive distributions. A remarkable difference is that the precision of estimation can be even up to one order of magnitude greater for hardly-informative priors than the one assuming low precision a priori.

---

[6]For comparison, more detailed results for the longer-sample case are available in Bijak (2008b).

On the other hand, the prior belief in a weak precision is a natural premise of migration research, which deals with phenomena that are by definition weakly predictable (see Chapter 1 for discussion). The a priori assumptions on a relatively small τ (large variance) seem much more important than beliefs concerning structural parameters of forecasting models or interrelations with potential predictors. The latter priors can be, more or less realistically, justified by various migration theories, regardless of the lack of generality thereof. In contrast, the assumptions of low precision concern the very nature of uncertain migratory processes.

The 80% predictive intervals yielded by models based on diffuse priors, as shown in Figs. 8.2, 8.3, and 8.4, are in several cases too narrow to be plausible – narrower than in-sample variability. This conclusion is also supported by insights from the ex-post analysis for 2005–2007, where in five cases (random walk, AR(1) and constant for m_{DE-IT}, as well as AR(1) and constant for m_{PL-DE}) the observed values never fell into the 80% intervals yielded under diffuse priors, which almost exclusively relied on the sample information. In addition, some other forecasts yielded by random walk and AR(1) models were outside the intervals either for 2006 and 2007, or for 2007 alone. Clearly, in the cases under study, the average *ex-post* accuracy of forecasts from diffuse-prior models appears to decrease over time. By contrast, under informative prior distributions reflecting beliefs in low precision, almost all actual migration rates for 2005–2007 fell into the respective intervals for the said models, with the only exception being the the constant models.

In order to manage uncertainty in migration forecasts in a proper way, solutions that underestimate it should not be accepted. Paradoxically, especially in the case of international migration, the researchers should not strive for as small *ex-ante* prediction errors as possible, as they can likely prove to be incompatible with the very uncertain character of the migratory phenomena under study. A more detailed analysis of both the *ex-ante* and *ex-post* performance of selected models is offered in the next section, together with some examples of comparisons of the obtained Bayesian forecasts and their traditional, sampling-theory equivalents.

8.2 Comparison of Selected Bayesian and Frequentist Forecasts

8.2.1 *General Remarks on* Ex-Ante *and* Ex-Post *Prediction Errors*

A key issue in all forecasting tasks, including demographic applications, is the assessment of the accuracy of predictions, which is usually understood as the consistency between forecasts and the actual realisations of the forecasted variables (e.g., Jóźwiak, 1987, p. 2). The possible accuracy measures, which can be constructed in a number of variants, each having slightly different characteristics (e.g., Smith & Sincich, 1992, p. 500), in principle belong to two broader groups of *ex-ante* and *ex-post* errors. The former ones enable the assessment of the degree of accuracy prior to observing the actual realisations of the variables under study, which is done by the means of statistical inference, and thus informs about the *expected* accuracy of forecasts. In turn, the *ex-post* measures are based on the empirical differences between the predicted and observed values, estimating the *real* accuracy, which can be later used to adjust the assumptions of subsequent predictions (Jóźwiak, 1987). Clearly, for the calculation of the *ex-post* indicators, the actual observations for the forecast period under scrutiny are indispensable. Among the *ex-ante* accuracy indicators, the most popular ones are based on the dispersion characteristics of the predictive distributions of the variables under study, or on the confidence or credible regions thereof.

In either case, Cohen (1986) further distinguishes the 'model-based' approach, derived explicitly from the estimated predictive models, and the 'empirical' one, based on the actual distributions of forecast errors yielded by other, external predictions. In the current section, the model-based approach is followed, focusing on two measures of accuracy: the half-widths of the respective 80 and 95% predictive intervals (*HCI*-80 and *HCI*-95), expressed in relative terms towards the point forecasts. A comparison of these two indicators provides an additional information with respect to the distributions' tails.

Among the variety of measures of *ex-post* forecast errors, Smith and Sincich (1992: 500) propose to focus on several ones, in order to account not only for the accuracy, but also for the direction of the forecast bias. The former group of indicators can consist for example of the Mean Absolute Percentage Error (*MAPE*), Root Mean Square Percentage Error (*RMSPE*) and 90th Percentile Error (90-*PE*). *MAPE* is defined as the average of the absolute values of errors (observed less forecasted values), expressed in percentage terms towards the actual observations.[7] *RMSPE*, a square root of the average of squared error terms, more strongly penalises larger discrepancies between the observed and forecasted values. In turn, the 90-*PE* indicator, the 'absolute percent error larger than exactly 90% of all other absolute percent errors' (*idem*), approximates the heaviness of tails of the empirical error distributions. The proposed measures of bias are the Mean Algebraic Percentage Error (*MALPE*), which is similar to *MAPE*, but accounts for the direction of error, as well as the fraction of positive errors (*%POS*) (*idem*). In the current study, all five above-mentioned *ex-post* measures of the quality of forecasts are applied.

8.2.2 Likelihood-Based Estimation and Model Selection Framework

As already indicated in Chapter 2, the traditional sampling-theory (frequentist) statistical paradigm differs from the Bayesian one in several aspects. Firstly, in the

[7] *MAPE* and its non-percentage counterpart, the mean absolute error (*MAE*) were used in previous chapters to evaluate the *ex-post* forecast errors for 2005–2007.

former case probability is interpreted in relation to the frequency of events in question rather than in the subjectivist fashion. Secondly, no additional knowledge other than the sample alone (e.g., through a priori distributions), is allowed in the inferential framework. These differences are related to distinct philosophies of both approaches: unlike in the Bayesian paradigm, the sampling theory treats model parameters not as random variables, but as fixed, albeit unknown quantities.

The sampling-theory approach to estimation, taking into account these considerations, has been proposed in a seminal paper by R. A. Fisher (1922), where the Maximum Likelihood (ML) method is presented (*idem*, pp. 323–330). In this approach, likelihood L of the data $\mathbf{x}$ sampled from a probabilistic model described by a density $p(\mathbf{x} \mid \boldsymbol{\theta})$ with unknown parameters $\boldsymbol{\theta}$, is defined as a function of these parameters: $L(\boldsymbol{\theta} \mid \mathbf{x}) \equiv p(\mathbf{x} \mid \boldsymbol{\theta})$ (Bernardo & Smith, 2000, p. 450). The ML estimate of parameters $\boldsymbol{\theta}$ is thus found by maximising $L(\boldsymbol{\theta} \mid \mathbf{x})$, or for convenience its monotonous transformation such as a natural logarithm, $\ln L(\boldsymbol{\theta} \mid \mathbf{x})$, over all possible $\boldsymbol{\theta} \in \boldsymbol{\Theta}$:

$$\hat{\boldsymbol{\theta}}_{MLE} = \arg\max_{\boldsymbol{\theta} \in \boldsymbol{\Theta}} \ln \left[\prod_{x \in \mathbf{X}} L(\boldsymbol{\theta} \mid x) \right] = \arg\max_{\boldsymbol{\theta} \in \boldsymbol{\Theta}} \sum_{x \in \mathbf{X}} \ln L(\boldsymbol{\theta} \mid x). \tag{8.2}$$

The ML point estimate thus maximises the likelihood that a given sample $\mathbf{x}$ was obtained from a model characterised by the estimated set of parameters $\hat{\boldsymbol{\theta}}_{MLE}$ given by (8.2). The use of the product of likelihoods for particular sample elements, x, requires an additional assumption on their independence. On the basis of the ML estimates (8.2), the point predictions $\hat{\mathbf{x}}^{\mathbf{P}}$ can be derived arithmetically from the form of the forecasting model. Conventionally, this is performed by setting all other predictors and noise parameters equal to their assumed expected values.

The traditional, sampling-theory interval estimation and prediction makes wide use of confidence intervals or regions, which are theoretical constructs that are meant to cover the unknown, estimated parameters $\boldsymbol{\theta}$, or the future values of the variables under study, $\mathbf{x}^{\mathbf{P}}$, with a 'confidence' of $(1-\gamma)\cdot 100\%$. 'Confidence' here, however, is not exactly the same as probability in Bayesian credible regions for the reasons discussed further, and the construction of confidence regions is also slightly different. The frequentist region estimates of parameters typically make use of such probability distributions, which are functions of observations not depending on unknown parameters (so-called *pivotal quantities*). Location of these distributions is then determined by the respective point estimates $\hat{\boldsymbol{\theta}}_{MLE}$ or predictions $\hat{\mathbf{x}}^{\mathbf{P}}$.

For example, double-sided intervals from symmetric univariate distributions, such as Normal or Student's t, would generally involve a product of the standard error of estimation or prediction, and the relevant quantiles of a rank $\gamma/2$ and $1 - \gamma/2$ from these distributions, which themselves are centred around the appropriate $\hat{\theta}_{MLE}$ or $\hat{x}^P$. In particular, predictive intervals from models assuming Gaussian error terms with an unknown (estimated) variance, make use of Student's t distributions with as many degrees of freedom, as the total number of observations T, less the number of estimated structural parameters, including a constant. In large-sample studies (for instance, where $T > 100$), Normal approximations can be used instead.

Multidimensional confidence regions are generalisations of the univariate cases, calculated from the appropriate multivariate distributions (e.g., Hotelling's T^2 instead of Student's t; cf. Anderson, 2003).

Derivation of standard errors of estimation and prediction is model-specific and detailed formulae for particular models can be found in basic econometric textbooks (e.g., Greene, 2000). In general, the standard errors of the estimation of $\boldsymbol{\theta}$ can be calculated via Fisher's information matrix $\mathbf{H}(\boldsymbol{\theta})$, which is a negative of the expected value of a matrix composed of second-order partial derivatives of the log-transformed likelihood function, with respect to $\boldsymbol{\theta}$ (after Lynch, 2007: 40):

$$\mathbf{H}(\boldsymbol{\theta}) = -E\left[\frac{\partial^2 \ln L(\boldsymbol{\theta} \mid \mathbf{x})}{\partial\boldsymbol{\theta}\partial\boldsymbol{\theta}'}\right]. \tag{8.3}$$

The standard errors are square roots of the diagonal elements of the inverse of the estimated information matrix $\mathbf{H}(\hat{\boldsymbol{\theta}}_{MLE})^{-1}$, which is an estimate of a covariance matrix of $\hat{\boldsymbol{\theta}}_{MLE}$ (*idem*).

In all cases, however, a careless interpretation of sampling-theory confidence regions may lead to a trap. Namely, no confidence region derived from single-sample data guarantees that the unknown parameter θ lies within with a probability $1 - \gamma$, but rather that the long-run behaviour of the whole procedure is that on average in $(1 - \gamma) \cdot 100\%$ of samples, such a region would cover the true value of θ. Hence, for any particular confidence region it is entirely uncertain whether it covers θ, or not (cf. Bernardo & Smith, 2000, p. 466). This problem is largely due to ignoring the randomness of parameters, which are treated here as fixed. In order to overcome this, some more advanced methods have been proposed within the frequentist paradigm, for example the bootstrap method.[8] Nevertheless, such approaches are often criticised from the Bayesian point of view, as being 'stylised procedures, such as those for point or interval estimation, or hypothesis testing, designed to satisfy or optimise an ad hoc criterion' (Bernardo & Smith, 2000, p. 444), rather than offering probabilistically-coherent inferential paradigms, as it is the case within the axiom-based Bayesian statistics (*idem*).[9]

Formal model selection can be performed within the frequentist framework; for example on the basis of Likelihood Ratio (*LR*) tests, or using other asymptotically equivalent procedures, such as the Wald test (with akin construction to the Bayesian HPD test presented in Chapter 6), or the Lagrange Multiplier test (Greene, 2000, pp. 150–157). These procedures are valid for nested models, where the specific model is obtained from the general one by imposing r restrictions on the parameters of the latter. The significance of these restrictions is subject to testing. In particular, the *LR* test is constructed as follows: let $\hat{L}_R$ denote the estimated maximum

[8] For a demographic application of a non-parametric bootstrap, consisting in resampling from the same set of observations, see e.g. Keilman and Pham (2006).

[9] In general, for a more detailed critical treatment of various inferential aspects the traditional sampling-theory approach made from the Bayesian perspective, see Bernardo and Smith (2000, pp. 443–488).

value of the likelihood for the restricted model, and $\hat{L}_U$ – for the unrestricted one. Then, under the null hypothesis of significance of the restrictions, as well as under 'suitable regularity conditions', the following asymptotic relationship holds (*idem*, p. 152):

$$LR = -2 \cdot \ln(\hat{L}_R/\hat{L}_U) \dot{\sim} \chi^2_{(r)} \, , \tag{8.4}$$

where the dot over the tilde (~) reflects an asymptotic property, valid for large samples. The null hypothesis is then rejected if the estimated *LR* statistic is greater than an appropriate critical value from the chi-squared distribution with as many degrees of freedom, as there are restrictions (r).

There are two major obstacles as to why the formal likelihood ratio tests cannot be directly applied in the migration forecasting examples presented in the current section. Firstly, the *LR* test statistic follows a known $\chi^2_{(r)}$ distribution only asymptotically and hence the sampling-theory results obtained for samples of ten-odd elements would be at least dubious. Secondly, the *LR* tests are suited only for comparing nested hypotheses, which is not directly the case for all models in the three considered spaces, $\mathbf{M_1}$–$\mathbf{M_3}$ (cf., Greene, 2000, pp. 153). Therefore, in the current section, only a simplified model selection procedure is offered, consisting of a choice of a forecasting model characterised by the highest (log-)likelihood.

8.2.3 Selected Bayesian and Frequentist Migration Forecasts for 2000–2007

For the sake of comparison of the outcome of various Bayesian forecasting models with their traditional sampling-theory counterparts, in the current section a simple empirical analysis is offered. The forecasts of migration both-way flows between Germany, and Switzerland and Italy, have been computed in each case on the basis of 15-element samples covering the period 1985–1999. For forecasting, five models from the model space $\mathbf{M_1}$ defined in (5.5) have been used, all being sub-models of the ARMA(1,1) class. In order to account for discontinuity in the series of emigration rates from Germany observed in 1991 (cf. Chapter 2), the respective models have been additionally equipped with dummy variables, set to equal one for flows only from the former West Germany (the then-Federal Republic) prior to unification, in the period 1985–1990. All migration to and from Poland was excluded from this exercise, due to the necessity to truncate the sample before the system transformation period, additionally removing the years 1985–1990. Forecasting on the basis of the nine-element remainder would yield too doubtful results, especially within the frequentist framework.

The forecasts for the four above-mentioned flows (m_{DE-IT}, m_{IT-DE}, m_{DE-CH} and m_{CH-DE}) are prepared for 8 years, for 2000–2007. The calculations of frequentist forecasts have been performed using the built-in functions of the statistical environment R, which is a free software[10] based on the GNU license. The predictions have been computed from the extrapolation of relevant time series with a particular model. The *ex-ante* error characteristics, such as the half-widths of confidence intervals (*HCI*-80, *HCI*-95) have been obtained from appropriate Student's t distributions with $T-k$ degrees of freedom, T being the sample size and k – the number of structural parameters, including constants. The t distributions were centred at the point forecasts.

In addition, the reference Bayesian models were estimated using the same procedure and with the same model priors as described in Chapter 5. The *ex-ante* uncertainty measures have been obtained using quantile-based credible intervals from the predictive distributions. The results of forecasts prepared on the basis of truncated time series, both within the Bayesian and frequentist approaches, are presented in Fig. 8.5, alongside the actual observations available for the forecast horizon (2000–2007). Like before, the figures show point forecasts and 80% predictive intervals, treated as the basic uncertainty assessments. Bayesian predictive medians and frequentist point forecasts of the respective emigration rates are shown in more detail in Annex C, Table C.7, while their *ex-ante* errors are discussed in the next subsection.

In all cases shown in Fig. 8.5, the Bayesian median and frequentist point forecasts are very close to each other, slight differences being due to the impact of additional Bayesian prior information on the parameters. Within the 8-year horizon, both types of 'central' forecasts also seem plausible from the demographic point of view. However, as it might be expected from the outcomes of the robustness analysis offered in Section 8.1, the main difference is in the *ex-ante* precision assessments. The 80% Bayesian predictive intervals in general seem more reasonable than the frequentist ones, devoid of a priori assumptions on low precision and in many cases rather implausibly narrow. This conclusion is additionally supported by an empirical analysis of the frequency of actual observations for 2000–2007 falling into respective predictive intervals, offered in the following subsection together with a more detailed treatment of *ex-ante* and *ex-post* error measures.

In turn, the outcomes of the Bayesian and sampling-theory model selection procedures are shown in Table 8.2, the former prepared under the assumptions of uniform priors over the model space. In the Bayesian case, the results for truncated samples are roughly consistent with the outcomes presented in Chapter 5, generally pointing to simple models: constant (M_1) and random walks with drift (M_2), with relatively high probability also for AR(1) models (M_3). On the other hand, the frequentist log-likelihood values indicate either random walks, in these cases, where Bayesian criteria opted for constant models (m_{DE-IT} and m_{DE-CH}),

[10]The whole R environment is available from the R Development Core Team (2008), *R: A language and environment for statistical computing*, R Foundation for Statistical Computing, Vienna; www.r-project.org (accessed on 10 June 2007). In computations, the built-in functions arima and predict.arima from the *stats* package of R 2.4.1 were used. For more information on the applications of R to Bayesian computations, see Albert (2007), Lynch (2007), and Chapter 9 of this book.

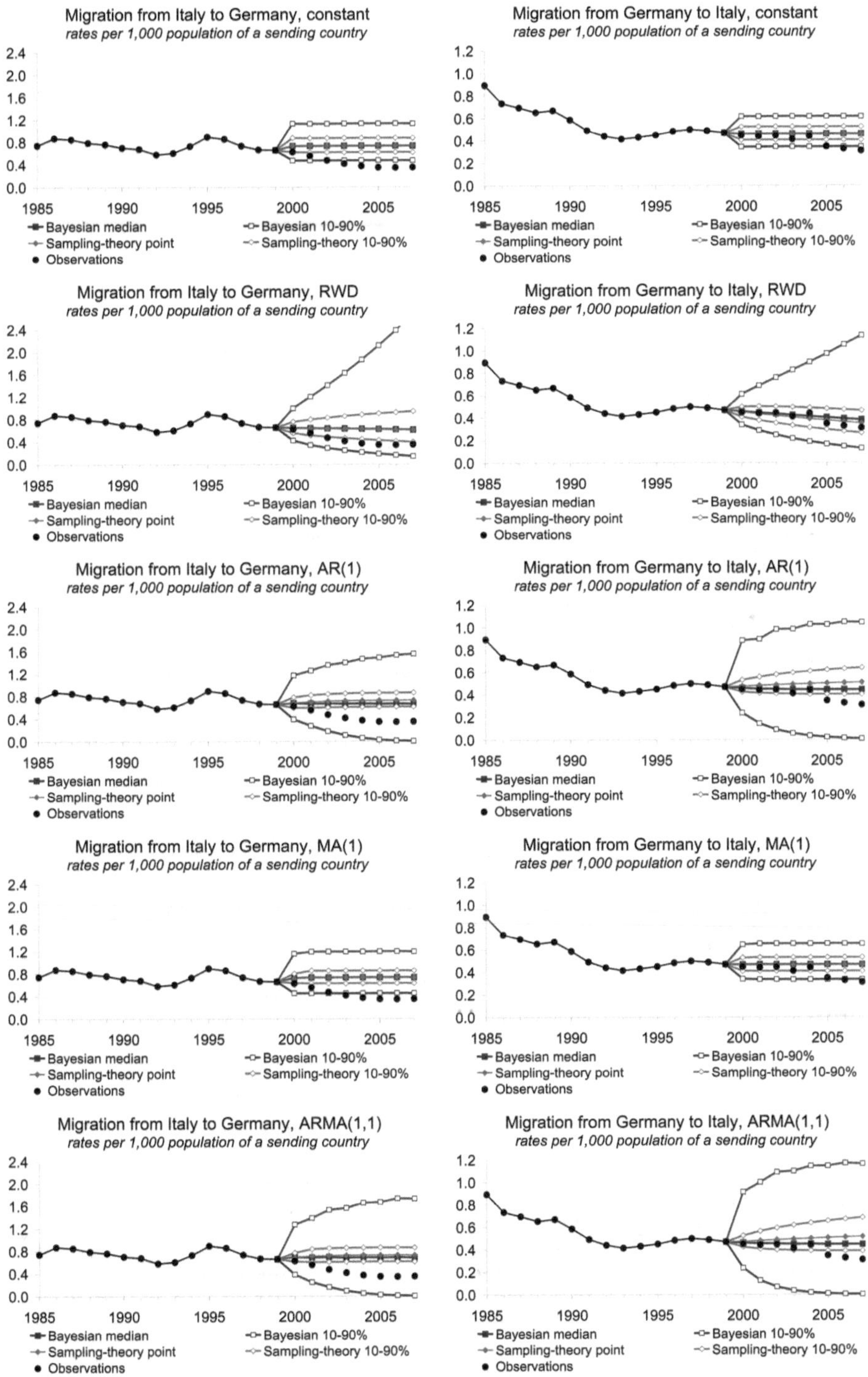

Fig. 8.5 Comparison of Bayesian and frequentist forecasts of migration rates, 2000–2007. Source: own elaboration in WinBUGS and R. Data until 2007: Eurostat/NewCronos; DESTATIS (see Annex A)

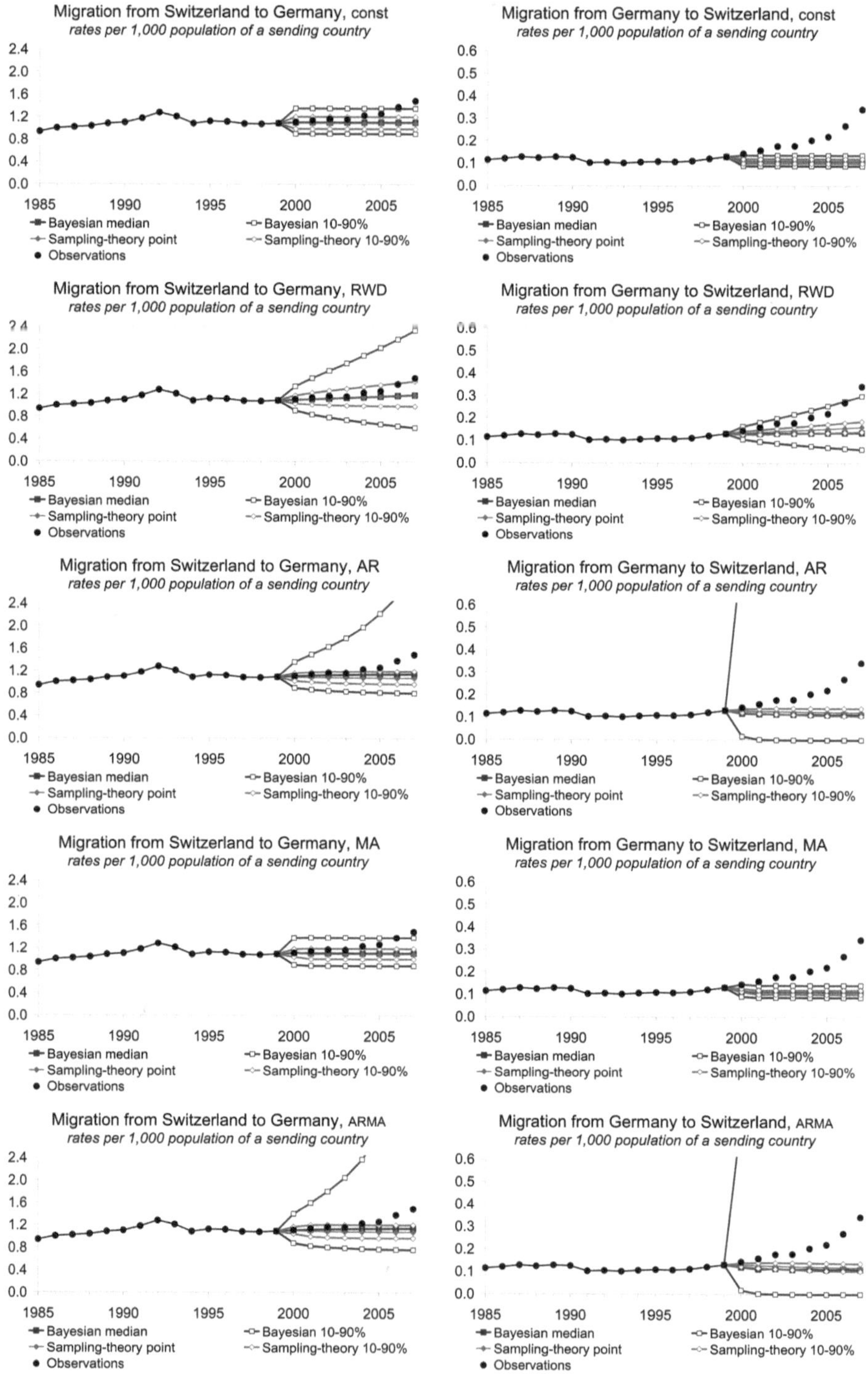

Fig. 8.5 (continued)

Table 8.2 Posterior probabilities and log-likelihood values for the estimated models M_i

Model (M_i)	M_1	M_2	M_3	M_4	M_5
Migration from Italy to Germany					
Posterior model probabilities $p(M_i \mid \mathbf{x})$: Uniform prior assumed	0.231	**0.325**	0.205	0.130	0.109
Log-likelihood of data $\propto \ln[L(\mathbf{x} \mid \boldsymbol{\theta})]$*	9.547	10.637	13.003	14.666	**15.603**
Migration from Germany to Italy					
Posterior model probabilities $p(M_i \mid \mathbf{x})$: Uniform prior assumed	**0.307**	0.244	0.208	0.143	0.099
Log-likelihood of data $\propto \ln[L(\mathbf{x} \mid \boldsymbol{\theta})]$*	13.992	**16.714**	15.653	14.507	15.940
Migration from Switzerland to Germany					
Posterior model probabilities $p(M_i \mid \mathbf{x})$: Uniform prior assumed	0.214	**0.354**	0.207	0.118	0.106
Log-likelihood of data $\propto \ln[L(\mathbf{x} \mid \boldsymbol{\theta})]$*	18.051	22.276	23.761	23.978	**25.783**
Migration from Germany to Switzerland					
Posterior model probabilities $p(M_i \mid \mathbf{x})$: Uniform prior assumed	**0.291**	0.233	0.206	0.172	0.098
Log-likelihood of data $\propto \ln[L(\mathbf{x} \mid \boldsymbol{\theta})]$*	19.822	**26.117**	24.356	24.203	24.792

Notes: *Log-likelihood values as provided in the R output (`arima$loglik`) and exact to proportionality. **Boldface** indicates highest-posterior probability, respectively highest-likelihood models for particular flows.
Source: own elaboration in WinBUGS and R

or ARMA(1,1) models, where the posterior probabilities point to random walks (m_{IT-DE} and m_{CH-DE}).

Clearly, such 'inverse' outcomes are an empirical issue, but nevertheless they confirm that the results of both procedures need not be equivalent, even under uniform model priors. Nevertheless, it is worth emphasising that the advantage of the Bayesian approach lies here in offering a probabilistically-coherent model selection framework, in which the uncertainty of observations, parameters and models are treated jointly, rather than separately and only partially, as is the case of the frequentist approach.

8.2.4 Comparison of Ex-Ante *and* Ex-Post *Errors for Various Predictions*

The indicators of forecast accuracy and bias presented in the current subsection are framework-specific (Bayesian or frequentist), model-specific (models M_1 through M_5), and flow-specific (four migration flows under study), in each case being calculated as means over the six-year forecast period, 2000–2007. Additionally, average measures are also computed as arithmetic means across four flows, thus for 32 observations. Notably, the 90th-Percentile Error (90-*PE*) due to its construction was calculated only for such a joint cross-flow universe. Hence 90% · 32 observations is 28.8, the 90-*PE* was derived from an interpolative formula, as a weighted average of 0.8 of the fourth-highest error and 0.2 of the fifth-highest one.

Detailed results in terms of various *ex-ante* and *ex-post* indicators are shown in Table 8.3. They indicate that, on average, the sampling-theory forecasts yielded

Table 8.3 Comparison of selected *ex-ante* and *ex-post* error measures for Bayesian and frequentist forecasts of migration rates, 2000–2007

Approach and models	Bayesian (median) forecasts					Frequentist (point) forecasts				
Indicators*	M_1	M_2	M_3	M_4	M_5	M_1	M_2	M_3	M_4	M_5
Migration from Italy to Germany										
ex-ante: *HCI*-80	1.39	2.14	4.02	1.54	4.57	0.55	0.67	0.48	0.47	0.48
ex-ante: *HCI*-95	2.19	3.36	16.06	2.49	20.15	0.88	1.07	0.77	0.75	0.78
ex-post: *MAPE*	63.6%	46.6%	53.0%	63.5%	56.2%	63.9%	46.0%	60.6%	63.0%	61.2%
ex-post: *RSMPE*	63.4%	48.2%	54.2%	63.5%	57.0%	63.7%	47.7%	61.7%	63.3%	62.1%
ex-post: *MALPE*	−63.6%	−46.6%	−53.0%	−63.5%	−56.2%	−63.9%	−46.0%	−60.6%	−63.0%	−61.2%
ex-post: *%POS*	0.0%	0.0%	0.0%	0.0%	0.0%	0.0%	0.0%	0.0%	0.0%	0.0%
Migration from Germany to Italy										
ex-ante: *HCI*-80	0.37	0.82	1.92	0.44	2.41	0.17	0.22	0.27	0.17	0.31
ex-ante: *HCI*-95	0.59	1.27	6.08	0.70	7.99	0.27	0.35	0.43	0.27	0.50
ex-post: *MAPE*	17.4%	9.2%	14.3%	18.8%	14.9%	17.4%	6.9%	24.5%	18.5%	24.8%
ex-post: *RSMPE*	22.5%	11.9%	19.8%	23.5%	20.1%	22.5%	8.3%	29.2%	23.3%	29.7%
ex-post: *MALPE*	−17.4%	−6.9%	−14.3%	−18.8%	−14.9%	−17.4%	−1.7%	−24.5%	−18.5%	−24.8%
ex-post: *%POS*	0.0%	25.0%	0.0%	0.0%	0.0%	0.0%	50.0%	0.0%	0.0%	0.0%
Migration from Switzerland to Germany										
ex-ante: *HCI*-80	1.96	3.33	3.46	2.21	4.28	1.06	1.00	1.26	0.92	1.18
ex-ante: *HCI*-95	3.07	5.19	9.98	3.54	14.78	1.69	1.59	2.02	1.48	1.91
ex-post: *MAPE*	52.3%	37.0%	44.0%	53.0%	42.7%	57.5%	37.0%	65.1%	58.5%	61.5%
ex-post: *RSMPE*	31.8%	23.2%	27.7%	32.1%	27.3%	33.6%	23.3%	37.1%	34.4%	35.9%
ex-post: *MALPE*	52.3%	37.0%	44.0%	53.0%	42.7%	57.5%	37.0%	65.1%	58.5%	61.5%
ex-post: *%POS*	100.0%	100.0%	100.0%	100.0%	100.0%	100.0%	100.0%	100.0%	100.0%	100.0%
Migration from Germany to Switzerland										
ex-ante: *HCI*-80	0.10	0.26	2.43	0.11	2.32	0.04	0.05	0.05	0.04	0.06
ex-ante: *HCI*-95	0.15	0.40	9.68	0.17	8.79	0.06	0.09	0.08	0.06	0.09
ex-post: *MAPE*	38.6%	26.8%	36.6%	38.2%	37.6%	38.6%	20.8%	30.8%	38.3%	31.8%

Table 8.3 (continued)

Approach and models	Bayesian (median) forecasts					Frequentist (point) forecasts				
Indicators*	M_1	M_2	M_3	M_4	M_5	M_1	M_2	M_3	M_4	M_5
ex-post: RSMPE	53.0%	39.5%	51.3%	52.9%	52.6%	53.0%	31.1%	45.0%	53.1%	46.6%
ex-post: MALPE	38.6%	26.8%	36.6%	38.2%	37.6%	38.6%	20.8%	30.8%	38.3%	31.8%
ex-post: %POS	100.0%	100.0%	100.0%	100.0%	100.0%	100.0%	100.0%	100.0%	100.0%	100.0%
Cross-flow averages**										
ex-ante: HCI-80	0.95	1.64	2.96	1.07	3.40	0.46	0.48	0.51	0.40	0.51
ex-ante: HCI-95	1.50	2.55	10.45	1.73	12.93	0.73	0.78	0.83	0.64	0.82
ex-post: MAPE	42.9%	29.9%	37.0%	43.4%	37.9%	44.3%	27.7%	45.3%	44.6%	44.8%
ex-post: RSMPE	42.7%	30.7%	38.2%	43.0%	39.3%	43.2%	27.6%	43.2%	43.5%	43.6%
ex-post: 90-*PE****	136.3%	124.7%	134.8%	136.3%	135.5%	136.3%	118.8%	128.9%	136.7%	130.2%
ex-post: MALPE	2.5%	2.6%	3.3%	2.2%	2.3%	3.7%	2.5%	2.7%	3.8%	1.8%
ex-post: %POS	50.0%	56.3%	50.0%	50.0%	50.0%	50.0%	62.5%	50.0%	50.0%	50.0%

* For definitions of all *ex-ante* and *ex-post* error measures used, see the beginning of the current section.
** Non-weighted means across four flows and eight forecast years (32 data points). ***Calculated for averages only.
Notes: *Ex-ante* indicators are expressed in terms of multitudes of respective point forecasts, while *ex-post* measures as percentages of the observed values. Grey shading indicates the measures of bias, and white one – of accuracy. All measures are related to log-transformed emigration rates, which were originally subject to forecasting.
Source: own elaboration in WinBUGS and R. Data until 2007: Eurostat/NewCronos; DESTATIS (see Annex A)

slightly smaller *ex-post* errors than their Bayesian equivalents. Remarks on the essential impact of (a lack of) prior assumptions on low precision on the *ex-ante* errors, made in the previous subsection, remain in force. With respect to particular migration flows, forecasts of emigration rates from Germany to Italy proved to be the most accurate in terms of relative error measures, while the predictions of flows in the opposite direction were the most imprecise ones.

As to the comparison of particular models, the lowest average *ex-ante* errors were yielded either by the Bayesian constant model M_1, or by the frequentist MA(1) model M_4. This outcome is not surprising, given that these two models are the ones that produce constant (non-increasing) predictive intervals, starting from the second forecast period (M_4) at the latest. In turn, the highest (on average) *ex-post* accuracy, according to all indicators applied, was obtained under either statistical inferential paradigm for the random walk models (M_2). The random walks yielded forecasts that were visibly closer to the empirical observations than the remaining models; on average by about one-third in the Bayesian case and by about 60% under the frequentist approach. In terms of forecast bias, the M_2-based error values were (again, on average) slightly more often positive than the ones of other models.

Further to synthetic measures of accuracy and bias, Table 8.4 presents the frequencies of empirical observations falling within various predictive intervals: 80% and 95%. The highest shares were obtained for the random walk and autoregressive models M_2, M_3 and M_5 – this time, in the Bayesian case, they were universally too high (100%), and in the frequentist case, too low and again unrealistic. Although the empirical frequencies of 'hits' of observations into predictive intervals differ for various models and migration flows, their cross-model and cross-flow averages reveal an interesting property. Namely, the shares obtained for the Bayesian forecasts (78.8 and 90.0%) are much closer to the nominal credibility or confidence levels $(1 - \gamma)$ than their sampling-theory equivalents (35 and 47%, respectively).

Despite such an interesting outcome, it has to be stressed that numerical results presented in the current section are mostly an empirical issue, which needs not hold in all particular cases. However, the presented general findings still seem very important both from the theoretical, as well as practical points of view. Firstly, as observed in Chapter 2, the existing migration theories are very difficult to be empirically verified, for the reasons including, but not limited to, the well-known deficiencies of available statistical data. Secondly, with respect to migration processes, there are many factors in play, which can be barely, if at all, possible to operationalise in the forecasting practice. These two justifications may help explain the clear support of the data for the random walk models in a vast majority of the presented empirical examples.

Whether migrations are indeed purely random processes (aleatory uncertainty), or whether such results as the ones presented above stem from imperfect knowledge concerning the determinants of population flows (epistemic uncertainty), is a matter of individual belief. This problem is thus in a way similar to the general discussion on the sources of uncertainty, presented in Chapter 2. Nevertheless, whichever the cause, such outcomes seem to ask for a proper assessment of the uncertainty of forecasts in the hardly-predictable migration reality.

Table 8.4 Percentage of empirical observations falling into various predictive intervals

Model	M_1	M_2	M_3	M_4	M_5	Average (cross-model)
80% predictive intervals, period 2000–2007						
Bayesian forecasts						
Italy to Germany	25.0%	100.0%	100.0%	37.5%	100.0%	72.5%
Germany to Italy	75.0%	100.0%	100.0%	75.0%	100.0%	90.0%
Switzerland to Germany	75.0%	100.0%	100.0%	87.5%	100.0%	92.5%
Germany to Switzerland	0.0%	87.5%	100.0%	12.5%	100.0%	60.0%
Average (cross-flow)	**43.8%**	**96.9%**	**100.0%**	**53.1%**	**100.0%**	**78.8%**
Frequentist forecasts						
Italy to Germany	12.5%	25.0%	12.5%	0.0%	12.5%	12.5%
Germany to Italy	62.5%	100.0%	62.5%	62.5%	62.5%	70.0%
Switzerland to Germany	50.0%	87.5%	50.0%	50.0%	50.0%	57.5%
Germany to Switzerland	0.0%	0.0%	0.0%	0.0%	0.0%	0.0%
Average (cross-flow)	**31.3%**	**53.1%**	**31.3%**	**28.1%**	**31.3%**	**35.0%**
95% predictive intervals, period 2000–2007						
Bayesian forecasts						
Italy to Germany	62.5%	100.0%	100.0%	100.0%	100.0%	92.5%
Germany to Italy	100.0%	100.0%	100.0%	100.0%	100.0%	100.0%
Switzerland to Germany	100.0%	100.0%	100.0%	100.0%	100.0%	100.0%
Germany to Switzerland	12.5%	100.0%	100.0%	25.0%	100.0%	67.5%
Average (cross-flow)	**68.8%**	**100.0%**	**100.0%**	**81.3%**	**100.0%**	**90.0%**
Frequentist forecasts						
Italy to Germany	12.5%	100.0%	25.0%	12.5%	25.0%	35.0%
Germany to Italy	62.5%	100.0%	62.5%	62.5%	75.0%	72.5%
Switzerland to Germany	75.0%	100.0%	75.0%	62.5%	75.0%	77.5%
Germany to Switzerland	0.0%	12.5%	0.0%	0.0%	0.0%	2.5%
Average (cross-flow)	**37.5%**	**78.1%**	**40.6%**	**34.4%**	**43.8%**	**46.9%**

Source: own elaboration in WinBUGS and R. Data until 2007: Eurostat/NewCronos; DESTATIS (see Annex A)

For the reasons mentioned above, the prior assumptions on high variability of migration flows appear to be just a convenient and honest way of expressing the researchers' uncertainty concerning the behaviour of the processes under study. The value added of using the Bayesian approach is the formality of inference, coupled with the explicitly-admitted subjectivism, which allows for the possibility to include the expert-based reasoning in the forecasts. These features can contribute to obtaining more realistic predictions than the weakly-reliable data alone, as demonstrated in the empirical examples of the current study.

Chapter 9
Bayesian Computing in Practice

This chapter starts with a short survey of available software for the Bayesian analysis (Section 9.1), with the emphasis laid on high-level programming languages and free, open-source software issued under the GNU General Public License.[1] Further, in Section 9.2, the WinBUGS environment for Bayesian analysis, applied to the analyses presented in this book, is described in more detail. In Section 9.3 an example of forecasting immigration flows using Bayesian approach, consisting of the estimation of the posterior distributions by means of the Gibbs algorithm, is demonstrated using the R environment, along with the Carlin–Chib (1995) model selection procedure (presented in Chapter 5). Finally, Section 9.4 offers general reflections on the advantages and limitations of particular software environments in practical applications.

9.1 A Short Survey of Available Bayesian Software

9.1.1 R programming Language

R is a free, open-source, high-level language and environment for statistical computing, available from www.r-project.org. It provides a command line interface and contains a tool for creating high-quality figures, graphs and charts. R can be seen as an open-source implementation of the S language, developed by Becker, Chambers, and Wilks (1988). One of the most important advantages of R is the availability of a great number of packages, which comprise of sets of functions and procedures that allow for handling particular models and applications. Among them, there are several easily downloadable packages for the analysis of the statistical and econometric Bayesian models. Table 9.1 presents the summary of the most common packages for Bayesian analysis that can be potentially useful for the migration modelling and forecasting.

Contributed by Arkadiusz Wiśniowski

[1] For more details see www.gnu.org.

J. Bijak, *Forecasting International Migration in Europe: A Bayesian View*,
The Springer Series on Demographic Methods and Population Analysis 24,
DOI 10.1007/978-90-481-8897-0_9,

Table 9.1 Selected R packages for Bayesian analysis

Package	Description
Arm	Bayesian inference in linear, generalised linear, ordered logit and probit models
bayesm	Analysis of linear regression models, multinomial logit, multinomial probit, multivariate probit, multivariate mixture of Normal distributions (including clustering), density estimation using finite mixtures of Normal distributions, Dirichlet process priors, hierarchical linear models, hierarchical multinomial logit, hierarchical negative binomial regression models, and linear instrumental variable models
DPpackage	R functions for Bayesian nonparametric and semi-parametric models. The package includes, among others, semi-parametric models for density estimation, censored data, binary regression models and generalised linear mixed models.
MCMCpack	Model-specific MCMC algorithms for inference in regression models (linear regression, logit, ordinal probit, probit, Poisson regression, etc.), measurement models (item response theory and factor models), changepoint models (binary and Poisson), and models for ecological inference. It contains a generic Metropolis sampler.
LearnBayes	R functions and sample data for the Albert (2007) book on R in the Bayesian computations
BMA	Bayesian averaging of linear, generalised linear and survival models
BAYSTAR	Bayesian analysis of threshold autoregressive models
MSBVAR	Estimation of Bayesian VAR and structural VAR models
BRugs	Provides R interface to the OpenBUGS environment (for Windows)
R2WinBUGS	Provides R interface to the WinBUGS environment (for Windows)
Rbugs	Provides R interface to OpenBUGS (for Linux)
Boa	Post-estimation tool for diagnostics, summarising and visualising the MCMC output, also useful for importing from the BUGS format
Coda	The *Convergence Diagnosis and Output Analysis* (CODA) package – a set of functions to summarise, plot, and diagnose convergence from MCMC samples, including a possibility of import from the BUGS format.

Source: Park, Martin, and Quinn (2009); cran.r-project.org/web/views/Bayesian.html.

A general introduction to the Bayesian analysis using R language was provided by Albert (2007). His book also contains information on the applications of various packages for performing MCMC simulations in BUGS (such as *boa*, *R2WinBUGS* and *Brugs*), and on the analysis of the output. An example of the Bayesian analysis of the simple time series models is presented in Section 9.3.

9.1.2 Octave

Octave is a high-level language for numerical computations, available from www.gnu.org/software/octave. Like the R language, the computations are handled by a command line interface and the tools for creating high-quality graphics are provided. Octave may be also used as a batch language, e.g. under UNIX systems. The language is similar to, yet not fully compatible with the MATLAB environment. The related project is Octave-Forge (octave.sourceforge.net), which contains some additional scripts and extensions for the language, although with no packages specifically aimed at Bayesian computations. Nevertheless the built-in sampling procedures from most of the probability distributions allow for easy implementation of the MCMC algorithms in Octave.

9.1.3 BUGS

BUGS (*Bayesian inference Using Gibbs Sampling*) is a software project aimed at providing an environment specifically devoted to the Bayesian analysis of complex statistical models using MCMC methods. Currently (as of November 2009), there are three main platforms with implementations of the BUGS: WinBUGS, OpenBUGS and JAGS (*Just Another Gibbs Sampler*). Of those, a discontinued WinBUGS project (www.mrc-bsu.cam.ac.uk/bugs/welcome.shtml; Lunn, Thomas, Best, & Spiegelhalter, 2000) is a proprietary software package developed for the Windows operating system. Its open-source continuation is OpenBUGS (mathstat.helsinki.fi/openbugs; Thomas et al., 2006), a release which in principle contains the BUGS implementations for Windows, DOS and Linux platforms. OpenBUGS also contains a DoodleBUGS, a graphical editor for model specification. In computational terms, OpenBUGS uses in fact three types of MCMC algorithms: Metropolis-Hastings, Gibbs (for description see Chapter 2) and slice sampling (Neal, 2003).

Finally, JAGS (www-fis.iarc.fr/~martyn/software/jags) is 'just another' command-line implementation of BUGS that was aimed at providing the engine of the BUGS language on UNIX systems, as well as to be a platform for experimenting and writing own functions and extensions. All implementations have similar syntax to the one used in R.

In the simplest way the analysis of a model in BUGS consists of the following stages: model and data specification, compilation, initialisation, sampling, monitoring and finally convergence diagnostics and analysis of the output. Detailed instructions on how to use WinBUGS (and DoodleBUGS) are provided in the manual for the software[2] (OpenBUGS Manual 2009). Section 9.2 presents in more detail the subsequent stages of programming in the WinBUGS environment.

9.2 Bayesian Computation in WinBUGS

9.2.1 Model and Data Specification

The first stage of working with BUGS consists of the model specification. In the case of WinBUGS, a model is specified within the graphical interface of the software, in an *.odc file. This format is used for WinBUGS to store all information, from model specification to the estimated statistics and related figures. In turn, JAGS requires

[2] Available at mathstat.helsinki.fi/openbugs/Manuals/Manual.html.

definition of the model to be written in a text file, which is then compiled using the command line. Specification of the model consists of a series of commands between the {and} brackets with a `model` heading. The example model for an AR(1) process is given below.

```
#Model specification in WinBUGS
model {
#Prior distributions specification 1
       tau     ~           dgamma(1,1);
       b[1]    ~           dnorm(0,0.0001);
       b[2]    ~           dnorm(0,0.0001);
#Prior distributions specification 2 (alternative)
#      tau     ~           dgamma(1,1);
#      bmu[1]<-0;          bmu[2]<-0;
#      bTau[2,2]<-2;       bTau[1,1]<-0.0001;
#      bTau[2,1]<-0;       bTau[1,2]<-0;
#      b[1:2]     ~        dmnorm(bmu[1:2],bTau[1:2,1:2]);
#Data transformation
      for (t in 1:n) {de[t]      <-      log(DE[t]) }
#Model
     for (i in 2:n) {
          mu[t]      <-      b0+b1*de[t-1]
          de[t]      ~       dnorm(mu[t], tau)
     }
}
#Data specification
list(n=12, DE=c(1599,1462,1693,1590,1773,1656,1744,2315,
  1921,1580,1781,2187))
```

In this code, the prior distributions for precision (`tau`) are defined as gamma distributions with expected value and variance equal to 1. Priors for the constant (`b0`) and autoregressive coefficient (`b1`) are assumed to be normal with zero mean and precision equal 10^{-4}. It should be stressed that different versions of BUGS are sensitive to the specification of priors. In WinBUGS 1.4 sampler the interdependencies between the structural parameters are assumed, hence the specifications 1 and 2 are equivalent. In WinBUGS 3.0.3 algorithm, in turn, specification 1 implies no correlation structure for the structural parameters of the conditional distributions, which may result in poor convergence. Therefore, in the latter version, specification 2 should be preferred unless there is rationale for assuming no correlation between the parameters.

The next step consists of the transformation of the data – natural logarithms are taken in order to assure the non-negativity. Subsequently, the 'model adaptation' phase follows whereby the likelihood of the observations is defined to be Normal with expected value `mu[t]` and precision `tau`. Finally, after the model brackets, the data are specified. The list of data items consists of the number of observations ($n = 12$) and the vector of observations, in this case denoting total immigration flows from Germany to Denmark in 1994–2005, as reported by the Statistics Denmark (see Section 9.3).

In WinBUGS data can be specified either before or after the model, while JAGS requires the data to be prepared in a separate text file created by means of the `dump()` command in the R environment. An example of such a data file is given below.

```
#Data specification in JAGS
'n' <- 12
'DE'<- c(1599,1462,1693,1590,1773,1656,1744,2315,1921,
   1580,1781,2187)
```

9.2.2 Model Compilation, Initialisation and Updating

Having specified a model, the programmer should check its syntactical correctness. After loading the data and setting the number of simulation chains to be simulated, the model is then compiled. In WinBUGS, these operations are handled by highlighting the headings model and list (preceding the data) and selecting the appropriate commands from the Specification Tool (menu option *Model*: *Specification*). In JAGS this is done by invoking a set of commands.[3] Subsequently, the model is initialised by providing or generating the initial values for the random quantities, which can vary between the chains. For generation, a random sample from the prior distributions is taken wherever possible; otherwise the initial values have to be specified by the user in a similar way (using lists) as the data are.

In the next step of the WinBUGS procedure, the 'stochastic nodes' depicting the variables of interest, such as model parameters or forecasts, are chosen for inference by entering them into the box in the *Sample Monitor Tool* (menu option *Inference*: *Samples*). In our example, the stochastic nodes of interest are `b0`, `b1` and `tau`. Then, after the number of iterations of the chain is provided, by selecting *Model*: *Update* from the menu, the simulation can begin. It can be monitored by 'real-time' trace plots of parameters that allow a visual assessment, whether the convergence has been achieved or not.

There are also several other options that can enhance the updating with respect to the convergence achievement and well-mixing[4] of the chain, which include

[3] For details, see the JAGS manual: www-fis.iarc.fr/~martyn/software/jags/jags_user_manual.pdf.

[4] A chain is said to be 'well-mixing' if it explores (almost) the whole parameter space. As opposed to a well-mixed chain, the poorly-mixed one stays in a small region for a long time.

thinning, over-relaxing and adapting.[5] Thinning is a technique that chooses every k-th iteration of the running chain. It can be applied during the simulation (which saves the disk space and memory), or after the simulation. It is advised to apply it when there is evidence of autocorrelation in the generated path of the node. Over-relaxing (Neal, 1998) is a technique which also helps to reduce the autocorrelation and aids the mixing of the chain; it can also reduce the number of iterations needed for achieving convergence but simultaneously increases the time required for the simulation. According to the OpenBUGS manual (2009) it consists in generating '... multiple samples at each iteration and then select[ing] one that is negatively correlated with the current value.' The authors of the Manual warn however that 'this method is not always effective and should be used with caution.' Finally, the adapting technique can be chosen only when the Metropolis or slice-sampling algorithms are used, in order to improve their performance.

9.2.3 Convergence Diagnostics and Inference

After the simulation has exceeded the pre-defined size of the burn-in sample (see Chapter 2), convergence can be assessed and characteristics of the distributions of quantities of interest can be obtained. In particular, the smoothed estimate of posterior distributions can be plotted and the summary statistics can be calculated, such as the means, standard deviations, main quantiles and the Monte Carlo standard error. The latter is calculated using a method proposed and described in more detail in Roberts (1995).

For convergence assessment a 'running mean'[6] with 95% confidence intervals ('running quantiles') may be plotted and checked for stability. Moreover, the stationarity of the autocorrelation functions[7] of samples can be verified. There are also other diagnostic tools available, described in more detail in the OpenBUGS Manual (2009). The whole sample history can be viewed and saved for more systematic and in-depth convergence diagnostics available in the R package *coda*. Finally, a tool for calculating the Deviance Information Criterion (DIC), described in Section 5.1, can be used for model comparisons.

In JAGS, various monitors for nodes, such as a simple trace of the sample, its deviance or output readable for the *coda* R package, are set by means of commands.

Examples of the WinBUGS code for migration forecasts to the problems discussed in Chapters 5, 6, 7, and 8 are provided in Appendix B. The following section, in turn, contains an illustration of similar models programmed within the R environment.

---

[5] Description of the techniques is based on the OpenBUGS Manual (2009).

[6] 'Running mean' or 'running quantile' denotes a series of means (or quantiles) calculated at each k-th iteration using the whole backward sample. It can be calculated recursively, e.g, a formula for a recursive mean is $\bar{x}_{n+1} = (n\bar{x}_n + x_{n+1})(n+1)^{-1}$.

[7] Autocorrelation function (ACF) depicts correlations between sample values at different moments in time.

9.3 Example of Bayesian Computation in R Language

9.3.1 Forecasting Migration Using R

In this section an R version of the sample code used for the analysis of forecasting models described in Chapter 5 (and Appendix B) is presented. The example consists of the Gibbs algorithm employed to obtain the posterior characteristics of the model parameters and forecasts, as well as for the model selection using the Carlin and Chib (1995) method. The empirical illustration, and therefore also the code, is based on the example of total (gross) immigration flows of foreigners from Germany to Denmark in the period 1994–2005.[8] Hereafter, the models employed for the analysis are presented.

9.3.2 The Model of Immigration Flows

Let Y_t denote the total immigration flows and $y_t = \ln(Y_t)$.[9] There are $t = 1, \ldots, T$ observations. Two underlying models are considered: autoregressive process of order 1 (denoted by k_1) and a random walk with a drift (denoted by k_2). Errors are assumed to be Gaussian white noise. For the purpose of formulating the pseudo-prior distributions in the Carlin–Chib (1995) model selection procedure (described in detail in Chapter 5), the following notation for the hyper-parameters (i.e. parameters of the prior distributions) is introduced: let $\alpha_{ij} = \alpha_i \left| k_j \right.$, thus when $i = j$ the hyper-parameter relates to the true prior, and when $i \neq j$ it depicts the linking distribution ('pseudo-prior'). The models are given below.

$$k_1: \quad y_t = \beta_{10} + \beta_{11} y_{t-1} + \varepsilon_t, \quad \varepsilon_t \sim \text{iid N}(0,\ \tau_1^{-1}); \tag{9.1a}$$

$$k_2: \quad y_t = \beta_2 + y_{t-1} + \varepsilon_t, \varepsilon_t \sim \text{ iid N}(0,\ \tau_2^{-1}). \tag{9.1b}$$

Hence, the corresponding parameter sets are $\boldsymbol{\theta}_1 \left| k_1 \right. = [\boldsymbol{\beta}_1{}',\ \tau_1]\,'$ with $\boldsymbol{\beta}_1 = [\beta_{10},\ \beta_{11}]'$ for the first model, and $\boldsymbol{\theta}_2 \left| k_2 \right. = (\beta_2,\ \tau_2)$ for the second. The likelihood functions can be written as:

$$L(\boldsymbol{\beta}_1,\ \mathbf{X}_1,\ \mathbf{y} \left| k_1 \right.) = (2\pi)^{-(N-1)/2}\ \tau_1^{(N-1)/2} \exp[-\tau_1/2(\mathbf{y} - \mathbf{X}_1\boldsymbol{\beta}_1)'(\mathbf{y} - \mathbf{X}_1\beta_1)], \tag{9.2a}$$

$$L(\beta_2,\ \mathbf{x_2},\ \mathbf{y} \left| k_2 \right.) = (2\pi)^{-(N-1)/2}\ \tau_2^{(N-1)/2} \exp[-\tau_2/2(\Delta\mathbf{y} - \mathbf{x_2}\beta_2)'(\Delta\mathbf{y} - \mathbf{x_2}\beta_2)], \tag{9.2b}$$

where $\mathbf{y}$ denotes a $(T-1)$-element data vector, $\mathbf{X_1}$ is a $(T-1) \times 2$ matrix of ones and lagged values of y, and $\mathbf{x_2}$ is a $(T-1)$-element vector of ones.

[8] Source: www.statbank.dk, accessed on 23 October 2009.

[9] Note that these are different flows than emigration introduced in Chapter 2 and denoted by $M_{i-j(t)}$.

Relatively vague Normal priors have been assumed for β_{10} and β_2 parameters, together with an informative Normal prior for the autoregressive parameter β_{11} in the first model. Hence, $\boldsymbol{\beta}_1 \sim \mathrm{N}(\mathbf{b}_{11}, \boldsymbol{\Sigma}_{11}^{-1})$, where $\mathbf{b}_{11} = [0, 0.5]'$, and Σ_{11} is a covariance matrix, here assumed to have 10^4 and 4.0 on the diagonal, and zeros outside of it. For the second model it is assumed that the parameter $\beta_2 \sim \mathrm{N}(b_{22}, \sigma_{22}^2)$, where $\sigma_{22}^2 = 10^4$. For precision (inverse variance), the Gamma[10] priors have been imposed: $\tau_1 \sim \Gamma(a_{11}, s_{11})$ and $\tau_2 \sim \Gamma(a_{22}, s_{22})$. Moreover it has been assumed that $a_{11} = s_{11} = a_{22} = s_{22} = 1$.

The resulting posterior distribution does not have a form of any known density; hence no direct inference can be made. Nevertheless, the full conditionals of the posterior distribution can be written (for details see Koop, 2003). For the model k_1, these are:

$$p(\boldsymbol{\beta}_1 \,|\tau_1; \mathbf{X}_1,\ \mathbf{y},\ k_1) \sim \mathrm{N}(\boldsymbol{\Omega}_1^{-1}(\boldsymbol{\Sigma}_{11}^{-1}\mathbf{b}_{11} + \tau_1\mathbf{X}_1'\mathbf{y}),\ \boldsymbol{\Omega}_1^{-1}), \tag{9.3a}$$

$$p(\tau_1 \,|\boldsymbol{\beta}_1;\ \mathbf{X}_1\ \mathbf{y}_1,\ k_1) \sim \Gamma((T-1)/2 + a_{11},\ [0.5(\mathbf{y}-\mathbf{X}_1\boldsymbol{\beta}_1)'(\mathbf{y}-\mathbf{X}_1\boldsymbol{\beta}_1) + 1/s_{11}]^{-1}), \tag{9.3b}$$

where $\boldsymbol{\Omega}_1 = (\boldsymbol{\Sigma}_{11}^{-1} + \tau_1\mathbf{X}_1'\mathbf{X}_1)$. Similarly, for k_2:

$$\mathrm{p}(\beta_2 \,|\tau_2;\ \mathbf{x}_2,\ \mathbf{y},\ k_2) \sim \mathrm{N}(\omega_2^{-1}(\sigma_{22}^{-2}b_2 + \tau_2\mathbf{x}_2'\boldsymbol{\Delta}\mathbf{y}),\ \omega_2^{-1}), \tag{9.4a}$$

$$\mathrm{p}(\tau_2 \,|\beta_2;\ \mathbf{x}_2,\ \mathbf{y},\ k_2) \sim \Gamma(0.5(T-1) + a_{22}, [0.5(\boldsymbol{\Delta}\mathbf{y}-\mathbf{x}_2\beta_2)'(\boldsymbol{\Delta}\mathbf{y}-\mathbf{x}_2\beta_2) + 1/s_{22}]^{-1}), \tag{9.4b}$$

where $\omega_2 = [\sigma_{22}^{-2} + \tau_2(T-1)]$ and the second element of this sum results from the fact that $\mathbf{x}_2$ is a vector of ones.

Forecasting in the Bayesian approach consists of sampling from the posterior predictive distribution; hence it requires the evaluation of the integral in Equation (2.4) (see Chapter 2). Within the Gibbs sampler this can be achieved by sampling from the conditional model for future observations given predictors. In both models the predictor is the lagged value of the observed immigration flows. Hence, sampling from the predictive distribution for the P future immigration flows, $y_{T+1}, \ldots, y_{T+P}$, consists of a recursive sampling using the conditional distribution (for details see e.g. Koop, 2003):

$$p(y_p \,|\boldsymbol{\theta}_1, y_{p-1},\ k_1) = (2\pi)^{-0.5}\tau_1^{0.5}\exp[-\tau_1/2(y_p - \beta_{10} - y_{p-1}\beta_{11})^2], \quad p = T+1, \ldots, T+P, \tag{9.5a}$$

$$p(y_p \,|\boldsymbol{\theta}_2,\ y_{p-1}, k_2) = (2\pi)^{-0.5}\tau_2^{0.5}\exp[-\tau_2/2(\Delta y_p - \beta_2)^2], \quad p = T+1, \ldots, T+P. \tag{9.5b}$$

For forecasting y_{T+1}, the lagged value is anchored in the last observation y_T.

---

[10]Here, Gamma distribution is parameterised as $p(x|a, s) = s^{-a}\Gamma(s)^{-1}x^{s-1}\exp(-x/s)$ with mean as and variance as^2.

9.3.3 Sampling

The Gibbs sampling algorithm, as described in Chapter 2, allows for obtaining the characteristics of the posterior distribution by a sequential sampling from the full conditionals (9.3) and (9.4). Below, the R code for the example of forecasting gross immigration from Germany to Denmark is presented. It should be noted that no attempts at code optimisation were made, since the main purpose of the example was to provide a transparent application of the Gibbs algorithm for the immigration forecasting model. In the code provided throughout this chapter, the hash signs (#) denote lines with comments.

As a first step of the computations, the raw data have to be loaded and the general transformations for the model formulation are subsequently required.

```
#DE->DK total immigration
#1599,1462,1693,1590,1773,1656,1744,2315,1921,1580,
  1781,2187
#data transformations
yraw<-log(c(1599,1462,1693,1590,1773,1656,1744,2315,1921,
  1580,1781,2187));
T  <- length(yraw);
y1 <- matrix(yraw[2:T],T-1);
X1 <- matrix(0,T-1,2); X1[1:T-1,1]<-rep(1,T-1);
  X1[,2]<-yraw[1:T-1];
y2 <- matrix(yraw[2:T]-yraw[1:T-1],T-1);
X2 <- matrix(0,T-1,1); X2[1:T-1,1]<-rep(1,T-1);
```

Here, `y1` and `y2` respectively denote the vectors of logarithms and first differences of logarithms of the immigration flows, in the autoregressive (in the code denoted as `M1`) and the random walk (`M2`) models. Besides, `X1` is a matrix with the first column of ones and the second – of lagged values of y, while `X2` is a vector of ones.

The next step consists of the specification of the priors (`#Priors specification` in the code-box below). Parameters `b1_1 or sig2_2` denote the hyper-parameters b_{11} and ${\sigma_{22}}^2$ respectively and this convention is followed hereafter. Next, recalculations required for the conditionals (`#calculations for M1` and `M2`) are carried out. After specifying the priors, the number of replications of the chain (`rp`) and the burn-in sample length (`nburn`), as well as the initial values of the models' parameters are also defined (`#Initials`). The initial values of the elements of $\boldsymbol{\beta}_1$ and β_2 are set to zero, and values of τ_1 and τ_2 to unity. Since the ultimate goal is to produce model-specific predictions, the vectors of forecasts need to be defined and the forecasting horizon set: in this case, being the 4-year period 2006–2009 (`#Forecasts specification`).

```
#Priors specification
#priors: theta1|M1
b1_1      <- matrix(c(0, 0.5),2,1);          #E(beta1)
sig1_1    <- matrix(c(10000, 0, 0, 4),2,2);  #Var(beta1)
a1_1=1; s1_1=1;                              #tau1 scale
                                             and shape
#priors: theta2|M2
sig2_2    <- 10000;                          #E(beta2)
b2_2      <- matrix(c(0),1,1);               #Var(beta2)
a2_2=1; s2_2=1;                              #tau2 scale
                                             and shape
#calculations for M1
isig1_1       <- solve(sig1_1);
isig1_1b1_1   <- isig1_1%*%b1_1;
y1y1=t(y1)%*%1; X1y1=t(X1)%*%y1; X1X1=t(X1)%*%X1;
is1_1         <- 1/s1_1;
a1new         <- 0.5*(T-1) + a1_1;

#calculations for M2
isig2_2       <- (1/sig2_2);
isig2_2b2_2   <- isig2_2*b2_2;
y2y2=t(y2)%*%y2; X2y2=sum(y2); X2X2=sum(X2);
is2_2         <- 1/s2_2;
a2new         <- 0.5*(T-1) + a2_2;
#Initials
rp<-100000;                                  #replications of
                                              the chain
nburn<-75001;                                #burn-in sample
#model parameters and initial values - theta1
tau1    <- c(1);
length(tau1)    <- rp+1;
beta1   <- matrix(c(0, 0),rp+1,2);
#model parameters and initial values - theta2
tau2    <- c(1);
length(tau2)    <- rp+1;
beta2   <- matrix(c(0),rp+1,1);

#Forecasts specification
yfM1 <- matrix(0,rp+1,for_l);
yfM2 <- matrix(0,rp+1,for_l);
for_l <- 5;    #forecast length
```

The next step of the procedure consists of a loop with the update steps of the Gibbs algorithm for both models, together with the respective predictive distributions. In this way, immigration forecasts for 2006–2009 are obtained.

```
for (i in 1:rp)
{
#sampling parameters

#theta1|M1
iom1 <- solve(isig1_1 + tau1[i]*X1X1); #auxiliary
                                              calculation
beta1[i+1,1:2]<- iom1%*%(isig1_1b1_1 + tau1[i]*X1y1)
  + t(chol(iom1))%*%rnorm(2);

tau1[i+1] <- rgamma(1, a1new, 0.5*y1y1 - t(beta1[i+1,])
  %*%X1y1 + 0.5*t(beta1[i+1,])%*%X1X1
  %*%beta1[i+1,] + is1_1);

#theta2|M2
iom2 <- 1/(isig2_2 + tau2[i]*X2X2);     #auxiliary
                                              calculation
beta2[i+1] <- iom2*(isig2_2b2_2 + tau2[i]*X2y2)
  + sqrt(iom2)*rnorm(1);
tau2[i+1] <- rgamma(1, a2new, 0.5*y2y2 - beta2[i+1]*X2y2
  + 0.5*beta2[i+1]*X2X2*beta2[i+1] + is2_2);

#Sampling model-specific forecasts
#M1
yfM1[i+1,1] <- rnorm(1, mean=beta1[i+1,1]+y1[9,1]
  *beta1[i+1,2], sd=sqrt(1/tau1[i+1]));
yfM1[i+1,2] <- rnorm(1, mean=beta1[i+1,1]+yfM1[i+1,1]
  *beta1[i+1,2], sd=sqrt(1/tau1[i+1]));
yfM1[i+1,3] <- rnorm(1, mean=beta1[i+1,1]+yfM1[i+1,2]
  *beta1[i+1,2], sd=sqrt(1/tau1[i+1]));
yfM1[i+1,4] <- rnorm(1, mean=beta1[i+1,1]+yfM1[i+1,3]
  *beta1[i+1,2], sd=sqrt(1/tau1[i+1]));
#M2
yfM2[i+1,1] <- rnorm(1, mean=beta2[i+1]+y1[9,1],
  sd=sqrt(1/tau2[i+1]));
yfM2[i+1,2] <- rnorm(1, mean=beta2[i+1]+yfM2[i+1,1],
  sd=sqrt(1/tau2[i+1]));
yfM2[i+1,3] <- rnorm(1, mean=beta2[i+1]+yfM2[i+1,2],
  sd=sqrt(1/tau2[i+1]));
yfM2[i+1,4] <- rnorm(1, mean=beta2[i+1]+yfM2[i+1,3],
  sd=sqrt(1/tau2[i+1]));
}
```

Table 9.2 Summary statistics of the posterior distributions of the model parameters

Parameter	Mean	Variance	q10%	q25%	Median	q75%	q90%
beta10	5.0228	60.6676	−4.4563	0.1404	4.9756	10.1157	14.9500
beta11	0.3269	1.0949	−1.0346	−0.3643	0.3365	0.9846	1.6170
beta2	0.0250	0.0185	−0.1372	−0.0506	0.0263	0.1080	0.1872
tau1	5.2518	4.6411	2.7066	3.6830	5.0569	6.2467	7.7596
tau2	5.2262	4.9831	2.6649	3.5792	4.7856	6.5568	8.2605

Note: qa% denotes the a-th percentile of the distribution. Sample: 25,000, burn-in sample: 75,000.
Source: own calculations in R

In the first part of this code, in order to sample from a multivariate Normal distribution with variance matrix $\mathbf{\Omega}_1^{-1}$ or $\mathbf{\Omega}_2^{-1}$, the transposition of the upper triangular factor of the Choleski decomposition[11] of these matrices (denoted as `iom1` and `iom2` respectively) has been used. In the function generating random samples from the Gamma distribution (`rgamma`) the inverse of the shape parameter, s, is provided.

Subsequently, the summary statistics of the posterior distributions and the histograms are produced, together with the autocorrelation functions, using the code provided below. Table 9.2 presents the summary statistics (`par_stat`) of the model parameters. In Figs. 9.1 and 9.2, the histograms and autocorrelation functions of the parameters are respectively presented. Some indication of autocorrelation

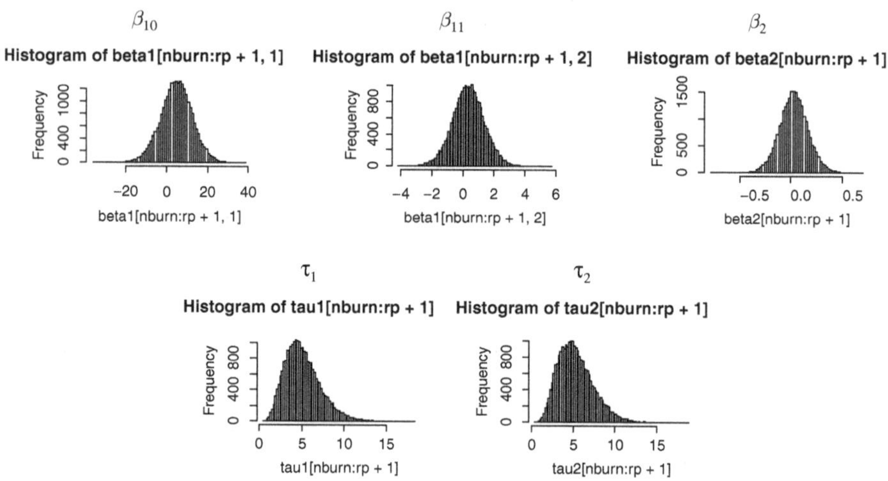

Fig. 9.1 Histograms of the posterior distributions of model parameters.
Source: own calculations in R

[11] If $\mathbf{A}$ is a symmetric, positive-definite matrix, then the upper triangular factor of the Choleski decomposition is a matrix $\mathbf{U}$, with strictly positive diagonal elements, such that $\mathbf{U}'\mathbf{U} = \mathbf{A}$ (after: MathWorld, mathworld.wolfram.com/CholeskyDecomposition.html, accessed on 10 December 2009).

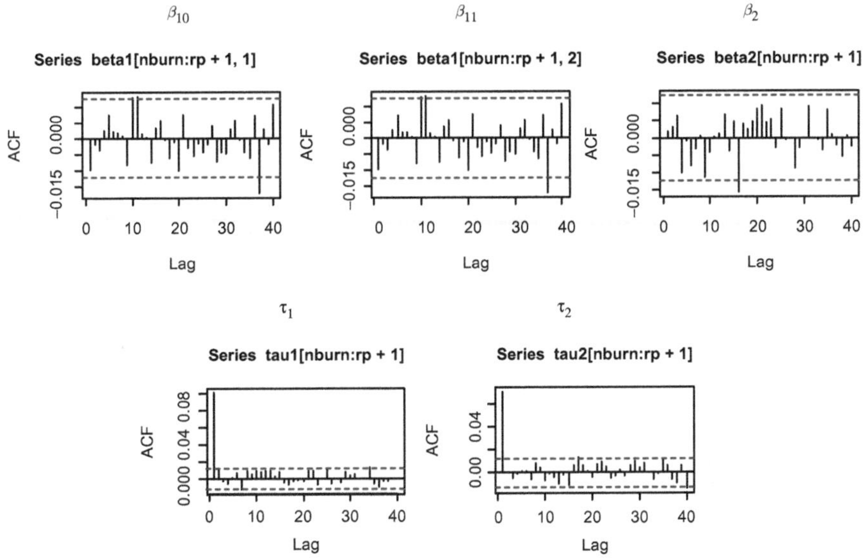

Fig. 9.2 Autocorrelation functions of the parameters.
Source: own calculations in R

of the first order can be observed for precision parameters τ_1 and τ_2. However, there is no indication of non-stationarity of the samples. It should be noted that an inspection of cross-correlations[12] can also be performed, omitted here for clarity of presentation.

```
#results
par_stat <- matrix(NA,5,7, dimnames=list(c("beta10",
  "beta11","beta2","tau1","tau2"), c("mean","var","q10%",
  "q25%","median","q75%","q90%")));
par_stat[1:5,1] <- c(mean(beta1[nburn:rp+1,1]),
  mean(beta1[nburn:rp+1,2]),mean(beta2[nburn:rp+1]),
  mean(tau1[nburn:rp+1]),mean(tau2[nburn:rp+1]));
par_stat[1:5,2] <- c(var(beta1[nburn:rp+1,])[1,1],
  var(beta1[nburn:rp+1,])[2,2],var(beta2[nburn:rp+1,]),
  var(tau1[nburn:rp+1]),var(tau2[nburn:rp+1]));
par_stat[1,3:7] <- quantile(beta1[nburn:rp+1,1],
  probs=c(.1,.25,.5,0.75,0.9));
par_stat[2,3:7] <- quantile(beta1[nburn:rp+1,2],
  probs=c(.1,.25,.5,0.75,0.9));
```

[12] Cross-correlation measures the correlation of different series (here, parameters) at different moments in time.

```
par_stat[3,3:7] <- quantile(beta2[nburn:rp+1],
  probs=c(.1,.25,.5,0.75,0.9));
par_stat[4,3:7] <- quantile(tau1[nburn:rp+1],
  probs=c(.1,.25,.5,0.75,0.9));
par_stat[5,3:7] <- quantile(tau2[nburn:rp+1],
  probs=c(.1,.25,.5,0.75,0.9));
par_stat;
var(beta1[nburn:rp+1,]);

#setting for histograms
wid=100; w=matrix(c(1:5),1,5,byrow=T);  #defining layout
                                         size
layout(w,1,1)                            #layout for figures
#producing histograms
hist(beta1[nburn:rp+1,1],wid); hist(beta1[nburn:rp+1,2],
  wid); hist(beta2[nburn:rp+1],wid);
hist(tau1[nburn:rp+1],wid); hist(tau2[nburn:rp+1],wid);

#autocorrelation functions (ACF)
par(mfrow = c(1,5));                    #ACF setting
                                        (layout for figures)
plot(acf(beta1[nburn:rp+1,1],plot=F)[1:40])
plot(acf(beta1[nburn:rp+1,2],plot=F)[1:40])
plot(acf(beta2[nburn:rp+1],plot=F)[1:40])
plot(acf(tau1[nburn:rp+1],plot=F)[1:40])
plot(acf(tau2[nburn:rp+1],plot=F)[1:40])
```

Finally, the execution of the following code returns the posterior characteristics of the predictive distributions – in this example, only the summary statistics.

```
yfM1_stat  <- matrix(NA,for_l,7, dimnames=list(c("yf1",
  "yf2","yf3","yf4"), c("mean","var",
  "q10%","q25%","median","q75%","q90%")));
yfM2_stat  <- matrix(NA,for_l,7, dimnames=list(c("yf1",
  "yf2","yf3","yf4"), c("mean","var","q10%","q25%",
  "median","q75%","q90%")));
#Forecasts
for (j in 1:for_l) {
yfM1_stat[j,1:2]=c(mean(yfM1[nburn:rp+1,j]),
  var(yfM1[nburn:rp+1,j]));
yfM1_stat[j,3:7]=c(quantile(yfM1[nburn:rp+1,j],
```

```
    probs=c(.1,.25,.5,0.75,0.9)));
  yfM2_stat[j,1:2]=c(mean(yfM2[nburn:rp+1,j]),
    var(yfM2[nburn:rp+1,j]));
  yfM2_stat[j,3:7]=c(quantile(yfM2[nburn:rp+1,j],
    probs=c(.1,.25,.5,0.75,0.9)));
  }
  yfM1_stat; yfM2_stat;
```

Tables 9.3 and 9.4, as well as histograms presented in Fig. 9.3 summarise the results of prediction. Summary statistics are given after exponentiation, since the modelled variables were the logarithms of immigration flows. After the simulations, the convergence of the chains was investigated. Figure 9.4 presents the respective autocorrelation functions. In this case, hardly any auto- or cross-correlation of the posterior samples was observed.

9.3.4 Carlin–Chib Model Selection Procedure

The calculations presented above provide the posterior characteristics of the parameters of the two models under consideration: autoregressive and random walk. However, since the ultimate goal is to produce forecasts of the immigration flows, the better-performing model has to be selected. Or, if an averaged forecast from both models under consideration is to be used, the weights for the average need to be computed. The Carlin–Chib procedure (Carlin & Chib, 1995), described in detail in Chapter 5, provides one of the possible solutions to this problem.

In the Carlin–Chib procedure the results of simulations described in the previous subsection serve as the first step, in that they are used to construct the *pseudo-priors*

Table 9.3 Forecasted immigration flows for 2006–2009, Model 1 (autoregressive)

	Mean	Variance	q10%	q25%	Median	q75%	q90%
yf1 (Y_{2006})	1,725	1.30	907	1,245	1,722	2,404	3,232
yf2 (Y_{2007})	1,631	1.90	657	1,100	1,692	2,539	3,847
yf3 (Y_{2008})	1,716	12.03	534	1,082	1,800	2,886	5,241
yf4 (Y_{2009})	1,288	$1.35 \cdot 10^7$	305	973	1,723	2,887	6,301

Note: qa% denotes the a-th percentile of the distribution. Sample: 25,000, burn-in sample: 75,000.
Source: own calculations in R

Table 9.4 Forecasted immigration flows for 2006–2009, Model 2 (random walk)

	Mean	Variance	q10%	q25%	Median	q75%	q90%
yf1 (Y_{2006})	1,620	1.28	872	1,183	1,618	2,228	2,990
yf2 (Y_{2007})	1,663	1.69	670	1,041	1,660	2,657	4,087
yf3 (Y_{2008})	1,714	2.31	541	952	1,714	3,069	5,363
yf4 (Y_{2009})	1,757	3.35	444	849	1,768	3,580	6,950

Note: qa% denotes the a-th percentile of the distribution. Sample: 25,000, burn-in sample: 75,000.
Source: own calculations in R

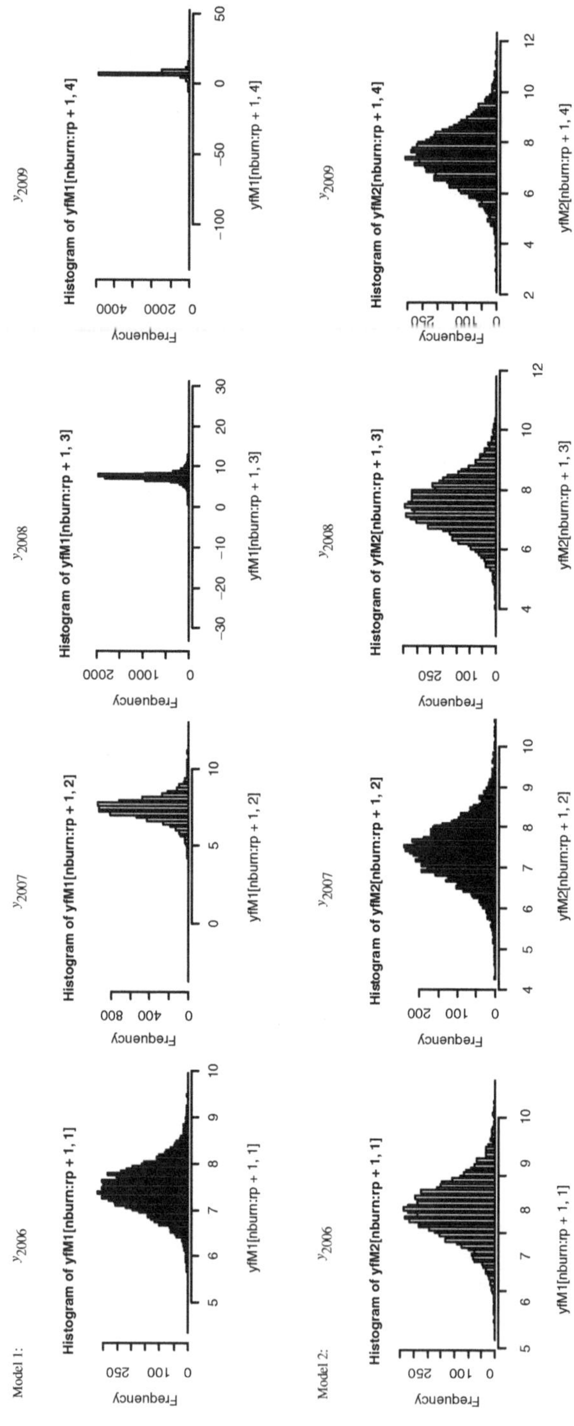

Fig. 9.3 Histograms of forecasts (logarithms) for 2006–2009, Models 1 and 2. Source: own calculations in R

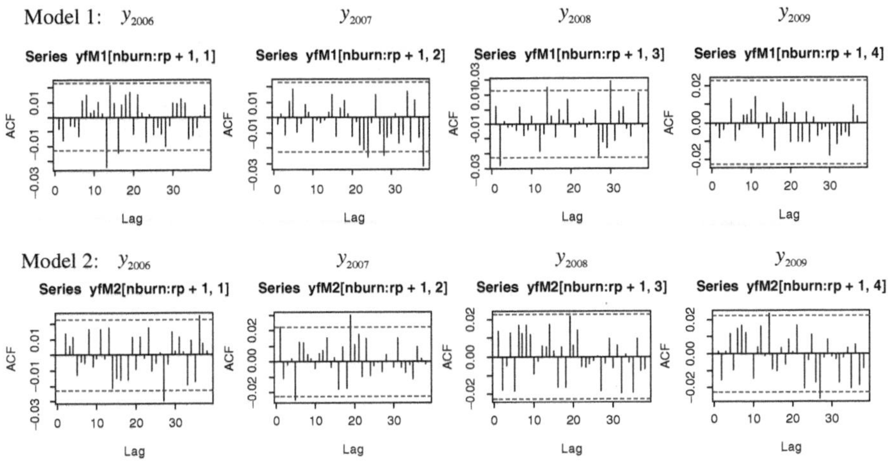

Fig. 9.4 Autocorrelation functions of sample forecasts.
Source: own calculations in R

for sampling from conditionals for model-specific parameters and model indices (see Chapter 5). Following the recommendations of Carlin and Chib (1995), the conjugate distributions have been assumed: Normal for parameters β and Gamma for τ. Here, their hyper-parameters reflect the first two moments (the mean and the variance) of posterior distributions obtained in the previous section. This is defined through the following code:

```
#Pseudo-prior hyper-parameters
b1_2   <- matrix(c(mean(beta1[nburn:rp+1,1]),
  mean(beta1[nburn:rp+1,2])),2,1)
b2_1   <- matrix(c(mean(beta2[nburn:rp+1])),1,1)
sig1_2 <- var(beta1[nburn:rp+1,]); sig2_1
  <- mean(beta2[nburn:rp+1]);
#transformation for sampling from normal distribution
sig1_2new      <- t(chol(sig1_2)); sig2_1new
  <- sqrt(sig2_1);
isig1_2         <- solve(sig1_2);
a1_2   <- mean(tau1[nburn:rp+1])^2/var(tau1[nburn:rp+1]);
s1_2   <- var(tau1[nburn:rp+1])/mean(tau1[nburn:rp+1]);
a2_1   <- mean(tau2[nburn:rp+1])^2/var(tau2[nburn:rp+1]);
s2_1   <- var(tau2[nburn:rp+1])/mean(tau2[nburn:rp+1]);

#Prior probabilities (Carlin-Chib)
pr10 <- c(0.5);
pr20 <- 1-pr10;
```

```
p1 <- c(0.5); p2 <- c(0.5);   #probabilities of initial
                                 values
mod <- 1;                                #initial model index
#Forecast horizon and specification of forecasts
for_l <- 5;                     #forecast length
yf <- matrix(0,rp+1,for_l);    #forecasts
```

Hence, the linking distributions (pseudo-priors) are specified as:

$$p(\boldsymbol{\theta}_1\,|k_2) = (2\pi)^{-1}\,|\boldsymbol{\Sigma}_{12}|^{-1/2}\exp[-0.5(\boldsymbol{\beta}_1-\mathbf{b}_{12})'\boldsymbol{\Sigma}_{12}^{-1}(\boldsymbol{\beta}_1-\mathbf{b}_{12})]\cdot \cdot[\Gamma(s_{12})s_{12}^{\;a_{12}}]^{-1}\tau_1^{a_{12}-1}\exp(-\tau_1/s_{12}), \tag{9.6a}$$

$$p(\boldsymbol{\theta}_2\,|k_1) = (2\pi)^{-1/2}\sigma_{21}^{\;-1}\exp[-0.5\;\sigma_{21}^{\;2}(\beta_2-b_{21})^2]\cdot \cdot[\Gamma(s_{21})s_{21}^{\;a_{21}}]^{-1}\;\tau_2^{a_{21}-1}\exp(-\tau_2/s_{21}). \tag{9.6b}$$

The conditionals for model indices can be derived analytically using the formula (5.6) provided in Chapter 5. In the code above, the hyper-parameters `sig1_2new` and `sig2_1new` denote the transformations of hyper-parameters, facilitating the sampling from the Normal distributions.

Prior probabilities need to be set for both models (`#Prior probabilities`), followed by the initial values for posterior probabilities, and the initial model index (`mod`). The forecast horizon and the resulting matrix of averaged forecasts also need to be defined. Then, the simulation is run in three phases, as discussed in Chapter 5, using the following algorithm:

```
for (i in 1:rp)

{

# Phase 1

#Model parameters
if (mod[i]==1) {
#theta1|M1
tau1[i+1]<-rgamma(1,a1new, 0.5*y1y1-t(beta1[i,]) %*% X1y1
  + 0.5* t(beta1[i,]) %*% X1X1 %*% beta1[i,] + is1_1);
iom1 <- solve(isig1_1+tau1[i+1]*X1X1);
beta1[i+1,1:2]<- iom1%*%(isig1_1b1_1 + tau1[i+1]*X1y1)
  + t(chol(iom1)) %*% rnorm(2);
#theta2|M1 - pseudo-prior
tau2[i+1]<-rgamma(1,a2_1, 1/s2_1);
```

```
beta2[i+1]<- b2_1 + sig2_1new*rnorm(1);
  } else {
#theta1|M2 - pseudo-prior
tau1[i+1]<-rgamma(1,a1_2, 1/s1_2);
beta1[i+1,1:2]<- b1_2 + sig1_2new%*%rnorm(2);
#theta2|M2
tau2[i+1]<-rgamma(1,a2new, 0.5*y2y2 - beta2[i]*X2y2
  + 0.5*beta2[i]*X2X2*beta2[i]+ is2_2);
iom2 <- 1/(isig2_2+tau2[i+1]*X2X2);
beta2[i+1]<- iom2*(isig2_2b2_2 + tau2[i+1]*X2y2)
  + sqrt(iom2)*rnorm(1);
  }

# Phase 2

#Model probabilities
p_aux1 <- tau1[i+1]^(T/2+a1_1-1.5) * tau2[i+1]^(a2_1-1)
  * det(sig1_1)^(-0.5)/(gamma(a1_1)*gamma(a2_1)
  *s1_1^a1_1*s2_1^a2_1*sqrt(sig2_1))* exp(-0.5*tau1[i+1]
  *sum((y1-X1%*%beta1[i+1,])^2) - 0.5*t(beta1[i+1,]
  -b1_1) %*%isig1_1 %*% (beta1[i+1,]-b1_1)
  - tau1[i+1]/s1_1 - tau2[i+1]/s2_1 - 0.5*(beta2[i+1]
  -b2_1)^2/sig2_1)*pr10;
p_aux2 <- tau2[i+1]^(T/2+a2_2-1.5) * tau1[i+1]^(a1_2-1)
  * det(sig1_2)^(-0.5)/(gamma(a2_2)*gamma(a1_2)
  *s2_2^a2_2*s1_2^a1_2*sqrt(sig2_2))* exp(-0.5*tau2[i+1]
  *sum((y2-X2*beta2[i+1])^2) - 0.5*(beta2[i+1]
  -b2_2)^2/sig2_2- tau2[i+1]/s2_2-tau1[i+1]/s1_2
  - 0.5*t(beta1[i+1,]-b1_2) %*% isig1_2 %*% (beta1[i+1,]
  -b1_2))*pr20;
#model probabilities
p1[i+1] <- p_aux1/(p_aux1+p_aux2);
p2[i+1] <- p_aux2/(p_aux1+p_aux2);

#drawing model index
mod[i+1]<-sample(1:2,1,prob=c(p1[i],p2[i]));

# Phase 3

#Averaged forecast
if (mod[i+1]==1) {
yf[i+1,1] <- rnorm(1, mean=beta1[i+1,1]+y1[9,1]
  *beta1[i+1,2], sd=sqrt(1/tau1[i+1]));
```

```
yf[i+1,2] <- rnorm(1, mean=beta1[i+1,1]+yf[i+1,1]
  *beta1[i+1,2], sd=sqrt(1/tau1[i+1]));
yf[i+1,3] <- rnorm(1, mean=beta1[i+1,1]+yf[i+1,2]
  *beta1[i+1,2], sd=sqrt(1/tau1[i+1]));
yf[i+1,4] <- rnorm(1, mean=beta1[i+1,1]+yf[i+1,3]
  *beta1[i+1,2], sd=sqrt(1/tau1[i+1]));
yf[i+1,5] <- rnorm(1, mean=beta1[i+1,1]+yf[i+1,4]
  *beta1[i+1,2], sd=sqrt(1/tau1[i+1]));
  } else {
yf[i+1,1] <- rnorm(1, mean=beta2[i+1]+y1[9,1],
  sd=sqrt(1/tau2[i+1]));
yf[i+1,2] <- rnorm(1, mean=beta2[i+1]+yf[i+1,1],
  sd=sqrt(1/tau2[i+1]));
yf[i+1,3] <- rnorm(1, mean=beta2[i+1]+yf[i+1,2],
  sd=sqrt(1/tau2[i+1]));
yf[i+1,4] <- rnorm(1, mean=beta2[i+1]+yf[i+1,3],
  sd=sqrt(1/tau2[i+1]));
yf[i+1,5] <- rnorm(1, mean=beta2[i+1]+yf[i+1,4],
  sd=sqrt(1/tau2[i+1]));
      }
}
```

In the first phase (`#Model parameters`) samples for model parameters are drawn, conditional on the model index, `mod`. Subsequently, the second phase (`#Model probabilities`) consists of drawing samples of the model indices according to the specification of the conditional probabilities (5.6). Then, having obtained the posterior probabilities for particular models, averaged forecasts are produced (`#Averaged forecast`), conditional on the model parameters and the model index. The averaged forecasts are obtained by means of sampling from the posterior predictive distribution given the model index drawn in a given iteration. The resulting distribution depends directly on the frequency of the occurrence of a given model index in the generated chain.

The resulting model probabilities can be calculated using the approximation proposed by Carlin and Chib (1995), namely by calculating the means of model-specific indices generated. Execution of the following code:

```
arobab   <- c(1-mean(mod[nburn:rp+1]-1),
  mean(mod[nburn:rp+1]-1));

#Summary statistics for forecasts
yf_stat <- matrix(NA,for_1,7,dimnames=list(c("yf1","yf2",
  "yf3","yf4","yf5"),
c("mean","var","q10%","q25%","median","q75%","q90%")));
```

Table 9.5 Averaged forecasted immigration flows for 2006–2009, Model 1 and 2

	Mean	Variance	q10%	q25%	Median	q75%	q90%
yf1 (Y_{2006})	1,637	1.29	875	1,180	1,645	2,276	3,044
yf2 (Y_{2007})	1,629	1.77	648	1,035	1,648	2,605	4,041
yf3 (Y_{2008})	1,695	3.73	518	942	1,725	3,074	5,588
yf4 (Y_{2009})	1,541	89.10	372	870	1,721	3,285	6,933

Note: qa% denotes the a-th percentile of the distribution. Sample: 25,000, burn-in sample: 75,000.
Source: own calculations in R

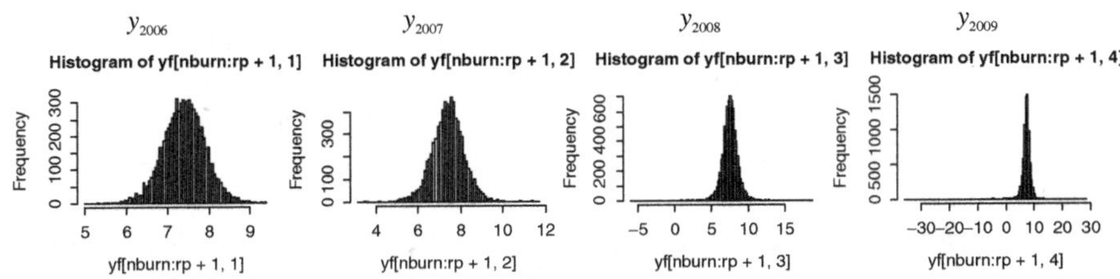

Fig. 9.5 Histograms of averaged forecasts for logarithms of immigration volumes.
Source: own calculations in R

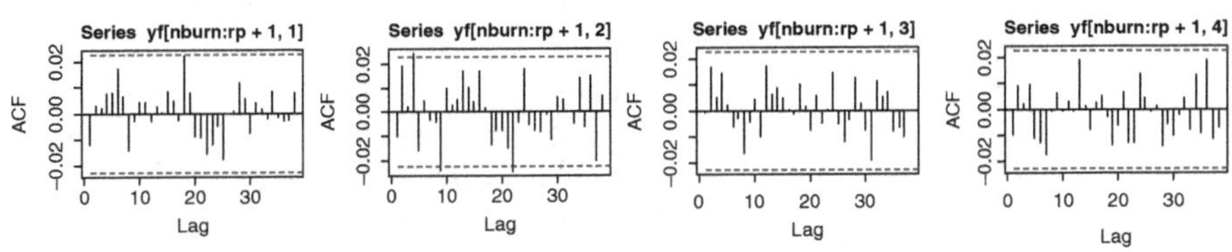

Fig. 9.6 Autocorrelation functions of the averaged forecasts.
Source: own calculations in R

returns the following estimates of the posterior probabilities: $p(k_1 \,|\boldsymbol{\theta}_1, \mathbf{x}_1, \mathbf{y}) = 0.3712$ and $p(k_2 \,|\boldsymbol{\theta}_2, \mathbf{x}_2, \mathbf{y}) = 0.6288$. Summary statistics for the averaged forecasts are also produced. Results, in the form of selected summary statistics and histograms of the averaged forecasts, are presented respectively in Table 9.5 and Fig. 9.5. The autocorrelation functions, indicating good convergence of the algorithms, are shown in Fig. 9.6.

9.4 Conclusions

In Section 9.3, a forecasting example using the R environment and programming language was presented, whereby the convergence of the algorithm was achieved relatively fast. In fact, a substantially smaller number of iterations were required to achieve the stability of the results, as compared to the computations carried out in WinBUGS. On the other hand, relatively more time was required to pursue the estimations in R. This may stem from the fact that in the latter case an explicit Gibbs algorithm for a linear model was used, whereas, as noted by the authors of the OpenBUGS manual (2009, Introduction),

> OpenBUGS tries to block update groups of nodes that are likely from the structure of the model to be correlated. If OpenBUGS s unable to set up block updaters, [...] [t]his can make convergence very slow and the program very inefficient for models with strongly related parameters.

In general, the R language provides a convenient framework for handling Bayesian computations. Techniques that facilitate the convergence of chains used in WinBUGS can be easily implemented and handled by means of R. In our example the ACF plots showed almost no trace of autocorrelation of the samples, hence neither thinning, nor over-relaxed sampling was required.[13] Finally, the histograms of the posterior distributions of the parameters, as well as of the predictive distributions were smooth, which also confirms a good convergence of the chains. It should be stressed, however, that this visual convergence evaluation is only preliminary and could be followed by more thorough analysis of MCMC errors, running quantiles or by means of formal statistical methods, such as those provided in the *boa* or *coda* R packages.

In contrast, various BUGS environments (WinBUGS, OpenBUGS and JAGS) provide a higher-level language for translating models into code for easy calculations. The packages also come with many examples on how to use the language for analysis of various models. Available literature (e.g. Congdon, 2001a, 2003, with further references therein) also provides a wide variety of illustrations. One of the greatest advantages of BUGS is that strenuous derivations of the exact or approximate formulae for the algorithm are not required. Hence, BUGS software seems to provide a very convenient tool for analysis of a wide range of statistical problems in the Bayesian setting, and may be especially useful for the practitioners. On the other hand, it seems that there is a trade-off between the simplicity of using BUGS and the efficiency of the code, especially in comparison to lower-level languages, such as R. An explicitly written MCMC sampler is more effective, but in the case of complicated models it is usually case-specific and requires much more in-depth analytical study. It should be also noted that a specification of the model in BUGS may influence its efficiency (see Section 9.2).

Notwithstanding the above, software from the BUGS family can be seen as a very useful computational tool, highly recommended especially for Bayesian practitioners. However, for academic enquiries, especially those with more theoretical focus, it might be also desirable to use explicit algorithms, if feasible. In such cases, BUGS may serve either as a benchmark tool, or as a tool for preliminary analysis of the problem at hand.

---

[13]Technically, thinning can be introduced by means of the 'modulo' operation, and over-relaxed sampling for example by using a double loop and a function which returns a minimal value from a given set.

Part IV
Perspectives of Forecast Makers and Users

Chapter 10
Extensions and Limitations of Migration Forecasts

The current chapter focuses on migration, and to some extent population forecasts from the point of view of forecast makers: official statisticians and individual researchers. Three possible pathways of extending the modelling framework are discussed in more detail, following the recent advancements both in statistical modelling and theoretical economics. The first one, discussed in Section 10.1, attempts at reconciling quantitative and quantitative research traditions, by combining data-based and theory-based approaches within the Bayesian paradigm. The second one, briefly presented in Section 10.2, is the 'Bayesian melding' framework of Poole and Raftery (2000), where statistical inference is performed on models exhibiting both deterministic and stochastic features. Subsequently, in Section 10.3, the 'imperfect knowledge' approach to macroeconomic models with plausible microfoundations, proposed by Frydman and Goldberg (2007), is discussed in the context of migration. Finally, Section 10.4 attempts to sketch the limits of migration forecasting, and propose a tentative agenda for future migration research, which can help improve the understanding of migratory processes.

10.1 Data, Theories and Judgement: Towards a Synthesis?

10.1.1 Theory in an Atheoretical Setting: Prior Distributions in Multivariate Models

In general, forecasting methods can be seen as either *data-driven*, as in the case of time series models presented in the preceding chapter, *theory-driven*, as in structural models, or *expectations-driven*, for example based on the results of surveys (Willekens, 1990). However, these possibilities need not to be mutually exclusive, as in the case of Bayesian VAR models discussed in Chapter 6. A general, unstructured VAR can be seen as a primarily data-driven model, but a migration theory could be potentially reflected in structurising the model by setting appropriate prior distributions on its parameters. For good migration theories, if they could really be trusted, such distributions could be potentially very concentrated around certain values. In such cases, point distributions would indicate upper limits of belief in

J. Bijak, *Forecasting International Migration in Europe: A Bayesian View*,
The Springer Series on Demographic Methods and Population Analysis 24,
DOI 10.1007/978-90-481-8897-0_10,

a particular theory.[1] In particular, good theories would enable the reduction of the complexity of VAR models by setting a priori point distributions concentrated in zero for parameters expected to link independent variables.

In Chapter 3 it was argued that, in general, the theory-based, nomothetic approach is not an appropriate option for migration forecasting due to a lack of comprehensive, sound and easy-to-operationalise theories of migration. However, as noted by Booth (2006), structural models and forecasts involving additional variables are inevitably theory-based. Summarising a literature review on this type of demographic models, Booth (*idem*: 552) concluded that the possible advantages of structural modelling include built-in feedback mechanisms and limiting factors. On the other hand, among the limitations of the approach, she listed the high risk of model misspecification owing to vague theoretical fundamentals of demography, and potential presence of spurious correlations between variables. Both factors seem to be very important especially in the case of migration. Notwithstanding, Booth (*idem*) underlined the usefulness of structural forecasts for policy purposes, for example through analysing various 'what-if' scenarios.

Expectations, in turn, can be included in the modelling process via the prior distributions, using information obtained for example from surveys. Such an attempt was undertaken by Bijak and Wiśniowski (2010), who used expectations elicited from migration experts, but due to data limitations, obtained ambiguous results (for more details, see the next subsection). Nevertheless, as suggested above, the data limitations or conflicting evidence could be at least to some degree mitigated by good theories, which would represent key prior information for forecasting models. Alternatively, should the available data become of better quality, including longer series; attempts could be made to deduct the structure of the model from the data in a similar way to tests for restrictions presented in Chapter 6.

To that end, one promising approach stems from the recent advances in the modelling of highly structured systems using graphical models. The statistical structure of such models – in particular, conditional independence – can be denoted using graphs, with variables depicted as nodes and their relationships as edges: directed (lagged) or undirected (contemporaneous). A summary of issues relevant to graphical modelling of time series using VAR models has been offered for instance by Dahlhaus and Eichler (2003). A worked-out example of a graph for a three-dimensional VAR is presented in Fig. 10.1.

The graphs in Fig. 10.1 illustrate both a general and a reduced VAR model, in the latter case with some of the insignificant edges removed. The left-hand panels present the 'time-series graphs', showing inter-temporal structure of the modelled processes, and the right-hand panels show abridged 'causal graphs', depicting the structures of causal relationships between variables. In all graphs, directed edges (with arrowheads) denote lagged impact, which can be interpreted

[1] Random variables following point distributions take a single value with probability one. As they have no dispersion (their variance equals zero by definition), they can be seen as reflecting 'absolute certainty'.

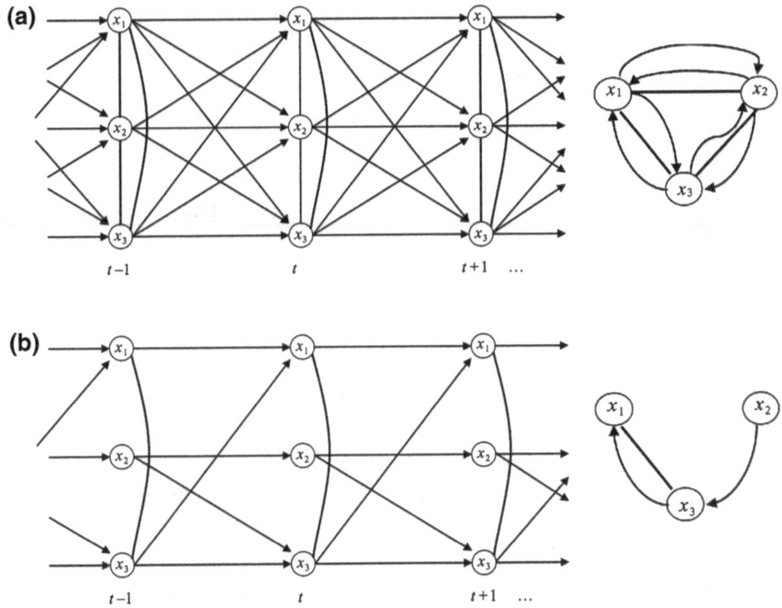

Fig. 10.1 Examples of graphs for general and reduced three-dimensional VAR(1) models. (**a**) General model with all possible relationships: time-series graph and causal graph. (**b**) Reduced model with selected relationships only: time-series graph and causal graph.
Source: Own elaboration, based on Figures 1 and 2 from Dahlhaus and Eichler (2003, pp. 118, 119)

in terms of Granger causality, whereas undirected edges indicate contemporaneous relationships.[2]

The particular examples shown in Fig. 10.1 can be read as follows. Let x_1 denote the variable of interest – migration, while x_2 and x_3 its covariates; to focus attention, respectively economic growth and unemployment, as in Chapter 6. Thus, in a general model (a), all variables influence each other both with a time lag, as well as instantaneously. In the presented example of a reduced model (b), the only significant inter-temporal causal relationship is $x_2 \rightarrow x_3 \rightarrow x_1$, coupled with a contemporaneous relationship between x_1 and x_3 (denoted as x_1—x_3). Of course, model (b) in Fig. 10.1 is just one of many possible examples of imposed inter-temporal and contemporaneous structures.

[2]As noted by V. Didelez in the discussion to Dahlhaus and Eichler (2003, pp. 139–141), such relationships can be seen as contemporaneous only exact to the measurement intervals, tacitly implying marginalisation over the time between the measurements. Charemza and Deadman (1992/1997, p. 157) also point to the fact that 'instant causality' and 'mutual contemporaneous causality' cannot exist, while the consequences has always to be preceded in time by their causes. In demographic context, causality was discussed in detail by Ní Bhrolcháin and Dyson (2007), who proposed ten criteria that would help to identify causal relationships, including, among others, time precedence, analysis of counterfactual outcomes and identification of underlying theoretical mechanisms.

In general, Dahlhaus and Eichler (2003) described a detailed algorithm for inferring a proper graph structure using the statistical properties of the sample. This approach, consisting of removing relationships (edges of the graph) rather than entire variables as in the 'from general to specific' approach is more subtle than simple techniques presented in Chapter 6. A long series of data is, however, a necessary prerequisite of any such analysis, which precludes its direct applications in migration studies given the current state of affairs with migration data in Europe.

Data problems notwithstanding, another very important obstacle in building theory-based VAR models is a lack of good and comprehensive migration theory, as alluded to in Chapter 3. Should such theory become available in the future (if this is at all possible, given the complexity of migration processes), Bayesian models potentially offer a coherent framework for theory-based modelling. Such an approach would retain all important advantages discussed before, in particular probabilistically-coherent description of predictive uncertainty. So far, given an imperfect theoretical framework and low-quality, incomparable short series of available migration data, the usefulness of theory-based models remain fairly limited, as indicated for example by the extremely wide predictive intervals obtained in Chapter 6. However, advances either in migration theorising or in data collection and harmonisation can bring about, although via different channels, better applications for multivariate Bayesian forecasting models, such as the VAR.

10.1.2 Data Versus Judgement: Elicitation of Expert Knowledge

As discussed above, solving the dilemma of data-driven versus theory-driven modelling can be to some extent facilitated by including theory-based prior information in the models. Given the imperfections of quantitative data, the role of any appropriate prior information, often qualitative, is not to be underestimated. In the current study, the prior information was based on the subjective beliefs of the author. However, to make full use of the potential of Bayesian statistics to incorporate expert knowledge, another idea would be to gather (elicit) prior information from experts in the field of demography and migration. This would extend the expert survey or Delphi framework, presented in Section 4.1, by incorporating its results into the full mechanism of statistical inference. Still, as discussed in Chapter 8, sensitivity of the results to changes in models, likelihoods and priors could be assessed in order to robustify the findings.

An attempt to take advantage of the subjectivist elements of the Bayesian paradigm, as well as of the knowledge of migration experts was undertaken by the author and contributor of this book in the framework of an EU-funded research project 'IDEA'[3] (Bijak & Wiśniowski, 2010). Within this project, Bayesian forecasts of gross immigration volumes into seven European countries were prepared.

---

[3]The project, entitled 'Mediterranean and Eastern European Countries as new immigration destinations in the European Union', was aimed at identifying common past, present and (likely) future patterns of migration in Western, Southern and Central-Eastern Europe, using an eclectic approach. The forecasting module constituted only one element in this very broad description of European migration reality. For more details, see the project website: www.idea6fp.uw.edu.pl (accessed on 4 August 2009).

The quantitative data came from the respective statistical offices, whereas prior distributions were elicited from between six and fourteen country-specific experts on various aspects of international migration. The models applied, akin to the ones presented in Chapter 5 of this book, included random walks and AR(1) processes, the latter additionally equipped with deterministic trends identified by the experts. Both types of models were analysed in two variants: either with constant or stochastic variance (SV). The elicitation was performed in a two-round Delphi survey, thus allowing for a correction of original opinions in extreme cases, which however scarcely took place. Similarly as in the current study, in the light of the evidence from data simple random walk models were found to be almost universally preferred, either alone, or mixed with AR(1) processes or stochastic-variance random walks (*idem*).

Additionally, within the framework of the same project, several VAR models were used to forecast the most important directions of inflows (again, identified by the experts), and to examine the impact of two macroeconomic and two demographic determinants of immigration. These were, respectively, GDP growth and unemployment rates, as well as natural population growth and share of population aged 15–64 years. Unfortunately, these ultimate results bore too high predictive uncertainty and in some cases were counterintuitive even with respect to the sign of relationship. This seems to confirm the observations made in the previous subsection and in Chapter 6, that given the shortness of data series on many European migrations, multivariate models such as VAR currently do not offer plausible solutions for forecasting purposes.

Regardless of empirical outcomes, the main lessons from the forecasting exercise in the IDEA project were learned with respect to the methodology of forecasting. In that regard, the Bayesian framework indeed provided an opportunity to combine quantitative and qualitative data and methods. However, even at this level of generality, some challenges can be identified. Firstly, the questionnaires should be prepared and questions formulated in such a way that they could be correctly understood by experts with various professional backgrounds. In IDEA the country-specific expert panels included the representatives of, among other disciplines, anthropology, law, sociology, economics, official statistics, demography, and geography. In general, the heterogeneity of experts is a desirable feature of survey and Delphi studies, since it allows for incorporating the uncertainty in expert judgement into the model. However, the main problems identified here were related to interpreting questions about probability and the stochastic nature of the processes in question. These issues are well described in the general literature on elicitation of prior distributions, for example in Kadane and Wolfson (1998) and O'Hagan et al. (2006).[4] In IDEA, some of these challenges were overcome using visualisations; for example presenting graphical patterns of sample random walk versus autoregressive processes, or of constant-variance versus stochastic-variance patterns. The questionnaires also purposefully avoided using formal terms ('stationarity', 'variance', 'quantiles', etc.), which might have been unclear for some respondents.

[4]For a review of selected literature on elicitation and cognitive biases especially with respect to interpreting probabilities, see Anderson (1998), as well as Bijak and Wiśniowski (2010).

Additionally, when designing the elicitation framework and interpreting the results, some general caveats have to be borne in mind. As summarised by Anderson (1998), insights from cognitive psychology allow for identifying at least several potential pitfalls of intuitive understanding of questions about probabilities. The most important are (*idem*, Section 6):

1. Interpretation of probability mainly in terms of relative frequencies of events, which does not correspond to the Bayesian subjectivist perspective of measures of belief and is not well suited for unrepeatable events, commonly encountered in social research.
2. Tendency to overestimate probabilities of single events and underestimate the uncertainty surrounding them. In this context, Kadane and Wolfson (1998) mentioned also the threat of 'anchoring' the answer close to some value perceived as highly likely.
3. The 'conjunction fallacy', whereby the probability of joint occurrence of two events is wrongly perceived as higher than probabilities for each of these events separately.
4. The 'illusion of control', leading to overestimating the probability of events, upon which the respondents perceive that they have influence.

The above-mentioned challenges do not preclude the usefulness of knowledge elicited from the experts. However, these issues have to be taken into account in the construction of a questionnaire for elicitation purposes in order to ensure that the forecasters and experts are using the same, or at least similar, mental frames and vocabulary (cf. Anderson, 1998).

Even despite the potential technical and interpretational difficulties, there seems to be a large unused potential in combining quantitative and qualitative methods in social research for example under the umbrella of Bayesian inference. Using more general terms, Bryman (1988) noted that, in spite of epistemological differences, the qualitative and quantitative approaches can be seen as to some degree complementary. Their interactions can potentially allow both for mutual corrections, as well as for a better understanding of respective research findings. In forecasting, this can be seen as another rationale for enhancing predictions with judgement, which provides important contextual knowledge that may be otherwise unavailable (Lawrence et al., 2006). To that end, taking a pragmatic rather than dogmatic approach allows for eclecticism, similarly as in Chatfield's (2002) stance on the frequentist-Bayesian debate (see Section 2.3).

An example of a qualitative study, the results of which are in broad agreement with the empirical findings presented in this book, includes the hypothesis of 'intentional unpredictability' amongst some economic migrants (Eade, Drinkwater, & Garapich, 2006), which seems to corroborate the quantitative hypothesis of a limited predictability of migration flows. Similar results and interpretation can be also found in a political science-based study of Pijpers (2008). He argued that the assumptions of forecasts of migration in Europe following the 2004 EU enlargement were completely unrealistic: they were too 'orderly' and 'aesthetic' (*idem*), which might be translated into statistical language for example as unreasonably stationary. The findings were then interpreted in terms of forecasts used as a political tool of influencing the public opinion in favour of unnecessarily restrictive migration policies. Indeed, as nicely summarised by Orrell (2007, pp. 182–183), 'scientific predictions sometimes say as much about politics and sociology as they do about the system under study'.

In practice, appropriate research paradigm – quantitative, qualitative or mixed – should be selected with the very purpose of research in mind. Bayesian statistics, given its subjectivist philosophical premises outlined in Chapter 2, can serve as a promising way of reconciling the views of both intellectual traditions of contemporary social research. Given that migration is an area where neither theories nor data are very reliable, combining quantitative and qualitative methods seems to be a promising way ahead. As indicated by the IDEA exercise discussed above, what is still needed is a common language that would ensure that the qualitative and quantitative researchers understand each other and relate to the same aspects of processes under study. This fully supports Bryman's (1988) caveat on safeguarding consistency in eclectic research by ensuring that both types of analysis (quantitative and qualitative) address the same issues.

As a final step, the results of a study combining different research traditions can be subject to various checks of plausibility and logical consistency. In particular, as suggested in Chapter 5, migration can be seen both as an input and as an output of the process of population change. At the core of the latter there is a deterministic mechanism of population dynamics embodied in the population balance equation and its cohort-component or multi-regional extensions. Hence, there is a need for procedures to control at the same time the plausibility of inputs and outputs of models of population dynamics involving migration. One example – the 'Bayesian melding' approach – is discussed in the next section.

10.2 Controlling Plausibility of Outcomes in Demographic Models

10.2.1 Combining Deterministic Population Models with Stochastic Forecasts

As mentioned in Chapter 5, the uncertain nature of population forecasts can be reflected for example by feeding the deterministic models of population change with stochastic inputs. It is worth noting that particular components of demographic change: fertility, mortality and migration – defined in terms of demographic rates or probabilities – are then controlled both by the outcomes of the model itself (e.g., number of potential mothers in the childbearing age range, number of prospective migrants in the most mobile age groups, etc.), as well as by individual rates or probabilities of experiencing certain demographic events. The latter can be possibly explained by a set of some socio-economic covariates on the basis of demographic theories.

In combining stochastic components of change in a deterministic model of population dynamics, one has also to control for correlations between the processes and between the spatial units, or, broadly speaking, between the states of the multi-state setting of analysis (for more details, see e.g. Alho, 2008). The final outcome is a probabilistic population forecast, whereby uncertainty assessments are correctly accounted for with respect to combining processes and aggregating outcomes over the state space. Two problems remain: firstly, our knowledge of mechanisms underlying particular components of demographic dynamics is imperfect and bears inevitable errors. Secondly, correlations between different processes are also uncertain.

As argued in this book, the imperfections of the researchers' knowledge can be at least partially incorporated in the Bayesian modelling framework via the prior distributions. However, as shown in Chapter 8, this can lead to very wide uncertainty assessments. If this is the case of forecasts of migration rates, the following problem can then arise: high predictive uncertainty of rates translates into high uncertainty of events, then (from the balance equation) into high uncertainty of the forecasted population size; which can then affect the rates through a feedback loop and possibly render the forecasts useless due to exploding uncertainty. Additionally, this effect can be propagated through inter-process correlations to other components of demographic change. In order to control not only the inputs, but at the same time also the outputs of population forecasts equipped with deterministic and stochastic components alike, additional mechanisms ensuring consistency are therefore required. Such mechanisms can also potentially account for 'exogenous' limiting factors of forecasts.[5] However, the problem of setting the limits to population growth seems to be equally difficult to address in a scientific enquiry, as are adequate predictions of demographic change.

The prevalence of deterministic models in many fields of science, coupled with an increasing need for coherent quantitative uncertainty assessments of their outcomes, called for a methodology that would allow for combining the features of deterministic and stochastic approaches in a comprehensive and coherent manner. Such a method was devised within a Bayesian framework by Poole and Raftery (2000), and was originally motivated by ecological modelling of populations of bowhead whales (*Balaena mysticetus*). Recently, this method, referred to as *Bayesian melding*, has been also applied in demography (to be more precise, in epidemiology) to model HIV prevalence in countries affected by AIDS epidemics (Alkema, Raftery, & Brown, 2009). In the latter study, prior information was added to the model used by the UN AIDS (*idem*). In both cases, the deterministic part of the model was accounted for by a system of dynamic differential equations. A brief introduction to Bayesian melding, following the seminal paper by Poole and Raftery (2000), is outlined in the next subsection.

[5]As an example, Orrell (2007, p. 282) suggested that global demographic forecasts should take into account the 'carrying capacity' of Earth, echoing earlier Malthusian and neo-Malthusian concerns.

10.2.2 The Bayesian Melding Approach: Outline and Discussion

The premises of the Bayesian melding approach are as follows. Firstly, a deterministic model $M : \theta \rightarrow \varphi$ is defined as a function transforming a set of input variables, $\theta \in \boldsymbol{\Theta}$, into output variables, $\varphi \in \boldsymbol{\Phi}$ (Poole & Raftery, 2000, p. 1245). The joint 'premodel' probabilistic distribution, $p(\theta, \varphi)$, is then defined over the Cartesian product $\boldsymbol{\Theta} \times \boldsymbol{\Phi}$, to reflect all information except their relationship defined by M. Assuming independence of inputs and outputs, the premodel distribution can be decomposed as (*idem*, p. 1246):

$$p(\theta, \varphi) = p^{[\theta]}(\theta) \cdot p^{[\varphi]}(\varphi) \propto q_1(\theta) \cdot q_2(\varphi) \cdot L_1(\theta) \cdot L_2(\varphi), \tag{10.1}$$

where q_i refer to the prior distributions and L_i – the likelihoods of both inputs (for $i = 1$) and outputs ($i = 2$). However, as $q_1(\theta)$ and M jointly also induce a prior distribution on φ, denoted as $q_1^*(\varphi)$, the likelihood L_2 can be also rewritten as: $L_2(\varphi) = L_2(M(\theta))$.

As suggested by the authors, the next step is then to combine two prior distributions for φ: one original and one induced by $M(\theta)$. As a solution, they propose a logarithmic pooling with a weight α, which yields the following proper joint prior distribution over $\boldsymbol{\Phi}$ (*idem*, p. 1247):

$$\tilde{q}^{[\varphi]}(\varphi) \propto q_1^*(\varphi)^{\alpha} \cdot q_2(\varphi)^{1-\alpha}, \tag{10.2}$$

where the prior distribution $q_1^*(\varphi)$ is the one induced on φ by $q_1(\theta)$ through model M. For single-expert prior judgements, the authors proposed using $\alpha = 0.5$, reflecting equal weight given to the inputs and outputs of the model. The posterior distribution of θ given the data is then obtained as being proportional to the product of the joint prior distribution over $\boldsymbol{\Theta}$, $\tilde{q}^{[\theta]}(\theta)$, and two likelihoods $L_1(\theta)$ and $L_2(M(\theta))$.

The only remaining problem consists now in deriving $\tilde{q}^{[\theta]}(\theta)$ knowing $\tilde{q}^{[\phi]}(\phi)$ given by equation (10.2). Its general solution depends on the invertibility of the model M (itself depending, among others, on the dimensionality of $\boldsymbol{\Theta}$ and $\boldsymbol{\Phi}$), and on the properties of the inverse, M^{-1}, if it exists. In that respect, the authors provide several examples, in particular for a simple discrete case with noninvertible M, a general solution for a discrete case, as well as for a continuous case. In the latter example, they propose the following solution (*idem*, p. 1249):

$$\tilde{q}^{[\theta]}(\theta) = \tilde{q}^{[\phi]}(M(\theta)) \cdot \left(\frac{q_1(\theta)}{q_1^*(M(\theta))}\right) = k_{\alpha} \cdot q_1 \cdot \left(\frac{q_2(M(\theta))}{q_1^*(M(\theta))}\right)^{1-\alpha}, \tag{10.3}$$

where k_{α} is a scaling constant ensuring that (10.3) is a proper probability distribution. The derivation of (10.3) made use of relationships between three sets: a subset $\mathbf{A}$ of the inputs space, $\mathbf{A} \subset \boldsymbol{\Theta}$, its transformation $\mathbf{B} = M(\mathbf{A})$, and inverse $\mathbf{C} = M^{-1}(\mathbf{B})$ (details: Poole & Raftery, *idem*).

In the empirical example presented by Poole and Raftery (2000), computations were performed using the Sampling-Importance-Resampling (SIR) algorithm. In terms of numerical outcomes, the melding approach appeared to be at least as efficient in terms of mean squared errors as a simple (non-melding) Bayesian estimation (Poole & Raftery, *idem*, p. 1252). In terms of possible paths for further research, the authors suggested, among others, accounting for model uncertainty by combining forecasts yielded by different models, and extending the analysis to a full Bayesian estimation of the pooling weight parameter α.

There are clear advantages of Bayesian melding for modelling complex systems of interactions. Most importantly, the models can retain their original deterministic features, which are crucial for demographic applications, while allowing for quantification of uncertainty in a coherent manner. However, their requirements for information are also high, especially in terms of prior distributions both on inputs and outputs. Hence, the role of adequately extracted and represented expert knowledge becomes even more important than in the case of simple models proposed in the current study. In general, the experts should not only be able to identify the characteristics of the process under study, such as migration, but also assess the plausibility of different outcomes of the whole model of demographic dynamics.

Some ideas to facilitate this issue through elicitation have been proposed in Section 10.1. In addition, macro-level models of demographic dynamics can be extended to include micro-level information about the behaviour of individuals ('demographic agents' or 'actors'); for example to simulate their reactions to policy changes. Some related issues are discussed in the next section.

10.3 Imperfect Knowledge Forecasting of Migration and Population

10.3.1 Micro-level Foundations in Macro-level Forecasting

In traditional econometrics, the policy-oriented modelling originally relied on structural sets of equations describing the macro-level relationships between various endogenous and exogenous variables. This approach gave grounds to specifying a demometric model of migration policy (Willekens, 1976), based on the paradigm set by the founding father of econometrics, Jan Tinbergen. In this paradigm, and thus in the resulting model, the endogenous variables are divided into two categories: target variables $\mathbf{y}_1$, of direct interest to the policy makers, and all other variables, $\mathbf{y}_2$, labelled as 'irrelevant' from the particular policy point of view. The exogenous variables, in turn, can be classified either as 'instrument variables' $\mathbf{z}_1$, on which the policy makers exert control, and 'data variables' $\mathbf{z}_2$, which are beyond any policy influence (Willekens, 1976, p. 4). The policy model in its simplest form can be then formulated as (*idem*):

$$\begin{bmatrix} \mathbf{y}_1 \\ \mathbf{y}_2 \end{bmatrix} = \begin{bmatrix} \mathbf{R} & \mathbf{S} \\ \mathbf{P} & \mathbf{Q} \end{bmatrix} \cdot \begin{bmatrix} \mathbf{z}_1 \\ \mathbf{z}_2 \end{bmatrix}, \tag{10.4}$$

where **R**, **S**, **P** and **Q** are matrix parameters resulting from the block partitions of exogenous and endogenous variables. The policy problem consists then in finding an optimal vector $\mathbf{y}_1$ (again, from a policy maker's point of view) with respect to $\mathbf{z}_1$, given $\mathbf{z}_2$ and all the matrix parameters.

The main limitations of the demometric approach to policy modelling, as sketched above, are threefold (Willekens, 1976, p. 108). Firstly, the underlying assumption is that the socio-economic reality could be adequately modelled. Secondly, linear relationships between the variables described by model parameters invariant to policies are assumed. Finally, the model requires the existence of a policy maker, whose goal function could be described in well-defined terms of target variables. These assumptions called for further research on theories of migration, which would help better understand the modelled relationships, and on the nature of the policy decision making process (*idem*). It appeared, however, that the assumption of invariance of parameters is of much more fundamental nature and importance.

The critique of traditional macroeconometric approach was undertaken by Lucas (1976). One of his main points of criticism of the then-prevalent policy-oriented models, such as the one given by (10.4), was that any policy intervention would likely change the model parameters, and this was not accounted for in the modelling process (after: Snowdon & Vane, 2005, p. 265). In other words, any forecast made for policy purposes is likely to be rendered invalid by the sole implementation of the policy, which the forecast was designed to support. In general, the parameters of models cannot thus be seen as invariant to policies. Models of international migration – a phenomenon highly impacted by policies – are no exception to this rule.

After the critique of Lucas (1976), the response of the mainstream economics was to incorporate microfoundations in the models. Thus, the macro-level models began to include micro-level representations of 'average individuals', who were able to anticipate and rationally respond to policy interventions. This assumption became known as the 'rational expectations hypothesis'. However, whilst solving the problem identified by Lucas, most of the models equipped with microfoundations created new ones, as noted by several authors in a discussion recently summarised by Frydman and Goldberg (2007). Several points from the debate are presented in the next subsection.

10.3.2 The Imperfect Knowledge Paradigm: Quantitative Versus Qualitative Predictions

The main point of the criticism of the 'rational expectations' models is that such models tacitly presume not only that individuals ('agents') are rational, but also that they are rational in a 'prespecified' manner, thus behaving exactly according to the model built by the analyst. According to Frydman and Goldberg (2007), this assumption is invalid for a variety of reasons. Firstly, it does not take into account the increasingly more dynamic and unstable character of modern economy, where individuals cannot be reasonably expected to be able to anticipate outcomes of a particular policy. As change itself is one of the core characteristics of our 'liquid times' (Bauman, 2007), exact predictions are not possible.[6] Moreover, as noted by Edmund Phelps in the preface to Frydman's and Goldberg's (2007) book, change is an endogenous element of any contemporary economy. In such a setting, even models with prespecified probabilistic features do not adequately describe the reality.

The second issue is that the individuals described in the model – rather than being purposeful agents – are in fact devoid of any free will and creativity and mechanically follow the pre-determined model with their expectations. Incidentally, with respect to migration forecasts this notion can be invalidated by some insights from qualitative research. For example, a recent anthropological study on Polish migrants in London (Eade et al., 2006) indicated that many respondents shared a common strategy of 'intentional unpredictability' with respect to migration-related decisions. This was interpreted as a way of adjusting to dynamically-changing economic and social conditions (*idem*). This idea is also mirrored in Bauman's (2007) notion of 'permanently temporary' migration.

Thirdly, the models as presented before seem to totally ignore the social context in which the individuals operate. Additional problems may also arise while trying to aggregate 'rational' decisions of individuals who do not necessarily share the same preferences, but may have different loss functions instead (see also Section 11.1). Recognition of this problem dates back at least to Arrow's (1950) theorem on the impossibility of a Pareto-efficient[7] aggregation of individual into social preferences in any other way than by imposing preferences on the individuals, dictatorially or otherwise, thus again excluding free will. Concluding the discussion, Frydman and Goldberg (2007) stated that economic models based on the rational expectations hypothesis are fundamentally flawed in that they lack sensible microfoundations.

Given the argumentation presented above, a question arises as to whether microfoundations can be at all helpful in predictions of macro-level processes, such as migration flows. The answer proposed by Frydman and Goldberg (2007) is positive, although the authors clearly state the limitations of such forecasts. Indeed, in their view microfoundations seem to be an indispensable element of any macro-level forecasts, which aim at being able to accommodate policy changes. At the same time, the models should be free of the pitfalls discussed before.

As a solution, Frydman and Goldberg (*idem*) proposed a new approach, termed as 'imperfect knowledge economics' (IKE). First of all, they posited that since all economic actors operate under the conditions of imperfect knowledge, they can at best make qualitative predictions about the direction of change of various policy-relevant

---

[6] See also the general discussion on the limitation of forecasts offered in Sections 10.4 and 11.2.

[7] According to the glossary of economic terms of the Economist, Pareto efficiency, named after an Italian economist V. Pareto, can be simply defined as a 'situation in which nobody can be made better off without making somebody else worse off [. . .]', in this context with respect to the aggregation of preferences (www.economist.com/research/Economics, accessed on 17 August 2009).

parameters. In this way, the models are only 'partially predetermined' with respect to individual expectations about the *signs* of the parameters. As a consequence, the model outputs in predictive terms are also qualitative. Such formulation of the problem implies admitting that there are regularities in economic life, but also that the analysts, not to mention the economic agents described by models, cannot specify them in an exact manner. As noted by the authors of IKE, their approach to uncertainty lies within the intellectual tradition of off-mainstream economists from both sides of the political spectrum, most notably of John Maynard Keynes, Friedrich von Hayek and Frank Knight (*idem*).[8] The IKE theory is thus fully consistent with von Hayek's suggestion that the economists can be only capable of making rough predictions instead of the 'pretence of exact knowledge' (Frydman & Goldberg, 2007, pp. 3). With this caveat in mind, the next section makes an attempt to outline the implications for the practice of modelling and forecasting demographic dynamics.

10.4 Implications for Forecast-Makers and Future Research Agenda

10.4.1 Limitations of Predictability and Plausible Horizon of Non-stationary Forecasts

In the context of migration and population change, implications of the lack of exact knowledge on the processes under study can be summarised as follows. To start with, unlike in many other economic or social settings, there exists a deterministic mechanism underlying population dynamics, consisting in the simplest form of the demographic balance equation (population change = births − deaths + net migration), or more generally, of the cohort-component or multi-regional population accounting. This mechanism can be thus seen as the only certain element underlying the process of demographic forecasting. All other elements, their interrelations and potential explanations are far from being deterministic or even 'prespecified', using the term of Frydman and Goldberg (2007).

As mentioned in Chapter 5, one of the recent attempts to tackle some of the limitations encountered in population forecasting consists of building bridges between the micro (individual-level) and macro (population-level) perspectives. As noted by Willekens (1990, pp. 31), migration can be 'subordinate to other life events, such as household formation, housing change and job change [and] should be predicted within a life course context'. The MicMac research project, mentioned in

[8] Interestingly, as noted by Skidelsky (2009, p. 162), Keynes and von Hayek shared similar liberal values grounded in the philosophy of Edmund Burke. The difference was that, while both of them acknowledged that the government possessed less knowledge than the individual economic actors, Keynes admitted the role of government in taking precautions against the uncertainty and its consequences (*idem*). For more discussion on the issues of uncertainty in macroeconomic thought, see also Snowdon and Vane (2005).

Section 5.2, is an example of such an attempt, although it does not deal with migration in a very detailed manner. However, even in very advanced models, such as the ones of MicMac, demographic forecasts are largely based on prespecified relationships embodied in the parameters of the underlying stochastic processes. The assumed relationships can be thus too precisely defined to reflect the real complexities of individual-level decisions and their translation into macro-level population outcomes. Hence, the limitations of knowledge on exact mechanisms underlying migration imply – once again – that its predictability is restricted to very general statements about the future.

The empirical findings of the current study suggest that migration processes can exhibit non-stationary unit root characteristics, such as in the case of random walk models. Of course, given the constraints imposed on migration for example by the population structure and the 'supply' of population in the mobile age, some values yielded by forecasts from such models can be implausible in the long term. Nonetheless, as shown in Chapter 8, in a short run, simple random walk models with appropriate prior information on precision of the random term can be seen as roughly adequate approximations of the irregular characteristics of the processes under study. Enforcing stationarity in modelling migration can lead to the 'pretence of exactness' and in consequence to large forecast errors *ex-post*. This was the case for example in the study of post-EU enlargement migration into the United Kingdom prepared for the UK Home Office by Dustmann et al. (2003). Their predictions were ultimately underestimated by more than an order of magnitude: instead of the expected five to thirteen thousand, in reality the annual inflow of migrants from the new EU member states likely exceeded 150 thousand persons (IPPR, 2008). Interestingly, according to Willekens (1990, pp. 24), the inappropriateness of assuming stationarity for demographic forecasts using the ARMA models was already pointed out by in the Ph.D. thesis of N. Baydar (1984) over 25 years ago.

The random walk character of migration, at least in a short run, does not seem to be unusual or unexpected. For many series of economic variables and aggregates, the unit-root hypotheses could not be rejected as well. For example, Mandelbrot (1963) and Fama (1965) suggested that stock prices can be modelled by heavy-tailed random walk processes. A summary of this line of reasoning, which led to seeing stock market prices as exhibiting fractal properties[9] has been recently provided by Mandelbrot and Hudson (2004). Furthermore, Nelson and Plosser (1982) could not reject a hypothesis on a random walk character for such macroeconomic aggregates as GDP, employment, prices and interest rates. Mankiw and Shapiro (1985) suggested that income and consumption, at least to a certain extent, may exhibit similar features.

Assuming a random walk character of such a process as migration has profound consequences both for policy makers and forecasters. If this assumption holds, the

[9]In application to time series, fractal features broadly correspond to self-similarity of the tails of the distributions, which in a simple case can be modelled for example by a Pareto-type power law: $p(X > x \,|x > x_0) \propto (x/x_0)^{-\alpha}$, where X is the random variable under study (Mandelbrot & Hudson, 2004, pp. 155–159, 295).

effects of policy interventions or other shocks, shifting the whole trajectory of the process, would be permanent, and not temporary, as it would be the case under stationarity.[10] The persistent character of shocks in the random walk processes and their 'unwillingness' to return to long-term trends result from the 'long memory' of the past propagated through time. From the practical point of view, as seen for the presented empirical results, increasing uncertainty in random walks puts limits on the plausible horizon of forecasts. The current findings suggest 10 years as a rule-of-thumb limit for migration predictions, which would be in line with many earlier suggestions of demographers. For example, Holzer (1959) suggested that migration should not be realistically predicted for more than 5–10 years ahead, whereas Keyfitz (1981) put the plausibility limits on 5–20 years for demographic projections in general.[11] Also Hajnal (1955, p. 312) noted that the prediction errors tend to increase over time and can be very large after 20 years. However, he also noted that the forecast users generally need relatively short-term forecasts under an assumption of continuation of past trends and no shocks. This would be in line with the famous statement of Keynes that 'in the long run, we are all dead' (Skidelsky, 2009, p. 150).[12]

These intuitions coincide with formal attempts to analyse the degrees of predictability. For example, Diebold and Kilian (2001) propose to measure predictability, P, based on a single-argument loss function L defined for the error of forecast made at time t for i periods ahead, $e_{t+i,t}$ (for a more formal definition of loss functions, see Chapter 11), conditional on the set of available information $\mathbf{\Omega}$. Predictability was defined as 'the difference between the conditionally expected loss of an optimal short-run forecast, $E[L(e_{t+j,t})]$, and that of an optimal long-run forecast, $E[L(e_{t+k,t})]$, [for] $j \ll k$', that is as $P(L, \mathbf{\Omega}, j, k) = 1 - E[L(e_{t+j,t})]/E[L(e_{t+k,t})]$ (*idem*: 659). In particular, for univariate random walks and quadratic loss function they show that $P(e^2, \mathbf{\Omega}, j, k) = 1 - j/k$, with predictability linearly decreasing along the forecast horizon (*idem*).

In demography, the general limits of predictability were discussed by Nico Keilman in his contribution to Willekens (1990, pp. 42–44). Keilman presented two conflicting views on the problem. The first one posits that the degree of predictability is related to the body of knowledge on the subject, while the second stance assumes that social change can never be adequately represented by deterministic mechanisms due to inherent unpredictability of human actions. Regardless of the possible outcomes of the debate, Keilman (*idem*) outlined two possible strategies of

---

[10]For an overview of economic implications, see for example Snowdon and Vane (2005, pp. 298–303, 335).

[11]For comparison, Orrell (2007, p. 151) suggested that for most weather forecasts plausible horizons do not exceed 2–3 days, except for precipitation, in the case of which this timeframe is even shorter (24 h). Clearly, this is mainly the inertia of demographic processes (the 'population momentum') that allows for such long horizons as suggested above.

[12]As a counter-example, however, one can think of long-term forecasts (or projections) indicating population ageing and decline in developed countries, which have direct, albeit rather long-term policy implications, especially in the area of social security. For more details, see also Section 11.2.

demographic forecasters for coping with unpredictable population processes: monitoring forecasts in the short term, and preparing various scenarios of change for the long-term horizon. His conclusions remain valid almost 20 years later. A tentative solution can be a combination of short-term forecast with appropriate uncertainty assessments and of long-term scenarios, clearly labelled as such, prepared on their basis.[13]

Although the predictability debate mirrors the one on the sources of uncertainty, outlined in Chapter 2, its consequences for researchers are much more severe. If the demographic phenomena are indeed inherently unpredictable, then any structural modelling of such complex phenomena as migration (or indeed any other social and demographic processes) for forecasting purposes is pointless. Willekens (1990, p. 49) noted that the question may be not 'whether', but 'to what extent' these processes can be seen as foreseeable. For this reason, he called for more research on the very issue of predictability of demographic processes in order to identify the 'best possible' forecasts, or, in other words, the true limits of prediction. The last part of this chapter proposes several topics, on which such research can focus.

10.4.2 Forecasting Migration and Population: Proposal for a Research Agenda

So far, three potential extensions of migration forecasting models have been proposed in this chapter: including judgement in forecasts, controlling for outcomes on the population modelling level and adding microfoundations to macro-level models in a plausible way. Other possibilities include going back to the ideas of Weidlich and Haag (1988) and making an attempt to frame migration and population forecasts using a complex systems approach. As noted by Orrell and McSharry (2009), one of the main characteristics of complex systems is that they cannot be reduced to simple mathematical models, which merely serve as local approximations of the true, complex relationships. Such systems can exhibit network-based non-linear features, high sensitivity to model specification, parameterisation and initial conditions (measurement), so that the corresponding mathematical models can be very difficult to compute.

One of the key properties of the complex systems is the presence of 'emergent phenomena', which can be understood as such that 'cannot be predicted in advance from a knowledge of the system components alone' (Orrell & McSharry, 2009, p. 736). In a similar context, Keilman (in Willekens, 1990, pp. 42–44) suggested looking at non-linear population models, which can potentially exhibit chaotic features and possible bifurcations under slightly altered initial conditions. However, according to Orrell (2007), these are not the differences in initial conditions, as in chaotic systems, but rather errors in model specification, that make complex systems extremely difficult to model and predict. Therefore, differences in initial conditions

---

[13] A similar idea was sketched by Marek Kupiszewski (personal communication).

can become irrelevant for forecast errors especially in the long term, as these errors are mostly due to model inaccuracies (*idem*, p. 295).

For migration and population forecasting, a perspective of applying a complex-systems approach is certainly very tempting, although it can be also very demanding in terms of required information. Especially for migration studies, involving many interactions between demographic and economic factors both on individual and aggregate levels, the data demands for complex-systems modelling are unlikely to be fully met under the current systems of collection of statistical information. Given that, the methods proposed in this chapter can at least supplement the necessary knowledge, enhancing it with judgement, consistency mechanisms and providing sensible linkages between the individual behaviour and aggregate outcomes. In that respect, next to the research on predictability, Lawrence et al. (2006) suggested further inquiries into when and to what extent adding judgement improves the outcomes of forecasting. Interestingly from the point of view of such a heterogeneous process as migration, Lawrence et al. (*idem*) also cited evidence that decomposing forecasts into sub-categories (such as particular groups of migrants, as suggested by Salt & Singleton, 1995; and de Beer, 2008) may not always be a way to increase accuracy, since judgements in such cases can be very difficult to elicit.

Even if decomposition of forecasted variables is not applied, in practical applications of migration forecasts attention should be paid at least to a proper disaggregation of the forecasted origin-destination migration rates by sex and age, in order to use them in the cohort-component or multiregional framework of population forecasting. Such disaggregation can be also made within a probabilistic framework consistent with the methodology used for forecasting the origin-destination emigration rates. In this context, such attempts have been made for example by Keilman et al. (2001) using the frequentist framework, and by Congdon (2008) in the Bayesian approach. In both cases, sex-specific age schedules of migration were estimated using the double-exponential curve of Rogers and Castro (1981), which in the work of Congdon (2008) was then additionally compared with a non-parametric approach.

Given the forecasted origin-destination matrices for subsequent years and the estimated sex-specific age schedules of migration, the detailed distribution of migration by four dimensions of interest (origin, destination, age and sex) could be obtained for example using the iterative proportional fitting procedure, as suggested by van Imhoff, van der Gaag, van Wissen, and Rees (1997). In general, the age-sex-origin-destination-specific migration rates can be available only a few population flows within Europe. In all other cases, the predictive uncertainty would be additionally enhanced by the errors of estimation of respective sex-and-age schedules.[14] An alternative probabilistic synthesis of forecasts of all three main components of demographic change, fertility, mortality and net migration, including decomposition

[14]For a recent overview of methods for modelling multi-dimensional migration flows involving the spatial aspects, and their subsequent extensions to correct for the missing values and differences in definitions, see Abel (2009). An example of joint migration modeling by origin, destination, age and sex, using the multiplicative component approach can be also found in Raymer (2008).

by sex and age, has been proposed by Hyndman and Booth (2008). Their 'functional data approach', successfully tested on the example of Australia, crucially depends on long time series of detailed data disaggregated by sex and age; and thus with respect to migration would probably not be very helpful in the European context.

Generally speaking, for the practice of migration and population forecasting, several plausibility and reliability criteria can be proposed. In the light of the discussion presented in this and previous chapters, demographic predictions should ideally:

1. Be based on an appropriate and clearly acknowledged mixture of data, theories and judgement (including expectations), as reliable as possible given the current state of knowledge,
2. Provide an assessment of the future uncertainty, together with underlying assumptions;
3. Provide a sensitivity analysis at least with respect to several key elements, such as for example the model formulation, prior distributions and the likelihood functions[15];
4. Include an explicit statement on the limits of predictability, expressed both in terms of the timeframe (forecast horizon), and interpretation of relationships.

In this way, paraphrasing Willekens (1990, p. 9), migration and population forecasters would acknowledge both 'the prospects of an ability to forecast and the impossibility of perfection'. Also Diebold and Kilian (2001, p. 658) noted that:

> poor forecast accuracy does not necessarily imply that the forecaster failed. The extent of a series' predictability in population depends on how much information the past conveys regarding the future; given an information set, some processes are inherently easy to forecast, and others are more difficult.

A message provided by forecasters to forecast users and to the general public should be thus explicit about which questions can the forecasts answer and in what ways do they attempt this. In this context, the next chapter deals with the consequences of uncertain forecasts of hardly-predictable variables for decision makers.

[15] In the Bayesian context, following the results obtained by Osiewalski and Steel (1993b, 1993c) and Kim and Mallick (2003), sensitivity analysis with respect to likelihood functions could be performed for example within the elliptical family encompassing distributions that can be either light-tailed (such as the Gaussian) or heavier-tailed (Student's t or Cauchy distributions).

Chapter 11
Dealing with Uncertain Forecasts: A Policy Perspective

The current chapter makes an attempt to deal with the uncertain forecasts from the policy-oriented perspective of forecast users – decision makers. Its first part, Section 11.1, presents a brief introduction to the decision analysis from the Bayesian perspective. In this section, selected insights into decision making and attitudes to uncertainty are briefly discussed, together with Bayesian estimation and prediction in the decision-analytic approach. The presentation is illustrated with stylised examples concerning migration forecasts. The second part, Section 11.2, contains an overview of literature and discussion on the generic limits of predictions from the point of view of forecast users. Although the discussion is conducted in very general terms, pertaining to all types of socio-economic forecasts, its conclusion can be applied in particular to migration and population predictions. Finally, after exploring which policy questions can be answered by the forecasts, and how, an interactive approach to demographic forecasting is proposed, based on an increased role of the dialogue between forecasters and users.

11.1 Preliminaries of the Decision Analysis: A Bayesian Perspective

11.1.1 Background: Selected Insights into Decisions and Attitudes Towards Uncertainty

Decision analysis and support constitutes a very broad area of science, with a range of methods, such as mathematical statistics and operational research, and with many applications. By necessity, this chapter sketches only selected issues, which potentially can be most relevant for the decision making in the areas of demography and migration under the prevalent conditions of uncertainty. The approach presented here follows the Bayesian perspective, not only for the sake of coherence with the rest of the book, but also since, as mentioned in Chapter 2 and thoroughly discussed for example by DeGroot (1970/1981), Bernardo and Smith (2000), or Robert (2001), the axiomatic construction of Bayesian statistics is firmly grounded in the decision analysis. A more general overview of various axiomatic approaches to the theory of

J. Bijak, *Forecasting International Migration in Europe: A Bayesian View*,
The Springer Series on Demographic Methods and Population Analysis 24,
DOI 10.1007/978-90-481-8897-0_11,

decision making under uncertainty, not limited to the Bayesian approach, has been recently offered by Gilboa (2009).

Within the general framework of statistical decision problems, such as estimation or prediction, the decisions often concern the choice of one value given probability distributions depicting the possible 'states of Nature' and their likelihoods. Given the scope of the book, the focus of this chapter will be on decisions concerning demographic and migration predictions. Noteworthy, in the context of population estimates, the use of formal decision analysis was suggested and discussed in more detail by Alho and Spencer (2005, p. 344*nn*).

In general, in the areas of public policy and planning related to demography and migration, decisions are made on a variety of levels: from local authorities, through national governments, to multi-national bodies, such as the European Union. A useful distinction can be also made between planning-related decisions, such as in the case of spatial organisation, and more general and strategic policy design, such as immigration policies of particular countries or of the whole EU. Although all are made under the conditions of uncertainty, decisions of the former type usually require some numerical input, whereas the latter ones are guided by more general, largely qualitative advice. The current section deals with a statistical decision analysis, suitable for the former, quantifiable, planning-related types of problems, whereas Section 11.2 provides some general guidance to more strategic policy challenges. In the latter case, recent advancements in forecasting and decision science are referred to.

As mentioned in Section 2.3, the Bayesian statistical inference is firmly grounded in the decision analysis. In particular, subjective probabilities can be derived from a set of axioms concerning the preference relations (weak preference, strong preference, and indifference), defined over a set of certain consequences **C** of different decisions **D**. These axioms include comparability of consequences with respect to the preference relations, as well as existence, transitivity and coherence of the preferences themselves (for details see e.g. Chapter 2 of Bernardo & Smith, 2000 or Chapter 2 of Robert, 2001).

This approach gave grounds to the definition of the (usually bounded) utility function u, being a measure of preference, defined over the space of possible outcomes $\mathbf{\Omega}$ of decisions **D**, such that $u : \mathbf{\Omega} \times \mathbf{D} \rightarrow \mathbf{R}$ (*idem*). As most persons, and even more so the policy makers, appear to be risk-averse, their utility functions are assumed to be concave and bounded from above, the convex functions being reserved for 'risk lovers' (Robert, 2001, p. 59). For public policy applications the cautious attitude to uncertainty is especially vital due to the possible large-scale consequences of wrong decisions, in particular involving taxpayers' money.

The axioms concerning preferences imply that the individuals are rational, have full information about the alternatives, and can make informed decisions by maximising their utility under the conditions of uncertainty. As noted by Robert (2001, p. 58), although the rationality stemming from the axioms can be questionable (see also discussion in Section 10.3), the presence of a coherent axiomatic framework for decision making under uncertainty cannot be undervalued. He proposes to view rationality not as a true description of the reality of decision making, but rather as a reductionist view inherent in all types of statistical analysis, and an ideal, against which the real-life decisions can be evaluated (*idem*).

With respect to the utility functions, there are several ways of eliciting them from the decision makers. The approach suggested by Bernardo and Smith (2000, p. 53) can be summarised as follows. For a selection of outcomes $w \in \mathbf{\Omega}$, the consequences $c \in \mathbf{C}$ of decisions $d \in \mathbf{D}$ can be specified and ranked from the least preferred ($c_{\min}$) to the most preferred ($c_{\max}$). Subsequently, for a series of different values of c, the decision maker can be asked whether she or he prefers (a) consequence c for sure, or (b) a gamble, where $c_{\max}$ is obtained with some probability p and $c_{\min}$ with probability $(1 - p)$. In that way, some value, c_p, can be found, for which both options (a) and (b) would be indifferent to the decision maker. Bernardo and Smith (*idem*) suggest that for an individual with coherent preferences, $u'(c_p) = pu'(c_{\max}) + (1 - p)u'(c_{\min}) = p$, where u' is a utility function defined over $\mathbf{C}$. Therefore, the series of pairs (c, p_c) can be used to approximate the utility function $u'(c)$ (*idem*). Finally, the function u' needs to be re-mapped into u defined over the space of decisions $\mathbf{D}$ and outcomes $\mathbf{\Omega}$.

Additional problems can arise in situations with more than just one decision maker and the absence of a shared set of preferences (utility function). As per Arrow's (1950) impossibility theorem, mentioned in Section 10.3, the aggregation of preferences can be problematic and the result may not be Pareto-optimal, or the consensus between decision makers may be impossible. Other difficulties are pointed out in the game theory, dating back to the works of von Neumann and Morgenstern (1944; after: Bernardo & Smith, 2000), where the outcomes of decisions of several independent subjects depend on each other. One of the possible decision strategies in such setting, known as 'minimax', consists of making such decisions that would minimise the maximum loss expected under all possible decisions of all other subjects.

The minimax approach can be also used in making statistical decisions concerning estimation or prediction. However, from the Bayesian perspective, this approach is not attractive for a number of reasons. Firstly, it assumes that in the 'game' involving a statistician trying to estimate or predict a particular state of Nature, the second 'player' – the Nature herself – always tries to come out with the worst possible outcome (Robert, 2001, pp. 67). Nature is thus presumed to be an enemy, rather than to be neutral towards the statistician or decision maker. Despite the built-in pessimism of such a viewpoint, it is also criticised for not taking into account all information available a priori and for ignoring the formal analysis of uncertainty (*idem*).

Secondly, as shown by Bernardo and Smith (2000, p. 449), minimax decisions may violate the transitivity of preferences and may be not invariant under transformation of the loss function. For example, the results may heavily depend on whether the loss function is defined in absolute terms or in relation towards other possibilities ('opportunity loss' or 'regret'; *idem*). Finally, as noted by DeGroot (1970/1981, p. 117), minimax decisions can often require randomisation; that is, the solution is not a single decision, but a combination of several, the choice among which has to be made at random. A Bayesian response to all these points of criticism is sketched in the next subsection, followed by some non-Bayesian and Bayesian ideas for assessing robustness of the decisions against uncertainty. Noteworthy, due to their independence on prior distributions, these are the minimax decision procedures that are often referred to as 'robust' (e.g. Adam, 2004).

11.1.2 Estimation and Prediction in the Bayesian Decision Framework

Within the Bayesian paradigm, the decision analysis offers a useful framework for calculating point estimates or forecasts from the relevant posterior or predictive distributions. As mentioned in Section 2.3, in order to do it, a non-negative *loss function* $L : \mathbf{\Omega} \times \mathbf{D} \rightarrow \mathbf{R}$, has to be defined over the space of possible states of Nature,[1] $w \in \mathbf{\Omega}$, and decisions $d \in \mathbf{D}$. The loss function can be defined either as negative utility, $L(w, d) = -u(w, d)$ (DeGroot, 1970/1981, p. 106; Robert, 2001, p. 60), or, more broadly as a difference between an arbitrary, fixed function f and utility: $L(w, d) = f(w) - u(w, d)$ (Bernardo & Smith, 2000, p. 256). The function $L(w, d)$ describes the loss (or cost) of making particular decisions d about w, in particular, the wrong ones.

In order to obtain optimal decisions in a Bayesian framework, let P denote the probability distribution defined over $w \in \mathbf{\Omega}$. For every possible decision $d \in \mathbf{D}$, the expected loss under this distribution, known as *risk* and denoted by $\rho(P, d)$ can be defined as (DeGroot, 1970/1981, p. 106):

$$\rho(P, d) = \int_{\mathbf{\Omega}} L(w, d)p(w)dw, \text{ for continuous } P, \tag{11.1a}$$

or:

$$\rho(P, d) = \sum_{w \in \mathbf{\Omega}} L(w, d)p(w), \text{ for discrete } P, \tag{11.1b}$$

where $p(w)$ respectively denotes the density or probability function of the distribution P. It is additionally assumed that such expected loss exists and is finite. The optimal Bayesian decision d^* is then such a decision d, for which the risk (11.1) is minimised (*idem*):

$$d^* = \arg\min_{\{d\}}\{\rho(P, d)\}. \tag{11.2}$$

This decision is specific to the probability distribution P and the loss function L. In practice, P is usually either a posterior distribution of a parameter, $p(\theta|x)$ in estimation problems, or a predictive distribution $p(x^P|x)$ in forecasting applications.

One of the important issues concerning the choice of a loss function is its symmetry. Symmetric functions can be useful to obtain point estimates of central

[1] In the general case $\mathbf{\Omega}$ needs not to be restricted to the space of model parameters $\mathbf{\Theta}$ and forecasted values $\{\mathbf{x}^{\mathbf{P}}\}$, but can depict any quantity of interest described using a respective probability distributions depicting uncertainty. Notation in this chapter follows DeGroot (1970/1981).

characteristics of the relevant distributions. It can be shown (e.g., Bernardo & Smith, 2000, p. 257) that the quadratic loss function $L(w,d) = a(w-d)^2$ yields the mean of the posterior or predictive distribution as the optimal solution to the decision problem, whereas the absolute value function $L(w,d) = a|w-d|$ yields the median. For the point function $L(w,d) = 1 - \mathbf{1}_{w=d}$, where $\mathbf{1}_X$ is the indicator function ($\mathbf{1}_X$ equals one if X holds and zero otherwise), the optimal solution is the mode of the distribution. In all cases it is assumed that the relevant characteristics exist (*idem*).

Nevertheless, as noted by Lawrence et al. (2006, p. 498), in many real-life forecasting situations, the shape of loss functions is asymmetric and decision makers can exhibit bias in the direction of asymmetry. Hence, asymmetric functions may be more appropriate for many types of decision settings. A simple example of an asymmetric loss function is the linear-linear (LinLin) one, whereby (e.g. Christoffersen & Diebold, 1997, p. 813; Bernardo & Smith, 2000, p. 257):

$$L(w,d) = a(w-d)\,\mathbf{1}_{w\leq d} + b\,(d-w)\,\mathbf{1}_{w>d}. \tag{11.3}$$

The optimal solution of the decision problem is then the quantile of rank $b/(a+b)$ from the relevant distribution (*idem*). For $a = b$ the problem thus reduces to the one under the absolute value function $a\,|w-d|$, and yields a median solution, as discussed above.

For more complex, possibly non-linear cases, Varian (1975) introduced the linear-exponential (LinEx) loss function, defined as (after: Zellner, 1986, p. 446):

$$L(w,d) = b\,\{\exp[a\,(w-d)] - a\,(w-d) - 1\}. \tag{11.4}$$

This function is 'almost linear' for $w < d$ and 'almost exponential' for $w > d$ when $a > 0$, and the opposite holds when $a < 0$ (*idem*). It can be also shown (Zellner, *idem*, p. 447) that the optimal decision is $d^* = (-1/a)\ln\{\mathrm{E}_w[\exp(-aw)]\}$, where $\mathrm{E}_w(\cdot)$ is the expected value with respect to the distribution of w over $\boldsymbol{\Omega}$. Noteworthy, $\mathrm{E}_w[\exp(-aw)]$ is the moment-generating function for the density of w (*idem*). It is assumed that this function exists, as in the case of Normal distributions.

For LinEx functions the optimal decisions may not exist under heavy tailed distributions; however, in some cases fairly accurate rough solutions can be obtained. For example, following Christoffersen and Diebold (1997), the LinEx function can be approximated using Maclaurin's series expansion. The function $\exp(-x) + x - 1$ can be approximated for example by the sixth-order polynomial: $x^2/2! - x^3/3! + x^4/4! - x^5/5! + x^6/6!$, where $a!$ is the factorial of a.

Figure 11.1 illustrates several examples of different loss functions. In the upper graph, the symmetric linear function (grey line) is sketched alongside the examples of LinLin (dashed) and LinEx (dotted line). Two asymmetric loss functions reflect underestimation being more costly, that is, generating higher losses; what implies preference towards overestimation ($d > w$). The lower graph in Fig. 11.1 presents a LinEx function and its approximation by Maclaurin's series expansion, this time indicating preference towards underestimation of w.

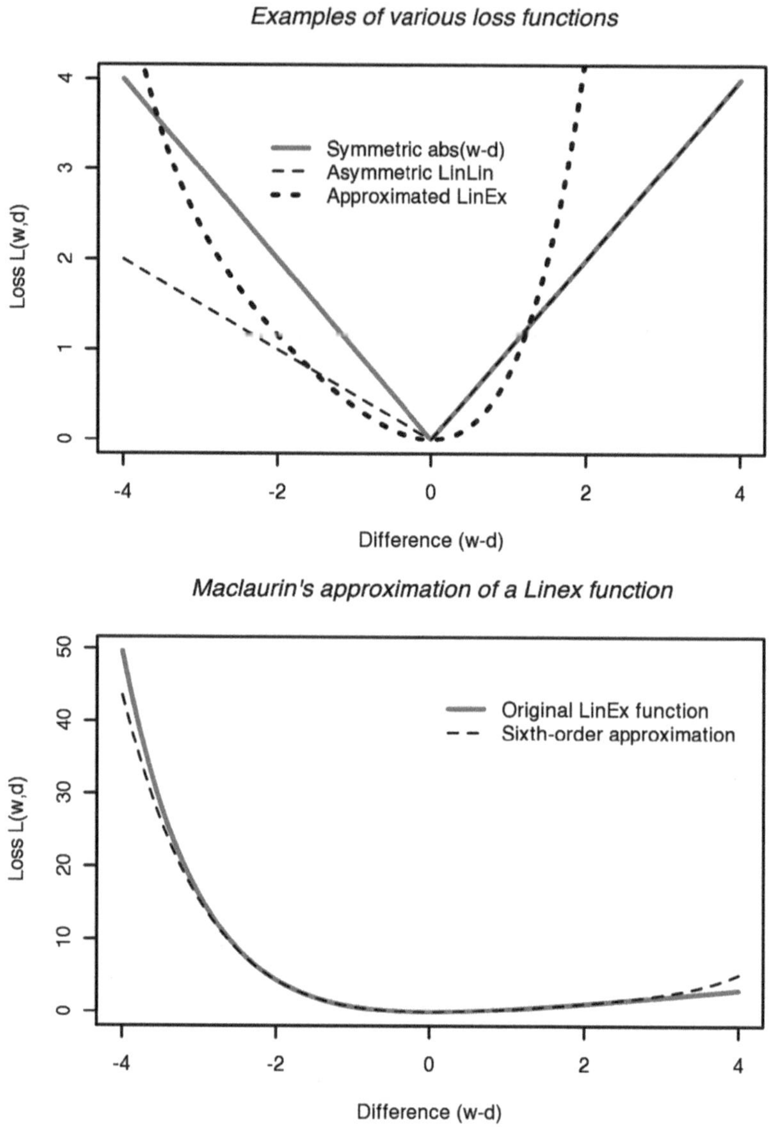

Fig. 11.1 Examples of various asymmetric loss functions and the approximation of LinEx. Source: own elaboration in R

To illustrate some optimal decisions, let $p(w)$ denote density of Student's t distribution with ten degrees of freedom. Consider the following five loss functions: (a) symmetric LinLin with $a = b = 1$; (b) asymmetric LinLin with $a = 1$ and $b = 2$; (c) asymmetric LinLin with $a = 2$ and $b = 1$, (d) approximated LinEx with $a = b = 1$; and finally (e) approximated LinEx with $a = -1$ and $b = 1$. The solutions of appropriate decision problems are illustrated in Fig. 11.2. Under (a) the optimal decision is a median, equal zero, while for (b) and (c) the respective solutions are the upper and lower tertiles (quantiles of rank 2/3 and 1/3), equal 0.445

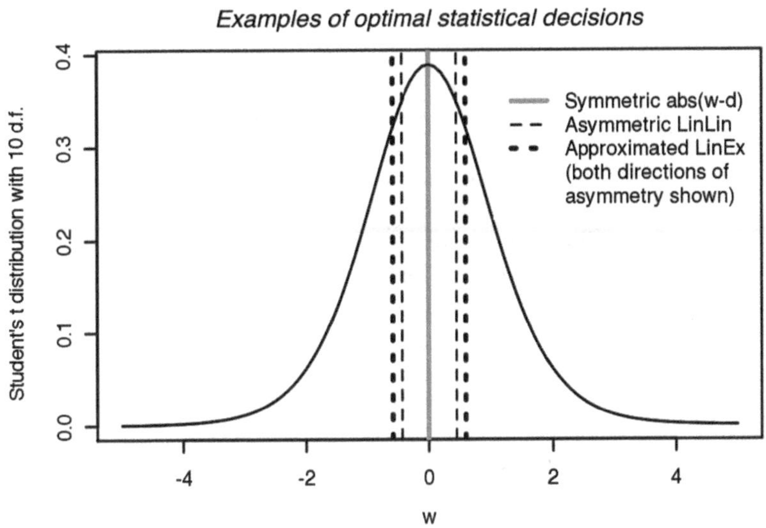

Fig. 11.2 Examples of optimal decisions for a Student's *t* distribution with ten d.f. Source: own elaboration in R

and −0.445. Under loss functions (d) and (e) the solutions for approximated LinEx functions, obtained numerically using the `integrate` function in R, equal respectively 0.594 and −0.594. Note that for Student's *t* distributions with more than one degree of freedom, zero is also an optimal decision under either quadratic or point loss functions, as it is also the median and the mode of the distribution.

At this point it is worth stressing the practical usefulness of simple LinLin functions in decision problems. From the statistical point of view, their quantile-based solutions have convenient properties. Firstly, quantiles are robust against the presence of outliers in the distribution. Secondly, they remain invariant under positively monotonous transformations (reparameterisations). Thirdly, alternative statistics may not always exist, as it is the case of moments in heavy-tailed distributions. Alternatively, for multi-modal distributions, there may be more than one unique solution. From the point of view of interactions between forecasters and forecast users, elaborated further in Section 11.2, such functions as LinLin can be relatively straightforward to elicit. This can be done by asking simple questions, such as: 'Assuming that the loss is proportional to the size of the forecast error, how many times more or less costly is overprediction than underprediction?' Alternatively, the approach for the elicitation of utilities presented before can be followed. The next subsection presents some worked-out migration-related examples, assuming that the loss functions are already known.

11.1.3 Bayesian Decision Analysis: Some Stylised Examples

Based on the theoretical foundations laid down before, consider the following stylised decision problems concerning migration forecasts. It has to be underlined that these examples are completely artificial and very simplistic, possibly far more than they would be in real-life policy situations, since they are meant merely for illustrative purposes.

Firstly, consider a civil servant in the Ministry for Immigration and Integration in Country X, who needs to budget the next year's expenditures for an information campaign. The campaign consists of producing brochures and booklets which form a 'welcome package' for all newly-arriving long-term immigrants who register at the local authorities. He knows that if too many materials were printed, the disposal and recycling of the unnecessary ones would cost 0.10 monetary units per item. However, when the influx of migrants is higher than expected, additional materials have to be printed on demand, at the price higher than would be normal, by 0.20 monetary units per item. The goal of the decision maker is to attempt to match the supply and the expected demand, which would be equivalent to the zero-loss situation, with all money optimally spent, at least according to the assumed policy aims.

A more senior colleague of the civil servant mentioned in the previous paragraph has to decide on the budget of the whole department, one of the elements of which, on the credit side, is a registration fee paid by each registered migrant. She knows that, should more migrants come than expected; the reserve money from extra fees collected can be invested at a simple interest rate of 2% per annum, with a one-off capitalisation at the end of the tax year. However, if the immigration levels are lower than expected, the money needed to balance the budget with respect to the revenue from fees needs to be borrowed at the annual rate of 4%, and the credit has to be paid back in one instalment, also at the end of the fiscal year. The goal here is to design the optimal budget size with respect to the fees.

In the meantime, the public relations department of the Ministry is planning to hold a press conference about the envisaged size of immigration to the country. At the conference, the journalists are expected to ask difficult questions about the 'exact' number of migrants forecasted to come the following year, and are almost certain to ignore all the caveats about uncertainty in their front-page stories the next day. From the point of view of public relations, overestimation is subjectively perceived to be as bad as underestimation; in both cases leading to a loss of credibility,[2] the loss being proportional to the size of the forecast error. Despite the department staff being well aware that precise forecasts are impossible, they would nonetheless like to provide the media with 'the one number', which would minimise the expected absolute error.

Finally, the analytical department of the Ministry is preparing a complex deterministic micro-macro model simulating the impact of immigration on various spheres of the economy. Among many inputs for such a model, they need a forecast of immigration one year ahead, exact to the nearest thousand persons. Since the model is really complex and highly non-linear, even small changes in the input

[2]For an argument, why, in order not to lose credibility, statisticians and demographers might prefer to avoid providing point estimates (or predictions) to the decision makers, see Alho and Spencer (2005, p. 354).

parameters beyond the prescribed tolerance can lead to completely different output trajectories, requiring the model to be re-calibrated after the true parameters become known. Again, given the complexity of the models, re-calibration is very resource-consuming in terms of computing power and time. The analysts are clearly aware that the probability of setting the right value of next year's immigration is very small, but would nevertheless like to set such a value that would minimise the prospects of re-calibration.

In order to support these decisions, the Ministry commissioned a Bayesian forecast of registered immigration based on relatively low-precision assumptions a priori, akin to the models presented in Chapter 5 of this book. As it happens, the available series of immigration data are short, even for simple models such as the random walk. Hence, let the forecasted number of immigrants, M, follow a log-t predictive distribution such that $3 \cdot [\ln(M) - 11] \sim t_{10}$, where t_{10}denotes a central t distribution with ten degrees of freedom, mean zero and precision one. The prediction was made available to all decision makers mentioned in the above examples.

Since the loss functions in the first three examples are linear (LinLin), the appropriate decisions are quantile-based and thus invariant under positively monotonous transformations. Hence, the appropriate quantiles q can be computed by transforming the relevant quantiles from the standard t_{10} distribution, q', using the formula $q = \exp(q'/3 + 11)$. The quantiles for t_{10} in this example can be found in statistical tables, but the general rule can be also applied to their numerical approximations, such as the ones presented in Chapters 5–8.[3] Hence, from the simple formulae for linear loss functions presented in the previous subsection, an optimal decision concerning the information campaign, with underestimation twice more costly than overestimation, is based on the upper tertile (quantile of the rank 0.667) of the predictive distribution. In this case, this forecasted number of immigrants equals about 69,448. With respect to the budget planning, the situation is the opposite and the decision is based on the lower tertile (quantile 0.333); that is, 51,620 immigrants. In turn, for the public relations department the optimal choice is the predictive median (59,874 immigrants).

In the fourth example, the loss function is (approximately) concentrated in a point. The resulting decision is the mode of the predictive distribution exact to the search grid (here defined in terms of full thousands of persons). The mode is usually not invariant under transformations, and therefore has to be derived numerically from the target log-t distribution.[4] For the analytical department, the decision on the model input would thus be approximately 54 thousand immigrants. All the decisions from the presented stylised examples are shown in Fig. 11.3 together with the predictive distribution, on the basis of which they were evaluated. Needless to say, all

[3]In these cases, the formula $q = \exp(q')$ should be used to obtain appropriate quantiles for migration rates from the ones estimated for the predictive distributions from particular forecasting models.

[4]For example, in R the appropriate code is `v = (which.max(dt((log(x)-11)*3,10)/x) -1)*step`, where `x` is a vector of values from the domain, from 0 to X_{max}, defined with intervals of `step` units.

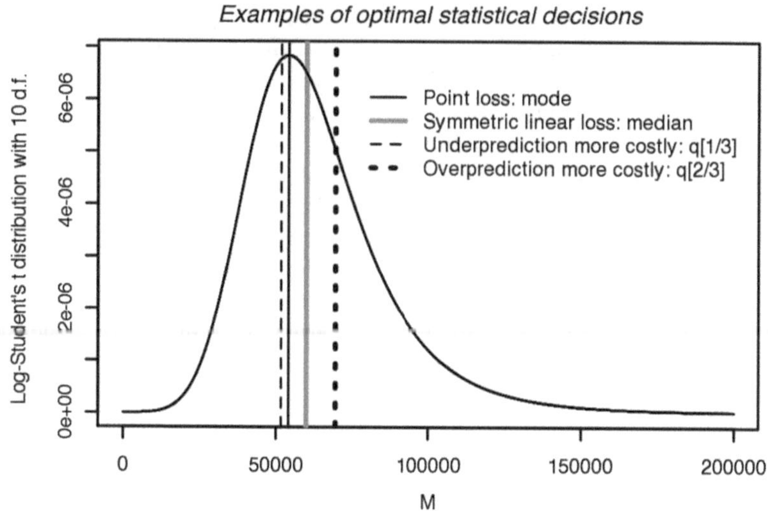

Fig. 11.3 Examples of optimal statistical decisions for stylised migration-related problems. Source: own elaboration in R

of them are different and depend both on the particular loss functions, which reflect the preferences of various decision makers, as well as on the prior distributions used to calculate the forecasts. However, it is worth noting that if the loss functions in all presented cases were of higher orders, for example quadratic or LinEx, optimal decisions would not exist, since due to heavy tails, the log-*t* distributions do not have positive moments.

11.1.4 Possible Extensions of the Decision Framework

Once a decision is reached, a further question arises, how susceptible is it to changes in our uncertainty about the world. To that end, there exist several non-Bayesian and Bayesian ideas for assessing the robustness of decisions against various types of uncertainty. One possible non-Bayesian extension of the decision making framework comes from the 'information gap' concept, originally proposed by Ben-Haim (2001, after: Regan et al., 2005). The idea of information gap aims at assessing 'how much uncertainty [in a decision problem] can be tolerated before [the] decision would change' (*idem*, p. 1471). Seen from that perspective, optimal decisions can reveal interesting properties. For example, an ecological study devoted to the conservation of Sumatran rhino (*Dicerorhinus Sumatranensis*) emphasises a trade-off between the 'immunity' of decisions against uncertainty and the level of aspirations of the decision maker (Regan et al., 2005). In other words, high demands of the decision maker with respect to the acceptable levels of utility (or loss) mean that the resulting decisions are very susceptible to uncertainty surrounding the decision problem. This particularly applies to decisions minimising the expected loss, owing to uncertainty in the loss or utility functions themselves (*idem*, p. 1476).

A Bayesian answer to caveats posed by the proponents of the information gap theory consists in treating loss functions as yet another source of uncertainty which may be subject to robustness analysis (e.g., Micheas & Dey, 2004). In that regard, a variety of numerical procedures for deriving a range of optimal decisions for a range of different utility (or loss) functions were proposed. For example, Abraham (2009) applied a Monte Carlo method based on the simulated annealing algorithm. Also in more general terms, various authors identified formal links between Bayesian and robust (e.g. minimax) decision procedures. At the conceptual level, Adam (2004) demonstrated that applying a sequence of k increasingly convex transformations T^k of the loss function L corresponds to an increase in the risk aversion of the decision maker.[5] As k increases, the resulting Bayesian optimal decisions (if exist), converge to the minimax solution of the decision problem (*idem*, p. 2112).

In addition, there is a vast statistical literature devoted to the robustness of Bayesian decisions against changes in prior distributions (see also the discussion in Section 8.1). Some options here consist in applying the minimax rule in Bayesian setting. In practice, this can be achieved by limiting the optimisation of the risk function $\rho(P,d)$ (11.1) to a certain class $\mathbf{\Gamma}$ of prior distributions P. The resulting solutions are thus referred to *conditional* $\mathbf{\Gamma}$*-minimax* estimates or predictions, conditionality referring to the dependence of estimates on a particular set of data, $\mathbf{x}$ (see e.g. Męczarski, 1998). Among different classes $\mathbf{\Gamma}$, one option consists in 'contaminating' the prior P with another distribution $Q \in \mathbf{Q}$, for example as a linear combination $(1-\varepsilon)\,P + \varepsilon\,Q$, where $0 \leq \varepsilon \leq 1$ (Berger & Berliner, 1986). An analysis of sensitivity can help verify, how much 'contamination' (ε) can be allowed before the decision changes. The analysis can be performed for different distribution classes $\mathbf{Q}$, such as all possible probability distributions, unimodal distributions, symmetric distributions, etc. (Męczarski, 1998).

The notion of conditional $\mathbf{\Gamma}$-minimax decision rules has also led to the development of a concept of *stable* estimates or predictions.[6] As defined by Męczarski (1998, p. 113), a stable decision $d^{\#}$ with respect to a parameter θ or prediction x^P is the one, for which oscillations of risk $\rho(P,d)$ are minimal for all distributions $P \in \mathbf{\Gamma}$:

$$\sup_{\{P\in\Gamma\}} \{\rho(P,d^{\#})\} - \inf_{\{P\in\Gamma\}} \{\rho(P,d^{\#})\} - \inf_{\{d\in\mathbf{D}\}} \{\sup_{\{P\in\Gamma\}} \{\rho(P,d)] \quad \inf_{\{P,\Gamma\}} \{\rho(P,d)\}\}. \quad (11.5)$$

Męczarski (1998) offered a number of analytical solutions for $\mathbf{\Gamma}$-minimax and stable decisions $d^{\#}$ for particular classes of prior distributions, whilst noting that a more general treatment of different statistical models seems hardly possible. Besides, general Bayesian caveats concerning the built-in pessimism of the minimax rules, mentioned earlier in this section, remain in force.

Nevertheless, from a policy perspective the notions presented above seem potentially appealing, since they can inform the policy makers, how robust are their

[5] An example of such transformation is $T^k[L(w,d)] = \exp[k \cdot L(w,d)]$ (Adam, 2004, p. 2111).

[6] Such estimates or predictions are sometimes also referred to as *most stable* or *most robust*.

decisions against different types of uncertainty depicted by prior distributions *P*. For statistical demographers, exploring these options further would additionally enrich the range of possibilities offered by the decision analysis to practical applications, mentioned by Alho and Spencer (2005), as well as in the current chapter. To that end, the next section briefly summarises a general discussion of other options of decision making under uncertainty, not limited to the statistical decision theory.

11.2 Limitations of Uses of Migration and Population Predictions

11.2.1 Alternatives to the Use of Optimal Forecasts

The optimal statistical decisions, such as the ones outlined in the precious section, can be useful in the context of planning, where the decisions should be based on certain numerical parameters, such as the expected numbers of migrants, associated costs, etc. However, there are many other decision settings in which, given the prevalent uncertainty, the room for manoeuvre is much more limited. These more difficult cases have been recently discussed in a special issue of *International Journal of Forecasting* (*IJF*), devoted to 'Decision making and planning under low levels of predictability'. The discussion and recommendations were made in general terms, but they also apply to population and migration forecasting in particular.

The problems with the traditional approach to forecasting, mentioned by the guest editors of the *IJF* special issue (Makridakis & Taleb, 2009), most notably include assumptions that the future will resemble the past and that events under study are independent, which need not hold in complex, network-based systems. Further, the use of overfit models was criticised, since they are predestined to yield wrong forecasts in any other way than by pure luck. Among other flaws of the traditional approach to forecasting, the use of tractable, thin-tailed error terms (such as Gaussian) and the assumption of the existence of a finite variance were mentioned.[7] Finally, the human tendency to underestimate uncertainty was stressed. These, as well as several other issues are briefly summarised below.

To start with, a paper by Taleb (2009) provided a synthetic typology of decision situations under uncertainty. He distinguished four classes according to (1) the type of uncertainty: thin-tailed versus all other types (including unknown); and (2) the type of payoffs: linear versus non-linear. The payoffs are here akin to loss functions introduced in Section 11.1, only with the opposite sign. Taleb's analysis further focused on the 'fourth quadrant' of hardly or completely unpredictable events carrying possibly non-linear payoffs. As discussed in the previous section, for the thin-tailed (e.g. Gaussian) uncertainty, optimal decisions exist even for many non-linear loss functions, such as the LinEx. Under linear loss functions, moment-based optimal decisions exist even for heavy-tailed distibutions, assuming that they

[7]Note that in the empirical examples presented in this study, the log-*t* predictive distributions for migration rates are heavy-tailed and their positive moments (including variance) do not exist.

can be properly estimated. However, in the 'fourth quadrant', the statistical decision theory fails and has to be replaced by general, common-sense strategies of the decision making, unless the problem can be reduced to the other types, for example by changing or bounding the loss function (*idem*, p. 755).[8] This is important especially in the migration context, where the relevant distributions can rarely be expected to be thin-tailed, due to the dynamic and changing nature of the process.

In that regard, Makridakis and Taleb (2009) summarised a number of such common-sense decision strategies emerging from various contributions to the *IJF* special issue. Their most important recommendations include: avoiding the 'illusion of control' (or – relevant for demographic and migration applications – illusion of having accurate predictions, which can bring about potentially dangerous consequences), adopting various protective strategies, including preparedness for difficulties, as well as setting up backup plans and additional reserves of resources (*idem*, pp. 842–843). Such reserves may seem redundant and unnecessary from the point of view of optimal decision making, although the latter, labelled by Taleb (2009, p. 757) as 'overoptimisation', was heavily criticised for making the complex systems in question much more dangerously vulnerable to unpredictable or hardly predictable events.

Another strategy suggested by Makridakis and Taleb (2009) was to apply the maximin approach, maximising the minimum payoffs. This can be seen as roughly equivalent to minimax (minimising the maximum losses), although, as noted by Gilboa (2009: 155), the equivalence is not perfect due to the gain-loss asymmetry stemming from risk-aversion. As discussed in Section 11.1, from the Bayesian point of view this strategy, if a solution uniquely exists (which needs not be the case), it has several drawbacks. As mentioned before, minimax decisions exhibit bias towards the worst-case scenarios (or the least favourable prior distributions), do not take into account all information available, and despite their cautious construction can sometimes lead to worse outcomes than the approaches that are less pessimistic with respect to the states of Nature (Robert, 2001, pp. 66–77). According to Bernardo and Smith (2000, p. 449), although some minimax solutions may be acceptable as optimal Bayesian decisions under certain pessimistic priors (as for example in the Γ-minimax framework), a general minimax rule 'seems entirely unreasonable'. Besides, in practical application, derivation of the least favourable distributions may pose a serious problem. As noted by Robert (2001, p. 73),

> the minimax principle is not necessarily appealing from a Bayesian point of view. Indeed, [...] the least favourable prior is often unrealistic, because it induces a strong bias towards a few points in the sample space.

Last, but not least, in precise terms the minimax (maximin) solutions also belong to the category of optimal decisions, with which they share many of the common drawbacks discussed before.

---

[8] As one of the practical ways of putting 'caps' on payoffs or loses, Taleb (2009, p. 755) proposes insurance, although he himself admits that this strategy may not work well under very heavy tails, such as in the case of catastrophe insurance (and reinsurance).

As to other alternatives to the use of optimal forecasts, Wright and Goodwin (2009) argued for making enhanced use of the scenario analysis in order to 'challenge the mental frames' and design various options of strategic responses to different states of the world using the crisis management approach (*idem*, p. 813). Also with respect to scenario-based analysis, Makridakis, Hogarth, and Gaba (2009) proposed preparing continuously-revised contingency plans for scenarios defined in terms of broader classes of events. In the case of probabilistic forecasts, such classes could be for example defined by intervals bounded by various predictive quantiles; e.g. assuming a symmetric loss, the 'base' class could lie between the fourth and sixth deciles, the 'high' class between the sixth decile and the third quartile, the 'very high' one between the third quartile and the ninth decile, etc. The decision scenarios could then focus on responses to events of particular types, rather than on the crude optimisation. As argued in Chapter 10, making use of the scenario approach for migration and population predictions in the longer horizon can be a useful way forward in the methodology of demographic forecasting.

As to forecasts, the authors concluded that in many instances their uncertainty cannot be assessed correctly, since for example for heavy-tailed distributions it takes much longer to reveal their properties than for thin-tailed ones (Taleb, 2009). A simple solution suggested by Makridakis et al. (2009) is to augment the initial uncertainty assessment; for example by a rule-of-thumb factor of two. At this point, however, it is worth reiterating the potential of Bayesian methods in a probabilistically coherent manner, as shown in the empirical examples throughout this book: a judgemental low-precision prior can provide a better (and more coherent) solution than a rule of thumb. Besides, the subjective interpretation of the results of Bayesian predictions, provided that it is explicitly stated and clear to the forecast users, allows bypassing the problem mentioned by Taleb (2009). If probability is interpreted as a degree of belief, measuring frequencies of events through long-run observations is not necessary.

In addition to these conclusions and recommendations, Orrell and McSharry (2009, p. 742) suggested that instead of futile attempts to predict and control, 'a more realistic goal for policy makers is to better sense the subtle dynamics of the complex systems in which we are embedded, and find ways to shape and direct them'. Under imperfect knowledge, such an approach would enable forecasters to produce general warnings rather than precise policy advice; however this may be the only truly reliable information that can be realistically expected from the forecasts.[9] In this case, subjective Bayesian interpretation of probabilities, if made explicit to the users, can be helpful in avoiding overconfidence in forecasts and admitting their inherent frailties. Some of such general limits of the usefulness of forecasts for policy makers are briefly addressed below.

[9] Recall also the argumentation of Frydman and Goldberg (2007), and Orrell (2007), presented in Chapter 10.

11.2.2 Which Questions Can the Forecasts Answer?

From the point of forecast users, a crucial question becomes: what types of decision problems can be answered or aided by forecasts, and what are the dangers or limitations of relying or not relying on them? In philosophical terms, issues related to predictability in a contemporary complex world have been recently addressed by Z. Bauman. In a recent book, Bauman (2007) suggested that inability to predict leads to an increasing anxiety and fear in the society. This in turn provokes putting in place mechanisms of illusory safety and control at the expense of the social life, as well as of human and civil liberties. On the other hand, Bauman warned against embracing the 'Utopia of predictability' and labelled such an approach as completely immature and unrealistic. In the context of migration, this warning seems especially vital. Bearing in mind the Pijpers' (2008) interpretation of migration forecasts as an instrument of political influence and control over the public opinion (see also Section 10.1), the caveats presented in this section appear to be of crucial importance for forecasters (demographers), forecast users (policy makers) and forecast readers (the general public) alike.

In a similar context, addressing the issues of policy responses, Orrell (2007) argued that risk assessments are especially relevant if the potential dangers (losses or negative payoffs) are large. At the same time, he warned against being too risk averse, which can lead to negative externalities in such situations, when policy responses can be more damaging than the problem they were trying to resolve. The example he quoted is the overreaction during the SARS pandemic in 2003, leading to multi-billion dollar losses worldwide (*idem*, p. 306). Similarly in the context of migration, one can easily imagine the negative externalities of immigration policies that were either too restrictive or too lax; in the former case leading to the loss of human capital of potential migrants, and in the latter – for example – to an increased financial strain on public services and possible problems with social cohesion. This issue provides yet another argument against an uncritical application of the minimax rule to decision problems.

In reality, one of the reasons behind extreme policy responses, as suggested by Lawrence et al. (2006, p. 504), can be the fact that forecast users tend to focus on extreme probabilities of events, close to 0 or 1, and thus prefer such $1 - \gamma$ predictive intervals, for which γ is very small, for example, $1 - \gamma = 0.95$ or 0.99. This is an additional argument for presenting and accepting predictive intervals with lower probabilities attached to them (e.g., with $1 - \gamma = 0.8$ or 0.667), as a way to avoid the illusion of control amongst the decision makers. Interestingly, Lawrence et al. (2006, p. 506) cite several studies that suggest that forecast users nonetheless tend to prefer interval forecasts to point forecasts, the former clearly providing more information.

Therefore, a tentative recommendation for both parties involved in the forecasting process, the forecasters and the forecast users alike, would be that interval forecasts are useful and can provide valuable information for the decision making, although the intervals should not be based on overly high probabilities in order to avoid overconfidence. Narrower probability ranges suggest additional caution, as the probability that the variables under study fall outside the predictive intervals cannot be in such cases seen as negligible and ignored. On the part of the users, as noted by Lawrence et al. (2006), an additional caveat would be that the performance and expertise of the forecasters should not be assessed on the basis of their ability to minimise the width of interval whilst maximising the probability. As indicated by the forecast of post-EU enlargement of migration to the United Kingdom (Dustmann et al., 2003), such forecasts are not only very likely to miss the real-life developments, but also to contribute to unjustified 'illusion of control' among the policy makers and ultimately to generate subsequent problems in addition to the ones that they were supposed to contribute to solving.

Apart from providing direct or indirect input to the decision making process, as sketched in this chapter, predictions have additional, very important roles to play. As noted by Orrell (2007, pp. 301, 349), mathematical models and forecasts, instead of being interpreted literally, should be rather seen as indispensable devices to 'provoke ideas and aid thinking about the future' among the policy makers, and to 'help understand the world and communicate [about it]'. Better understanding of socio-economic mechanisms underlying demographic change and population processes, such as migration, has also a very important educational role in the society. A prime example in that respect is the concern of demographers about the sustainability of pay-as-you-go pension systems. This is the result of current demographic developments, coupled with a general agreement on both implausibility and impracticability of any 'solutions' to population ageing through increased migration.[10] From the policy point of view, as suggested by Ledbury, Miller, Lee, Fairman, and Clifton (2006), the role of education and advice cannot be underestimated under the conditions of imperfect information, in which both the decision makers and individuals operate.

Several elements of the above argumentation, such as the need for appropriate reporting of forecast outcomes, suggests that a both-ways dialogue could be beneficial for a proper understanding of the requirements of forecasts and their decision background by the forecasters, as well as of the forecast outcomes and limitations by the users. The final part of this chapter sketches some key elements around which such interactions could develop.

11.2.3 Towards Interactive Demographic Forecasting?

In order to provide decision support in a way proposed in Section 11.1, the loss functions, together with additional constraints and other elements of the decision problem, should be elicited by forecasters from the decision makers who will be using the forecasts. However, as sketched in the stylised examples of Section 11.1, the loss functions in practical settings can be only specific to given real policy problems. Hence, elicitation of the loss functions by the forecasters from the forecast users should be also problem-specific. This would imply that the offered decision

---

[10] For a literature review and discussion of some related issues, see for example Bijak et al. (2007).

support will not be universal, but rather limited to solving a problem at hand. In this approach the 'centre of gravity' of the forecasting process would thus be gradually shifting from the provider towards the user of forecasts, with the overall aim of providing, if not more accurate, then at least more useful predictions.

As to the elicitation of information from forecast users, all remarks made in Section 10.1, such as ensuring a correct interpretation of the elicited quantities or functions, remain in force. Moreover, elicitation of loss functions can additionally require transforming the initial forecasted variables into the units applied by the forecast users in their cost analysis, for which the loss function would be defined. An example here is a transformation of migration rates into numbers of migrants. The simplest case here would consist of a positively monotonous transformation and a linear (LinLin) loss function, since the resulting optimal decisions would be based on the quantiles of the same rank in both distributions and the computations would be straightforward. In other cases, after ensuring that the relevant integrals are finite, numerical approximations can be offered. However, should the predictive distributions be too heavy-tailed for the optimal decision to exist under a given loss function, the forecast user should be explicitly made aware of this. It would allow either for the redesigning of the loss function in a simpler form (e.g. linear), if the nature of the problem would make it possible, or for switching from the optimal statistical decisions to broader, common-sense strategies, such as those discussed at the beginning of this section after the *IJF* special issue.

Speaking of plausible policies in general terms, Ledbury et al. (2006) singled out seven broad criteria to identify and short-list various policy options. These criteria included (*idem*, p. 2): the likely efficiency, costs, unintended externalities, impact on international competitiveness, possibility of redundancy of policy means, impact on particular society groups, and uncertainty. The pre-selected policy options should be subsequently evaluated by the decision makers according to various constraints, such as their legality, possibility of enforcing a given policy, and public acceptance, as well as budgetary and infrastructure limitations (Ledbury et al., 2006, p. 3). In this framework, the role of input from demographic forecasters in population or migration-related policy issues can be important in terms of informing the potential efficiency and uncertainty criteria, having been previously provided information about the costs (loss function) by the users.

The distinction between specific planning-related and more general policy-relevant decisions, introduced at the beginning of Chapter 11, will have an impact on what is possible in terms of decision support from the point of view of the forecasters. In the former case it can be a proper statistical decision analysis (such as the Bayesian one presented in Section 11.1), while in the latter, for example, it could be a set of scenarios with clear caveats about uncertainty. In any case, enhancing the dialogue between forecast providers and users seems to be one of important possible pathways of future developments of the demographic forecasting methodology, and, of forecasting methodology in general. According to Lawrence et al. (2006), interactive procedures encompassing various 'user-provider' interactions are especially important in the context of probabilistic predictions. The presence of feedback on forecasts' performance, with information flow from users back to providers, can be efficient in calibrating probabilistic forecasts, although with several caveats, concerning understandability, accuracy and the form of the feedback (*idem*).

The possible danger in this respect, again according to Lawrence et al. (2006), lies in relying exclusively on the 'output feedback' – observed values from the forecast period – rather than additional information that could allow for the adjustment of the judgement of the forecaster and an improvement in the forecasts. Also without wider contextual information, e.g. about the changes in the institutional or legal frameworks, the forecast providers are unlikely to be able to differentiate between random errors and systematic errors of their predictions (*idem*). If the aim of both parties involved in the process is to obtain better, more useful forecasts, constructive forms of feedback seem indispensable. Besides, in the same way that the forecasters should not promise the impossible and should clearly state the limits of predictability, the users should not expect the impossible from the providers of predictions. Therefore, confronting the users' expectations with what is actually possible from the scientific point of view should constitute the most important element of the dialogue between the forecasters and forecast users, much more so than the bare issues of loss functions or decision parameters.

Part V
Conclusion

Chapter 12
Summary and Conclusion: Beyond Migration Forecasting

In the evaluation of various approaches to migration forecasting discussed in Part II of this book it was argued that the predictions of population flows should ideally be model-based, probabilistic and explicitly allowing judgemental elements. They should be also tailor-suited for solving particular research or policy problems rather than aiming at providing general migration forecasting solutions. However, all modelling frameworks presented in Part III have their specific strengths and weaknesses, as well as possibilities for further extensions, signalled in Part IV. This chapter summarises the key findings in Section 12.1 and attempts to put them into the context of general demographic forecasting debates in Section 12.2. Finally, Section 12.3 attempts to sketch possible futures of both international migration and its forecasting in the twenty first century.

12.1 Summary of the Key Findings

12.1.1 Bayesian Model Selection and Forecast Averaging

Among various modelling frameworks and perspectives presented in this book, the Bayesian model selection described in Chapter 5 allows for incorporating the prior knowledge in the ultimate decision, as to which model to use for producing forecasts. One option is the a priori equiprobability of models, but the analysis can be designed so that the prior beliefs prefer simple models over more complex ones, according to the 'Occam's razor' principle. The Bayesian averaging provides an alternative, in which the forecasts from various models are combined.

According to Hoeting et al. (1999), the theoretical advantages of averaging are: better predictive performance, no need to justify the use of any particular model and a possibility to incorporate several competing models proposed by different researchers working on the topic under study. This is consistent with the claim of Ahlburg and Land (1992) specifically related to population forecasting, in which the authors argue that the forecast accuracy is expected to improve when judgements of several experts are averaged. Hoeting et al. (1999, p. 399) noted that

> averaging is more correct because it takes account of a source of uncertainty that analyses based on model selection ignore [and its] conclusions are more robust than those that depend upon the particular model that has been selected.

J. Bijak, *Forecasting International Migration in Europe: A Bayesian View*,
The Springer Series on Demographic Methods and Population Analysis 24,
DOI 10.1007/978-90-481-8897-0_12,

In the current study, the formal model selection procedures applied for simple stochastic processes showed preference for such models as random walks or, exceptionally, constant models – simple, unstructured and of an indeterminate character. These results confirm a rather unpredictable character of migratory processes; especially in that they appear to be rather robust against the changes in the prior distributions over the model space, at least with respect to the identification of the model with highest probability a posteriori.

With regard to potential future research on Bayesian model selection and averaging, Hoeting et al. (*idem*) suggested that the issues to be further explored include: robustness of the results on different prior information, performance of the methods if the 'true model' is not in the class under study, fine-tuning of the 'Occam's window' approach, and development of computational algorithms more efficient than the MCMC. In the field of applications of the presented methodology to forecasting international migration, the further work could for example concentrate on extending the model space to cover a wider variety of models.

12.1.2 Vector Autoregression Models and Their Reduction

Another methodological perspective, attempted in Chapter 6, is the analysis of models with additional socio-economic explanatory variables. The main advantage of the proposed application of Bayesian VAR modelling and the 'from general to specific' approach is the possibility to incorporate prior expert knowledge on the relationships between migration and its main determinants, following the indications of various theories.

The empirical results of VAR modelling in the presented examples indicated that two selected macroeconomic determinants of migration – income differentials and unemployment – had hardly any impact on population flows under study. In this way, the general VAR models could be reduced to marginal univariate autoregressive processes. However, even if such reductions were unjustified, which would indicate that forecasting models might include variables suggested by migration theories, the price to pay would be an extremely high predictive uncertainty. Moreover, in several cases the forecasts of both migration determinants, yielded by VAR models, were found implausible or inconsistent, additionally casting doubts on the usefulness of their application in forecasting population flows. Nonetheless, all results obtained in the current study are to a large extent a by-product of small sample sizes.

There are several possibilities of extending the framework presented in Chapter 6. First of all, within the VAR structure, more complex Bayesian tests can be performed, in order to examine not only the direct (although lagged) impact of various determinants on migration, but also the feedback effects, as well as the indirect (but instantaneous) impacts through the elements of the covariance matrix of the random component. Other possible extensions of the analysis, mentioned in Chapters 6 and 10, include: adding more structure to the models, including graphical modelling, incorporation of alternative variables, examination of VAR (or VECM) processes of higher orders, and formal studies of cointegration within the Bayesian framework. Nonetheless, the main limitation of all these extensions is the availability of generally short series of low-frequency data concerning migration in Europe.

Additionally, applying more complex structures of the prior distributions can be also considered. LeSage and Pan (1995) noted, for example, that the main rationale of using the 'multidimensional random walk' priors, like the Minnesota ones of Litterman (1979), is the heuristic observation that the macroeconomic time series are usually consistent with such a priori beliefs. As an alternative for use in regional studies, LeSage and Pan (*idem*) proposed priors based on the contiguity matrices for adjacent regions. The latter can be applied to international migration studies; for example based on VAR models comprising migration flows between a particular country and several competing origins or destinations. In such models, the contiguity matrices, including elements such as transportation costs and the presence of various barriers, can provide a valuable source of prior knowledge. It is worth noting that in the context of the EU enlargement, models of the latter type would potentially enable capturing the effect of substitution of various destinations or origins. The underlying notion is that migration is instigated by a gradual erosion of formal boundaries due to the liberalisation of the flow of persons in Europe, proceeding at a different pace in various countries.

12.1.3 Models Acknowledging Discontinuity in Trends

As mentioned in Chapter 7 on modelling discontinuity in trends or changes in variance, forecasts based on analogies to the past experience of other countries, as well as produced by the models characterised by stochastic volatility, seem to be generally well-suited to address migration issues. The former ones involve the explicit use of expert judgement in a form of cross-country analogies, which sometimes may be the only justification for the expectations on the outcome of such political events as, for example, the EU enlargement. This is especially important given the problems with migration theories, most of which do not address the problems of discontinuity at all (see the discussion in Chapter 3). In turn, the stochastic volatility models reflect not only the uncertainty of the processes under study, but also the variability of their own uncertainty characteristics, addressing the postulate made in that respect by Cohen (1986).

Nevertheless, for the historically highly variable East-West European migratory flows, such as those between Poland and Germany, forecasting by analogy to migration experience of other countries, although potentially promising, did not yield the expected results. However, despite very low posterior probabilities for models with linear analogies, the idea to accommodate effects of the expected structural changes which had their equivalents in the history, remains one of the very few ways of predicting the outcomes of political events, which otherwise would be very difficult to foresee. As to models allowing for changes in the conditional variance, high posterior probabilities obtained for the stochastic volatility processes indicate that not only migratory flows, but also their variability characteristics, are likely to follow random patterns.

Further ways of exploring the presented approaches; for example by including analogies of migration to the experience of other countries, or examining more complex forms of stochastic volatility, are again limited by numerous problems with the availability of comparable data on population flows in Europe. Another idea of potential future research may concern for example the 'early-warning systems' designed for forecasting the turning points of migration trends within the Bayesian framework, following the ideas proposed by LeSage (1990).

12.1.4 Sensitivity of the Results to Changes in Priors

Since the Bayesian approach often requires an assessment of robustness of the results against the changes in prior distributions, an example of such an analysis has been offered in Chapter 8 for selected types of models. Its general conclusion is that, given the availability of short time series on origin-to-destination migration, the forecasts are very sensitive to the selection of the priors, especially on the precision parameters. Therefore, the priors should be ideally properly justified by expert judgement. In particular, the empirical results proved to be visibly susceptible to changes in the dispersion of the prior distributions for the precision of the error terms. Notwithstanding, in all cases, attention has to be paid to uncertainty assessments. In sensible forecasts of population flows, the predictive intervals should be neither too wide, nor too narrow, in order to accommodate the unpredictable character of the future migration developments in a proper fashion.

Clearly, the presented basic sensitivity assessments do not exhaust all such analytical possibilities, which are worth addressing in a separate study. Such more general research could focus for example on the robustness of the outcomes of the model selection procedures against changes in the priors for parameters and models jointly. With respect to potential future research pathways, another idea is to broaden and include more formalism in the analysis; for example with respect to non-informative Jeffreys' priors or conjugate analysis (DeGroot, 1970/1981; Męczarski, 1998). Following the suggestions of Osiewalski and Steel (1993a, 1993b), sensitivity analysis can be performed not only with respect to the prior distributions, but also to the likelihood functions; for example within the wider class of elliptical distributions.

12.1.5 **Ex-ante** *and* **Ex-post** *Comparison of Forecasts: Implications for Users*

The empirical comparison of various *ex-ante* and *ex-post* measures of forecast accuracy and bias, presented in Chapter 8, emphasised a prominent role of simple, unstructured models, such as the random walks, in the forecasts of migration flows under study. The analysis also confirmed the previously mentioned intuition that the uncertainty span assessments, as measured by the predictive intervals, seem more reliable for Bayesian forecasts. The main reason is that their sampling-theory equivalents are devoid of prior assumptions on low precision of the error term, which is a natural premise of migration research. In particular, the frequency of observations falling within various Bayesian predictive intervals was on average much closer to the nominal credibility levels than was the case for the respective frequentist confidence intervals.

Although detailed numerical results remain an empirical issue, the general outcome of the comparison additionally reconfirms and strengthens the conclusions of the robustness analysis presented before. In particular, the assessments of forecast uncertainty yielded by the presented Bayesian models proved to be more realistic than the ones obtained using traditional, sampling-theory predictions. This is a direct consequence of including additional prior knowledge on low precision (high variability) of migration processes under study in the forecasting models, taking advantage of the possibility offered by the Bayesian approach and the subjectivist viewpoint on probability. Such assumptions coincide with another empirical finding of the study; that is, with the indications of an important role of random walks in modelling migration, which reflects the hardly predictable character of population flows.

As a by-product of the obtained empirical forecasts, a simple typology of migration flows can be proposed, consisting of the following three broad groups: (1) well-predictable, stationary processes; (2) random-behaviour flows, such as the unit root processes, in particular, the random walk, and (3) almost unpredictable flows, non-stationary, exploding stochastic processes. From the outcome of the empirical analysis, which favoured the random walk models according to formal model selection procedures (such as posterior odds) and several *ex-post* quality criteria, the second group of flows can be treated as a benchmark, or a 'normal situation' in the very uncertain world of migration forecasts. In general, from the point of view of users, although the first group of forecasts would be likely preferred due to stationarity and its associated features, it is mainly the second one that can be thought of as the migratory reality, at least in the short term. Nevertheless, it needs to be stressed that a hardly predictable character of randomly-behaving migration flows does not preclude the validity of any particular migration theories, but rather indicates their weak empirical verifiability, given incomplete and deficient statistical information.

Another impact of the proposed typology of migrations and their predictions concerns the feasible length of the forecast horizon: only the forecasts of well-predictable flows can be done for longer periods, until the assumption of unchanging stochastic properties can reasonably hold. The random-behaviour processes, as shown in the empirical analysis in Chapter 8, yielded plausible forecasts for a period of about a decade, but a further extension of the forecast horizon would be too risky, especially given the underlying problems with data at hand. At the other end, for the almost unpredictable flows, only very short-term predictions, for 2–3 years ahead can be seen as realistic and providing any meaningful information to the users. All these issues have been addressed in more detail in Part IV of this book.

12.1.6 General Conclusions

Empirical results obtained in the current study indicate that the models under study usually yielded median forecasts of migration rates plausible from a demographic point of view. However, this was not always the case with respect to the uncertainty estimates measured by the 80% predictive intervals. In this case, the plausibility of obtained forecast ranges heavily depended on the selected predictive model and its stochastic features, such as stationarity. On the other hand, as indicated by the sensitivity analysis, without assuming low precision a priori, the *ex-ante* prediction errors would be in many cases too low for such uncertain processes as those related to international migration.

The outcome of a crude analysis of *ex-post* errors for 2005–2007 indicates an interesting pattern. Among all models considered in the study, highest inaccuracies have been universally produced by the random-deviations constant model and the moving average MA(1) process, both stationary. The lowest errors, in turn, have been yielded either by random walks, or by general vector autoregressive models VAR(1). It seems that either the simplest, barely structured models, or models with additional theory-based insights were capable of producing relatively most accurate predictions in the short term (up to 3 years ahead), based on scarce sample information. With respect to random walk models, this conclusion was supported by research on the *ex-post* accuracy of forecasts, offered in Chapter 8, where the random walk models again were found the most accurate. With regard to forecasts from VAR processes, although including migration theory improved the point accuracy *ex post*, the problem consists in very high predictive uncertainty *ex ante*, as mentioned in Chapter 6.

Also, as already noted before (Bijak, 2008a), the Bayesian statistical paradigm has several major advantages for forecasting international migration. To start with, the Bayes theorem allows for updating the prior knowledge (for example, expert-based) on features of the processes under study using sample-based information. The priors are nothing else but marginal distributions of parameters of the joint probability distribution $p(\boldsymbol{\theta}, \mathbf{x})$; that is, of the full Bayesian model, averaged over the space of observations. Secondly, formal uncertainty assessments are immanent in Bayesian forecasts producing whole predictive distributions, which naturally follow from the sample likelihoods and posterior estimates of parameters. The obtained predictive intervals can be appropriate for producing variant forecasts, if required.

Finally, Bayesian inference, based on a subjective definition of probability, avoids several drawbacks of the traditional sampling-theory approach, where probability is related to the frequency of events under study. As observed by Withers (2002, p. 561) with respect to quantitative geographical studies, the repeatability assumption is so often violated by studying 'apparent populations' (non-replicable and not randomly-sampled data sets) that the Bayesian paradigm offers a more appropriate and more flexible approach, 'solving the actual problems' in the words of Robert (2001, p. 513). The same applies to other disciplines within social sciences, where repeatable samples do not form a natural research premise. The applications to demography and migration studies only confirm this general rule.

12.2 Bayesian Forecasts in the Population Forecasting Debates

12.2.1 Bayesian Methods in Perspective: Uncertainty, Judgement and Occam's Razor

As stated in Chapter 1, one of the aims of this book was to evaluate the existing methods of forecasting international migration and to propose Bayesian alternatives, which would formally approach the uncertainty issue and openly include expert judgement in the models. Such proclaimed research aim stems from the wider context of developing a population forecasting methodology based on a stochastic paradigm, where the problems related to uncertainty assessment are a central matter of investigation. In such studies, the focus is usually on two main sources of uncertainty, concerning randomness of the processes, which results in errors due to trends extrapolation, and the uncertainty of the estimation of parameters, which under the frequentist statistical paradigm can be assessed for example by using bootstrap techniques.[1] The Bayesian approach addresses both issues within a common probabilistic framework extending over the space of observations and parameters jointly.

The formal Bayesian model selection and averaging techniques provide valuable tools for reducing uncertainty emerging from another important source – the model specification – in a coherent probabilistic framework. Their outcome in the Bayesian framework allows for identifying models with the relatively highest data support. The issue is especially important in the case of international migration, a phenomenon that can possibly be explained by many competing hypotheses and theories, leading to a wide variety of predictive models. In that respect, the presented approaches address some of the methodological suggestions made with respect to demographic forecasting, focusing on the analysis of uncertainty. The results yielded either by forecasts from the selected models, or the averaged ones, enable the uncertainty assessment of the future values of variables under study in terms of predictive distributions and intervals.

The analysis of multivariate processes, for example within the VAR framework, allows for taking into account, in addition to the uncertainly of migration, the randomness of potential socio-economic explanatory variables and their mutual interrelations. Although, as shown in the empirical examples in the current study, even if such joint predictive uncertainty can be too high for any practical purposes and applications, VAR models still can be useful for deriving probabilistically coherent 'what-if' scenarios by means of conditional forecasting.

Model uncertainty is related to another issue present in the discussions on population forecasting, namely the simplicity versus complexity of tools used for predictions. Available literature overviews (Keyfitz, 1982; Ahlburg, 1995; Smith, 1997; Makridakis & Hibon, 2000) indicate that there is no firm evidence that more complex models perform better in terms of their predictive accuracy than

[1] For a demographic example concerning mortality forecasting, see Keilman and Pham (2006).

simple ones. If, for philosophical reasons, preference is given to more straightforward explanations of migratory phenomena, then the formal selection and averaging methods can be applied under the 'Occam's razor' assumptions a priori, as discussed in Chapter 5.

In addition to the formal and coherent assessment of uncertainty, the Bayesian paradigm allows the incorporation of prior expert knowledge into migration forecasts in a formal way, as suggested by Willekens (1994). With respect to the controversy about the subjective character of such an approach with respect to the selection of prior distributions, Withers (2002, p. 562) noted that '[f]or the opponents of the method, Bayesian probabilities seem arbitrary [and] proponents, in contrast, would claim they are honest'. Coherence of the Bayesian paradigm is also contrasted with the 'false sense of objectivity associated with classical inference' (*idem*).

As stated before, in the context of predicting international migration within Europe, allowing judgemental information seems to be another advantage of the proposed approach. The prior knowledge can supplement the short time series of available data carrying very limited sample-based statistical information. In this context, the advantages of the Bayesian paradigm have been summarised by Bauwens et al. (1999, p. 7), who stated that

> Bayes' theorem provides us with a coherent learning mechanism whereby a prior density is transformed into a posterior density. Explicit comparisons between prior and posterior densities constitute an invaluable device for calibrating and interpreting sample information, especially when the latter is deficient in some ways, e.g. in small-sample situations.

Besides, Bayesian statistics provides a formal language for the description of uncertainty, which enables quantification of the latter under various assumptions a priori.[2]

Notwithstanding the empirical results, the key methodological conclusion from the presented analysis provides full support for including in migration forecasts formal assessments of uncertainty on various levels, including model specification, to which the Bayesian approach provides probabilistically-coherent tools. Quoting one of the leading theorists of the Bayesian approach, D. V. Lindley (2000, p. 294), 'statistics is essentially a study of uncertainty'. Applications to population forecasting are here by no means an exception, which is important especially with respect to modelling and predicting migratory processes, given their hardly determinate nature.

12.2.2 Migration Forecasting as a Continuous Process

As already indicated in the Introduction, forecasts that inform policy decisions, as is the case with population and migration predictions, can have a direct impact on the reality they are trying to foresee through various feedback effects. In such

[2] Credit for this remark (cited *verbatim*) to Jacek Osiewalski.

cases there may exist several built-in mechanisms of interactions between the forecasts and the real sphere of social life. Former president of the American Economic Association, K. E. Boulding (1969, p. 3) used the Heisenberg uncertainty principle as a metaphor[3] to describe the social sciences reality: trying to obtain information about a system changes it, to the effect that science 'no longer merely investigates the world; it creates the world it is investigating' (*idem*). As mentioned in Chapter 10, a similar observation was made by Lucas (1976), who criticised economic policy recommendations based on structural macroeconometric models with parameters treated as invariant to policy changes (for more details, see Snowdon & Vane, 2005, and also Chapter 10 of this book).

For this reason, besides the methodology of simple forecasting, several elements of which are outlined in this book, more attention should be paid to perceiving prediction-making as a never-ending process of continuous adjustments to the changing world. Of key importance is the role of conditional inference ('what-if'), which can incorporate such a variable and dynamic setting into a formalised multi-stage forecasting. The Bayesian framework can potentially provide some useful tools in that respect, itself being largely based on probabilistic conditioning. Such an approach would be more coherent than the widely-used judgemental migration scenarios, at least for the short-term forecast horizon.

Hence, in addition to continuous research aimed at improving the accuracy of forecasts and at the proper management of associated uncertainties, efforts can be also made to generalise forecasting, this time seen as a continuous process, within a coherent probabilistic framework. In particular, as observed by Dawid (1984, pp. 278–279), the forecasting task has a sequential nature: once the outcomes of the forecasted processes under study become known, they can be incorporated in a subsequent forecast for more distant future. In order to accommodate this, the study of Dawid (*idem*) provides a complete Bayesian framework for a probabilistic, sequential (hence abbreviated in a *portmanteau* word 'prequential') forecasting. In demographic applications, this idea has been used in the 'conditional forecasting' approach proposed by Lutz et al. (2004), where the future trajectories of the processes under study are conditioned on their realisations at some point in time, all within the expert-based subjectivist framework outlined in Chapter 5.

Another possible analytical setting, in which the dynamics of migration processes can be transformed into sequential predictions once new information becomes available, is through the application of hierarchical models (alternatively, models with random parameters or models with latent variables). One of the classical examples is the Kalman filter (Kalman, 1960; after: Welch & Bishop, 2006),

---

[3]It has to be borne in mind that from the theoretical point of view, this metaphor is imperfect, as it concerns the macro world rather than the micro world of quantum physics. Nevertheless, its underlying meaning – the impact of the observation on the observed – remains valid. Notwithstanding, one has to be very careful when using the natural science metaphors to describe the social reality in a 'postmodern' fashion. I am grateful to Izabela Koryś to providing me with this citation and to Prof. Marek Męczarski for making the above caveats.

which itself has many built-in Bayesian features. Welch and Bishop (*idem*, p. 4) described a Kalman filter as a technique that

> estimates the process state at some time and then obtains feedback in the form of (noisy) measurements. [...] The time update equations are responsible for projecting forward (in time) the current state and error covariance estimates to obtain the *a priori* estimates for the next time step. The measurement update equations are responsible for the feedback – i.e. for incorporating a new measurement into the *a priori* estimate to obtain an improved *a posteriori* estimate.

The idea of using Kalman filters in the migration context seems especially appealing, as the technique explicitly acknowledges noise in the observations, as well as in the whole system; for example through the use of latent variables.[4] As mentioned in Chapter 2, measurement noise is a characteristic and indispensable feature of almost any available data on international migration.

Finally, especially when seeing migration forecasting as a process, one has to be aware of its possible ethical implications. Boulding (1969, p. 3) noted that 'as science moves from pure knowledge toward control, that is, toward creating what it knows, what it creates becomes a problem of ethical choice'. This is another, very important, challenge to the already sufficiently complex task of forecasting international migration. However, as mentioned in Chapter 11, when dealing with such a socially-delicate matter as migration, under no circumstances should the ethical issues related to the forecasting process, as well as the reporting of its outcomes, be neglected. Adherence to professional ethics in that respect is the ultimate responsibility of the forecaster, and is equally as important as – if not more than – the methodological correctness and sophistication of the applied methods. Of course, beside the researchers, the main responsibility for ethical implications of the forecasts ultimately remains with the forecast users, that is, primarily with the political decision makers.

C. R. Rao (1989/1994, p. 166) expressed a belief that in striving for a better understanding of the uncertain world, with the aim to make more informed and thus more efficient decisions, statistics is a key 'technology' allowing to shape the future of mankind. His conviction seems especially important in the long-term context of population forecasts: public policies designed on the basis of their outcomes can have significant impact on the life of subsequent generations. However, despite the best efforts of statisticians and forecasters, the future will always remain uncertain. An honest solution is therefore to admit the presence of uncertainty. What can be maximally attempted is its quantification in probabilistic terms, using available statistical tools. In that respect, as advocated in this book, the Bayesian approach provides an appropriate and internally-consistent paradigm of statistical inference.

---

[4] I am very grateful to Katarzyna Bijak for this suggestion.

12.2.3 *From Point Predictions to Decision Support: In Need of a Paradigm Shift?*

As argued before, the Bayesian approach can offer an umbrella framework for forecasting and decision making, providing a coherent mechanism of inference and decision support, and allowing for the inclusion of theory-based (or science-based) elements in the forecasts next to the quantitative data. However, following the argumentation in Chapter 11, unlike in other approaches to forecasting, the decision support requires a dialogue between forecasters and decision makers aimed at tackling a specific decision problem, which requires the forecasts as input. Besides forecasters and forecast users, the 'third party' to this Bayesian dialogue can comprise experts in the field providing prior information, especially vital in the absence of reliable data. Lawrence et al. (2006) argued that combining judgement with data leads to better forecasts than relying on either data or judgement alone. They also stated that (*idem*, p. 501):

> probability forecasts provide an elicitation format whereby subjective probabilities supply the communicative means for facilitating the users' understanding of 'vagueness' surrounding the presented forecasts, by enabling the forecast provider to give a more complete judgemental portrayal.

Needless to say, this is the very essence of the Bayesian approach in general.

The framework for interactive expert-based Bayesian migration or population forecasts can be summarised as follows. After the decision problem has been formulated by the forecast users, the researchers (forecasters) elicit from them the decision framework, including the loss functions, forecast horizon and other parameters. In order to make full use of the possibilities offered by the Bayesian approach, prior distributions of the parameters of the forecasting models can be additionally elicited from the domain experts. These elements are subsequently combined with data in the forecasting models, and the final outcomes – forecasts and decision guidance – are reported back to the users. As suggested by Lawrence et al. (2006), at this stage users could also provide feedback on the results to the forecasters, possibly resulting in iterative adjustments of the forecasts. In this way, the scheme of flow of information in the forecasting process could resemble the chart in Fig. 12.1. This procedure can be additionally enhanced by including feedback loops (omitted here for the sake of clarity of presentation) from researchers to experts and forecast users with respect to the elicited information.

The final outcome of the procedure is a set of user-specific forecasts enhanced by simple decision advice provided to the forecast users (decision makers). As mentioned in Chapter 10 such forecasts should ideally comprise other elements and caveats, most importantly including an explicit uncertainty assessment and a clear statement of the limits of predictability. In this way, the paradigm shift in demographic forecasting from deterministic point forecasts, through variant to stochastic predictions, as outlined in Chapter 2, would be continued towards the

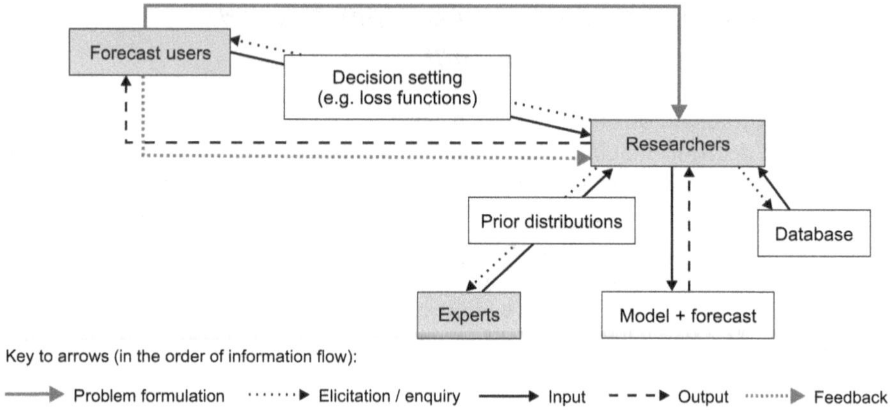

Fig. 12.1 Example of information flow in interactive expert-based Bayesian forecasts. Source: own elaboration

decision-analytic outcomes. Such forecasts would then become an explicit tool of well-defined decision support rather than merely a numerical exercise.

The price to pay is that interactive forecasts would inevitably lose generality, having to respond to specific problems faced by the decision makers. The researchers would also be no longer fully autonomous in preparing the forecasts and interpreting their outcomes, as these would emerge in a multi-stage process involving dialogue with forecast users and possibly also other experts in the domain. On the other hand, such procedures as outlined in Fig. 12.1 can be very time-consuming, costly and quite complex from the organisational point of view. Clearly, whenever decisions need to be taken quickly, or the budget for their analytical support is tight, not all paths indicated in Fig. 12.1 will be employed in the process. Hence, other forms of forecasts, including scenarios, variants or simple probabilistic predictions, are still very useful under the limited availability of resources. From the point of view of the forecasting methodology, they can be seen as special cases of a more general framework proposed above. Despite not having all the ideal properties, such predictions are also important decision support tools, although under the condition that their limitations are made explicit.

12.3 A Possible Future of Migration and Its Forecasts

The final part of this book, following Orrell's (2007) example, makes a futile attempt to predict the unpredictable by summarising some of the possible narratives for the future developments of both migration and its forecasts in the twenty first century. A rough sketch of a narrative about the possible future of migration worldwide was recently provided by Bauman (2007). He speculated that the 'migration potential' in many parts of the World is likely to remain very high, since emigration is a convenient way of 'crowding-out' of the surplus working-age populations of developing countries. The increased migratory pressure on the developed countries will then lead to social tensions and ever-tighter migration policies in response to them.

The 'crowding-out' mechanism seems to coincide with another demographic narrative for the future – the 'reproductive revolution' proposed by MacInnes and Pérez Díaz (2009) – which has quite different long-term implications. Behind the idea of 'reproductive revolution' there is a sequence in which countries undergo modernisation, coupled with the demographic transition from high to low fertility. As demographic transition implies emergence of large working-age cohorts, the developing countries will observe high pressure for emigration, exactly as did the developed countries decades or even centuries ago. However, if global modernisation continues, this effect will be limited in time: it will remain only until fertility levels become universally low worldwide, and the cohort effects of their decline, visible through the population momentum, level off. Needless to say, this mechanism describes global drivers of migration, while the local effects can be entirely different (see also Orrell, 2007). Importantly, in neither of these scenarios is migration expected to stop. Adversely, most likely the population flows will continue shaping – albeit at a different pace – the demographic and social structures of the future populations.

The future of migration forecasts seems to be less uncertain than the future of migration itself. The need for population forecasts for policy and planning purposes generates a demand for at least rough approximations of future migration trends and their uncertainty, such as the ones proposed in the current book. Besides, one could easily imagine various methodological extensions of the forecasting methodology, examples of which have been proposed in Chapters 10 and 11. On the other hand, there are inherent limits as to what can be reasonably predicted and with what accuracy. As noted by Orrell (2007, p. 241) specifically with respect to population flows, a combination of positive and negative feedback loops makes the mathematical models of migration very sensitive to parameterisation. In his view, the far-reaching implication is that 'self-regulating systems, which consist of a dynamic balance between the opposing forces, are hard to model and predict' (*idem*, p. 361). Migration, governed by various 'push, pull, retain and repel' factors is a model example of such a complex system.

As noted by King (2002), the dynamic – liquid – times require new analytical approaches to migration. According to him, globalisation, decline in travel costs, diversification of individual strategies and emergence of hybrid forms of migration result in multi-faceted migration processes, even more difficult to analyse using traditional, for example economic, theories and models. Besides, some traditional 'dichotomies' on which migration studies used to be based are no longer obvious and can only obscure the real picture. King (*passim*) discussed four examples:

(1) Internal versus international migration (less relevant within the EU/Schengen area),
(2) Voluntary versus forced migration (how to classify flows driven by extreme poverty?),
(3) Temporary versus permanent migration (no one would know it in advance); and

(4) Legal versus illegal migration (which distinction can be invalidated by regularisations).

In his view, all these issues call for true interdisciplinarity of migration studies in the future, which from today's point of view seems to be a quite likely scenario.

Regardless of general approaches to migration studies, the predictability of the related processes is likely to remain limited. The repercussions of this can be summarised by two general caveats. Firstly, the decision makers should not expect the impossible from the researchers. Secondly, the forecasters themselves should be more modest in presenting and interpreting the outcomes of their forecasts, and explicitly acknowledge their limitations. Admitting that the knowledge of the subject is imperfect can increase the chances for more cautious political decisions. In the case of government plans and programmes financed from the taxpayers' money, taking into account various scenarios, their probabilities and the costs of taking or not taking various courses of action is especially crucial. The economic and financial crisis of 2008–2009 showed that disregard for the implications of uncertainty can be dangerous. The same holds for population processes: uncertainty and complexity are natural elements of the social reality, and both of them can be subject to a scientific inquiry, but under no circumstances should be ignored. Paraphrasing Keynes, in order to avoid possibly disastrous consequences, any sensible policy advice should recommend 'prudence in the face of the unknown' (Skidelsky, 2009, p. 158).

In that respect, two caveats made by Hajnal (1955, p. 321) can be borne in mind: 'first that as little forecasting as possible should be done, and second that, if a forecast [. . .] is undertaken, it should involve less computation and more cognition than has generally been applied'. Given the prevailing uncertainty, they can serve as general guidance to population predictions, even more valid in the increasingly complex world of the twenty first century.

Annex A
Empirical Illustrations: Data Sources and Preparation

Annex A contains an inventory of data sources, a brief description of algorithms used for data preparation, and a listing of the time series of observations (Tables A.1 and A.2).

Migration Flows

For the reasons mentioned in Chapter 2 and to ensure data comparability for different flows, the current study universally employs the German time series of the population flows. The figures predominantly come from the NewCronos database of Eurostat, with the exception of auxiliary flows from/to Portugal and Spain used in the 'analogy' modelling in Chapter 7, which come from the Genesis database of the German Statistical Office (DESTATIS). The German data prior to 1991 concern West Germany (the then-Federal Republic). Further remarks are listed in the notes to Tables A.1 and A.2. Following the argumentation in Chapter 2, the numbers of migrants between the origin country i and the destination country j in year $t, M_{i-j}(t)$, as registered in the German official statistics, are divided by the population sizes of the origin countries, $P_i(t)$, in order to yield emigration rates according to the formula: $m_{i-j}(t) = M_{i-j}(t)/P_i(t) \cdot 1000$.

Population Stocks

Data on population stocks used in this study, $P_i(t)$, are also taken from the NewCronos database and from the statistical yearbook of the Council of Europe (2004). Detailed information is provided in notes to Tables A.1 and A.2, the former devoted to the four countries under study (Germany, Italy, Poland, and Switzerland), while the latter to the two countries of analogy considered in Chapter 7 (Portugal and Spain).

As noted in Chapter 2, population stocks in the countries with highly underestimated emigration can be corrected on the basis of the outcome of a population

J. Bijak, *Forecasting International Migration in Europe: A Bayesian View*,
The Springer Series on Demographic Methods and Population Analysis 24,
DOI 10.1007/978-90-481-8897-0,

Table A.1 Listing of data series used in the study and their sources – main variables

Variables, Countries*	Source(s)¶	Notes¶	1985	1986	1987	1988	1989	1990	1991	1992	1993	1994	1995	1996	1997	1998	1999	2000	2001	2002	2003	2004
Mid-year population (million persons)																						
$P_{DE}(t)$	A	1 2	61.03	61.08	61.19	61.47	62.19	63.20	80.01	80.62	81.15	81.44	81.68	81.91	82.03	82.05	32.10	82.21	82.35	82.49	82.53	82.52
$P_{CH}(t)$	A		6.47	6.50	6.55	6.59	6.65	6.72	6.80	6.88	6.94	6.99	7.04	7.07	7.09	7.11	7.14	7.18	7.23	7.28	7.34	7.39
$P_{IT}(t)$	A B		56.59	56.60	56.60	56.63	56.67	56.72	56.76	56.80	56.83	56.84	56.85	56.86	56.89	56.91	56.92	56.95	56.98	57.16	57.60	58.18
$P_{PL}(t)$	A	3	37.20	37.46	37.67	37.82	37.86	37.88	37.97	38.07	38.16	38.24	38.28	38.29	38.31	38.32	38.31	38.29	38.26	38.23	38.20	38.18
Migration flows (thousand persons)																						
$M_{CH\text{-}DE}(t)$	A C	1 4	6.1	6.6	6.7	6.9	7.2	7.4	8.0	8.8	8.4	7.6	7.9	7.9	7.7	7.7	7.8	8.0	8.3	8.5	8.5	9.1
$M_{IT\text{-}DE}(t)$	A C	1 4	41.9	49.2	48.0	44.5	43.1	39.7	38.4	32.8	34.2	41.2	50.6	48.5	41.6	37.7	37.2	35.4	31.6	26.9	23.7	21.4
$M_{PL\text{-}DE}(t)$	A C	1 4 5	89.7	105.4	158.2	309.2	450.5	296.1	141.1	139.1	81.7	88.1	99.7	91.3	85.6	82.0	90.2	94.1	100.5	101.0	104.9	139.3
$M_{DE\text{-}CH}(t)$	A C	1 4	7.1	7.4	7.9	7.7	8.1	8.0	8.3	8.5	8.3	8.7	9.0	8.9	9.2	10.0	10.8	11.9	13.1	14.7	14.8	16.9
$M_{DE\text{-}IT}(t)$	A C	1 4	54.6	44.8	42.5	40.0	41.6	37.0	39.2	35.4	33.5	35.0	36.6	39.4	40.8	39.9	38.4	36.7	36.1	36.5	33.8	36.3
$M_{DE\text{-}PL}(t)$	A C	1 4	58.1	62.8	71.5	101.4	145.9	162.1	118.0	112.1	104.8	70.3	77.0	78.9	79.1	70.6	69.5	71.4	76.0	78.7	82.9	104.5
Migration rates per 1,000 persons																						
$m_{CH\text{-}DE}(t)$	as above	as above	0.944	1.007	1.024	1.042	1.087	1.106	1.180	1.283	1.213	1.088	1.128	1.122	1.086	1.081	.093	1.115	1.146	1.171	1.165	1.235
$m_{IT\text{-}DE}(t)$	"	"	0.741	0.870	0.848	0.787	0.760	0.700	0.676	0.578	0.602	0.726	0.891	0.853	0.730	0.662	.654	0.621	0.554	0.470	0.411	0.368
$m_{PL\text{-}DE}(t)$	"	"	2.411	2.814	4.201	8.175	11.898	7.817	3.715	3.654	2.142	2.305	2.605	2.385	2.235	2.141	.354	2.458	2.627	2.641	2.746	3.648
$m_{DE\text{-}CH}(t)$	"	"	0.116	0.121	0.129	0.125	0.130	0.127	0.104	0.106	0.102	0.107	0.110	0.108	0.112	0.122	.131	0.145	0.160	0.178	0.179	0.204
$m_{DE\text{-}IT}(t)$	"	"	0.895	0.733	0.694	0.651	0.669	0.586	0.490	0.439	0.413	0.429	0.448	0.481	0.497	0.486	.467	0.446	0.438	0.443	0.410	0.440
$m_{DE\text{-}PL}(t)$	"	"	0.952	1.028	1.169	1.649	2.346	2.565	1.475	1.390	1.291	0.863	0.943	0.963	0.964	0.861	.847	0.869	0.923	0.955	1.005	1.267
GDP *per capita* (PPP), 2004 int. $, x 1,000																						
$Y_{DE}(t)$	D	1 6	20.8	21.1	21.6	22.3	23.1	24.2	25.3	25.4	25.0	25.6	26.0	26.1	26.4	26.8	27.4	28.2	28.5	28.1	28.1	28.3
$Y_{CH}(t)$	D	6	28.5	28.9	29.2	29.8	31.0	32.1	31.3	30.7	30.1	30.2	29.9	29.7	29.5	29.8	30.0	30.5	31.7	31.1	30.7	33.0
$Y_{IT}(t)$	D	6	20.6	21.0	21.7	22.6	23.4	24.0	24.3	24.2	23.9	24.3	25.0	25.0	25.4	25.6	26.0	27.0	27.5	27.4	27.5	28.2
$Y_{PL}(t)$	D	6	:	:	:	:	:	7.9	7.2	7.1	7.3	7.7	8.2	8.6	9.2	9.8	10.2	10.8	11.2	11.3	11.8	13.0
Auxiliary variable: US GDP deflator																						
Annual %	D	7	3.2	2.2	3.0	3.4	3.8	3.9	3.6	2.4	2.4	2.1	2.2	1.9	2.0	1.2	1.4	2.2	2.4	1.7	1.8	2.6
Index: 2004=100	own calculations		62.9	64.9	66.4	68.4	70.7	73.4	76.3	79.1	81.0	82.9	84.7	86.5	88.2	89.9	91.0	92.3	94.3	96.6	98.2	100.0
Unemployment rates (standardised, %)																						
$U_{DE}(t)$	E F	1 8	7.1	6.4	6.2	6.2	5.6	4.8	4.2	6.4	7.7	8.3	8.0	8.5	9.1	8.8	7.9	7.2	7.4	8.2	9.0	9.5
$U_{CH}(t)$	E F	8	0.8	0.7	0.7	0.6	0.5	0.5	1.9	3.0	3.9	3.9	3.5	3.9	4.2	3.6	3.0	2.7	2.6	3.2	4.2	4.4
$U_{IT}(t)$	E F	8	8.5	9.2	9.9	10.0	10.0	8.9	8.5	8.7	10.1	10.6	11.2	11.2	11.3	11.3	10.9	10.1	9.1	8.6	8.4	8.0
$U_{PL}(t)$	F	8 9	:	:	:	:	:	5.5	10.4	12.2	14.0	14.4	13.3	12.3	10.9	10.2	13.4	16.4	18.5	19.8	19.2	18.8

*For definitions of variables, see Chapter 2. Country codes: DE – Germany, CH – Switzerland, IT – Italy, PL – Poland.

¶Notes are indicated by Arabic numerals and data sources by capital letters, and are listed jointly for Tables A.1 and A.2 underneath the latter. Colons (:) indicate that reliable values are not available.

Table A.2 Listing of additional data series used in Chapter 7 of the study and of their sources

Variables, Countries*	Source(s)¶	Notes¶	1974	1975	1976	1977	1978	1979	1980	1981	1982	1983	1984	1985
Mid-year population (million persons)														
$P_{DE}(t)$	**G**	**1**	62.05	61.82	61.54	61.40	61.34	61.38	61.55	61.69	61.63	61.43	61.18	61.03
$P_{ES}(t)$	**G**		35.15	35.53	35.94	36.37	36.78	37.10	37.44	37.74	37.94	38.12	38.28	38.42
$P_{PT}(t)$	**G**		8.75	9.09	9.36	9.46	9.56	9.66	9.77	9.85	9.91	9.95	9.99	10.01
Migration flows (thousand persons)														
$M_{ES\text{-}DE}(t)$	**H**	**1 4**	15.7	9.8	8.4	7.9	7.4	7.2	7.7	7.8	5.9	5.0	5.5	5.7
$M_{PT\text{-}DE}(t)$	**H**	**1 4**	13.8	8.9	6.3	5.2	4.8	4.8	4.3	3.5	2.4	2.0	2.0	2.1
$M_{DE\text{-}ES}(t)$	**H**	**1 4**	49.7	42.1	34.7	26.5	19.5	14.0	11.8	10.8	12.7	12.8	11.8	10.6
$M_{DE\text{-}PT}(t)$	**H**	**1 4**	17.8	15.3	12.6	9.3	7.0	6.1	8.7	8.2	9.7	10.0	21.4	3.9
Migration rates per 1,000 persons														
$m_{ES\text{-}DE}(t)$	as above	as above	0.447	0.275	0.232	0.216	0.201	0.195	0.205	0.207	0.157	0.132	0.143	0.150
$m_{PT\text{-}DE}(t)$	"	"	1.580	0.984	0.677	0.545	0.507	0.495	0.438	0.354	0.241	0.204	0.199	0.212
$m_{DE\text{-}ES}(t)$	"	"	0.802	0.681	0.563	0.432	0.317	0.229	0.191	0.175	0.207	0.208	0.193	0.173
$m_{DE\text{-}PT}(t)$	"	"	0.287	0.247	0.205	0.152	0.114	0.099	0.141	0.133	0.157	0.163	0.350	0.063

Variables, Countries*	Source(s)¶	Notes¶	1986	1987	1988	1989	1990	1991	1992	1993	1994	1995	1996	1997
Mid-year population (million persons)														
$P_{DE}(t)$	**A**	**1 2**	... As in Table A.1											
$P_{ES}(t)$	**G**		38.54	38.63	38.72	38.79	38.85	38.93	39.05	39.17	39.26	39.34	39.43	39.52
$P_{PT}(t)$	**G**		10.01	9.99	9.97	9.94	9.90	9.92	9.97	9.98	10.00	10.03	10.06	10.09
Migration flows (thousand persons)														
$M_{ES\text{-}DE}(t)$	**H**	**1 4**	6.9	6.7	7.2	7.9	8.1	8.5	9.4	9.8	10.0	11.5	12.6	12.2
$M_{PT\text{-}DE}(t)$	**H**	**1 4**	3.8	3.7	4.2	6.1	7.8	11.5	11.0	14.9	30.8	34.0	37.1	27.2
$M_{DE\text{-}ES}(t)$	**H**	**1 4**	9.7	10.1	9.6	10.1	9.7	9.5	10.5	11.4	12.9	12.8	14.3	15.6
$M_{DE\text{-}PT}(t)$	**H**	**1 4**	2.7	2.5	2.6	3.7	3.8	4.9	5.7	7.4	16.7	23.2	28.8	27.4
Migration rates per 1,000 persons														
$m_{ES\text{-}DE}(t)$	as above	as above	0.178	0.174	0.186	0.205	0.208	0.219	0.241	0.250	0.254	0.292	0.318	0.308
$m_{PT\text{-}DE}(t)$	"	"	0.380	0.369	0.424	0.616	0.788	1.158	1.100	1.498	3.076	3.389	3.693	2.696
$m_{DE\text{-}ES}(t)$	"	"	0.158	0.166	0.156	0.163	0.154	0.119	0.130	0.141	0.159	0.156	0.175	0.190
$m_{DE\text{-}PT}(t)$	"	"	0.045	0.041	0.042	0.059	0.060	0.061	0.071	0.091	0.206	0.284	0.351	0.334

Table A.2 (continued)

*For definitions of variables, see Chapter 2. Country codes: DE – Germany, ES – Spain, PT – Portugal.
¶Notes are indicated by Arabic numerals and data sources by capital letters, as listed below.

Notes:

1 Prior to 1991, the data concern only West Germany (FRG).
2 German population size for 1989–1993 has been adjusted.
3 Polish population size for 1998–2002 has been adjusted.
4 Migration flows are shown according to the German statistics.
5 Polish-German migration flows for 1988–1992 have been adjusted.
6 The 2004-based international dollars are recalculated from current dollars using a US GDP deflator (2004 = 100).
7 For 2000–2004, GDP was calculated from the gross values from the World Bank Quick Reference Tables (see **D**).
8 For definitions, see the methodological part of the United Nations (2005b) report.
9 For 1990–1992, the estimates are shown, assuming the same dynamics for standardised rates and the registered unemployment, the latter following United Nations (2000b, p. 230).

Sources:

A Eurostat/NewCronos database, population figures downloaded in May 2005, migration figures – in July 2004;
B For 2003–2004, Italian Statistical Office ISTAT/Demo database: demo.istat.it (accessed on 10 August 2005);
C Data for 1994 supplemented from: Statistisches Bundesamt (1995), *Statistisches Jahrbuch 1995 für die BRD*, Stuttgart; and for 2003–2004 by the Joint UNSD-Eurostat-UN ECE-Council of Europe-ILO Migration Questionnaire.
D For 1985–2001, *World Development Indicators 2003* (World Bank, 2003); for 2002–2003 supplemented by the World Bank figures available from the Earthtrends website (www.earthtrends.org/searchable_db/index.php?theme=5, accessed on 5 April 2006), while for 2004 – calculated from the gross values presented in theWorld Bank Quick Reference Tables (www.worldbank.org/data/quickreference/quickref.html, accessed on 5 April 2006);
E For 1985–1989, United Nations (2000b, pp. 205, 230);
F For 1990–2004, United Nations (2005b, p. 68);
G Council of Europe (2004, country-specific Tables 8);
H German Statistical Office/Genesis database (www.destatis.de/genesis, accessed on 21 September 2006).

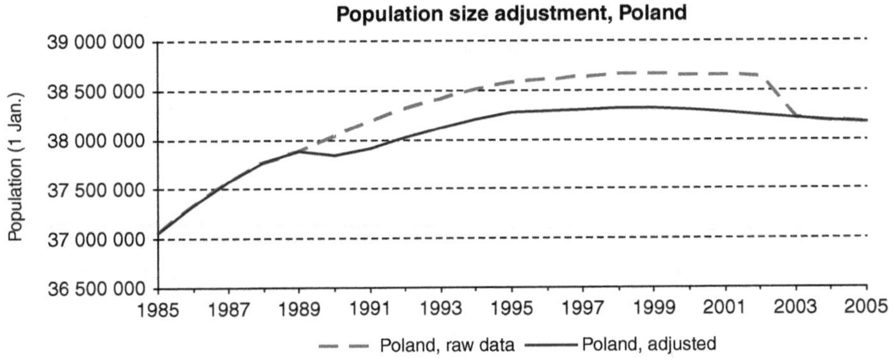

Fig. A.1 Population size of Poland, 1985–2004: official and corrected.
Source: Official figures and adjustment size – Council of Europe (2004, Table 8). Adjusted figures – own calculations

census[1]. Here, following Bijak and Koryś (2009), such correction was made for Poland for the recent inter-census period, between 6th December 1988 and 21st May 2002. The 'statistical adjustment' based on the results of the 2002 census was estimated as a net of −390,285 persons (Council of Europe, 2004, Table 8), who 'invisibly' emigrated in the period 1989–2002. This number has been distributed over the period between the censuses, proportionally to the size of net migration from Poland registered in Germany according to the German data. However, in the German statistics for 1993 there is an administrative adjustment of the Polish migrant stock by −23,049 persons. In the current study, the last number has been distributed equally over the period 1988–1992 in order to correct migration flows, and for the years 1989–1993 to adjust the German population size. The presented method involves an implicit assumption that registered and unregistered migration follow the same pattern over time, which in reality may not necessarily be true. The recalculation of Polish population stocks in the period under study is illustrated in Fig. A.1.

Economic Variables

The time series of the purchase power parity (PPP)-adjusted GDP *per capita*, $Y_i(t)$, were taken from the *World Development Indicators* database of the World Bank (2003), supplemented for the most recent years by the World Bank data from various Internet sources, as listed in the notes to Table A.1. The PPP concept makes the variable $Y_i(t)$ comparable between the economies with different costs of living,

[1]The calculations presented here are by necessity simplistic, as they do not take into account for example the errors of the census estimates, which themselves can be biased, especially for the most mobile population groups.

hence bring most appropriate as a proxy of a migration determinant. In order to ensure the temporal consistency, the 2004-based international dollars were recalculated from the current international dollars listed in the World Bank data, using a series on the US GDP deflator – a proxy of inflation in the US economy, and the depreciation of the dollar.

The standardised unemployment rates, $U_i(t)$, originate from the UN ECE Economic Surveys of Europe (United Nations, 2000b, pp. 205, 230; 2005b, p. 68). In the case of Poland, the rates for 1990–1992 were estimated under the assumption of the same dynamics for standardised rates and the registered unemployment, the latter also following United Nations (2000b, p. 230). Detailed remarks with respect to particular series of economic data are provided in notes to Table A.1.

Annex B
WinBUGS Code Used in the Forecasting Examples

Annex B contains the listing of WinBUGS 1.4 code underlying the calculations of Bayesian forecasts of migration rates presented in Chapters 5, 6 and 7.

B.1. Bayesian forecast selection and averaging, simple stochastic processes

```
model {
                                          # Model probabilities
mod ~ dcat(p[])                           # categorical distrib. over the model space (phases 2-3)
# for (k in 1:5) { p[k] <- 1/5 }          # uniform prior (phase 2)
# p[1] <- 0.30769; p[2] <- 0.30769; p[3] <- 0.15385; p[4] <- 0.15385; p[5] <- 0.07692
                                          # alternative Occam's razor prior (phase 2)
p[1] <- 0.232; p[2] <- 0.337; p[3] <- 0.212; p[4] <- 0.123; p[5] <- 0.096
                                          # p(M|x) estimated in phase 2, used in phase 3

# Definitions of data
for (t in 1:n) {    for (k in 1:5) { m[t,k] <- log(MR[t]) }
#                   mav[t] <- log(MR[t]) }          # phase 2 only
}

# Model update
for (t in 2:n) {    mu[t,1] <- c[1]
                    mu[t,2] <- c[2] + m[t-1,2]
                    mu[t,3] <- c[3] + phi[3] * m[t-1,3]
                    mu[t,4] <- c[4] - theta[4] * e[t-1,4]
                    mu[t,5] <- c[5] + phi[5] * m[t-1,5] - theta[5] * e[t-1,5]
                    for (k in 1:5) { m[t,k] ~ dnorm(mu[t,k], tau[k]) }
#                   mav[t] ~ dnorm(mu[t,mod], tau[mod])      # model probability update, phase 2
                    e[t,4] <- m[t,4] - mu[t,4]; e[t,5] <- m[t,5] - mu[t,5] }
for (k in 1:5) {    m.new[n,k] <- m[n,k] }

# Artificial MA error terms for the first period (t=1)
e[1,4] ~ dnorm(0,tau[4]); e[1,5] ~ dnorm(0,tau[5])

# Priors used in phases 1 and 3
for (k in 1:5) { c[k] ~ dnorm(0,0.0001) } # constants
phi[3] ~ dnorm(0.5,1); phi[5] ~ dnorm(0.5,1) # AR coefficients
theta[4] ~ dnorm(0.5,1); theta[5] ~ dnorm(0.5,1) # MA coefficients
for (k in 1:5) { tau[k] ~ dgamma(4,0.4) } # precision for the random component

# Priors used in phase 2
# for (k in 1:5) { c[k] ~ dnorm(mu.c[k,mod],tau.c[k,mod]) }
# phi[3] ~ dnorm(mu.phi[3,mod],tau.phi[3,mod]); phi[5] ~ dnorm(mu.phi[5,mod],tau.phi[5,mod])
# theta[4] ~ dnorm(mu.th[4,mod],tau.th[4,mod]); theta[5] ~ dnorm(mu.th[5,mod],tau.th[5,mod])
# for (k in 1:5) { tau[k] ~ dgamma(r.tau[k,mod],m.tau[k,mod]) }

# Input for the estimation priors, phase 2
for (k in 1:5) { mu.c[k,k] <- 0; tau.c[k,k] <- 0.0001; r.tau[k,k] <- 4;  m.tau[k,k] <- 0.4 }
mu.phi[3,3] <- 0.5; mu.phi[5,5] <- 0.5; tau.phi[3,3] <- 1; tau.phi[5,5] <- 1
mu.th[4,4] <- 0.5; mu.th[5,5] <- 0.5; tau.th[4,4] <- 1; tau.th[5,5] <- 1

# Input for pseudo-priors, phase 2 (obtained as posterior means in phase 1)
mu.c[1,2] <- 0.1159; mu.c[1,3] <- 0.1159; mu.c[1,4] <- 0.1159; mu.c[1,5] <- 0.1159
mu.c[2,1] <- 0.0141; mu.c[2,3] <- 0.0141; mu.c[2,4] <- 0.0141; mu.c[2,5] <- 0.0141
mu.c[3,1] <- 0.0488; mu.c[3,2] <- 0.0488; mu.c[3,4] <- 0.0488; mu.c[3,5] <- 0.0488
mu.c[4,1] <- 0.1162; mu.c[4,2] <- 0.1162; mu.c[4,3] <- 0.1162; mu.c[4,5] <- 0.1162
mu.c[5,1] <- 0.0543; mu.c[5,2] <- 0.0543; mu.c[5,3] <- 0.0543; mu.c[5,4] <- 0.0543

tau.c[1,2] <- 971.70; tau.c[1,3] <- 971.70; tau.c[1,4] <- 971.70; tau.c[1,5] <- 971.70
tau.c[2,1] <- 1042.6; tau.c[2,3] <- 1042.6; tau.c[2,4] <- 1042.6; tau.c[2,5] <- 1042.6
tau.c[3,1] <- 362.95; tau.c[3,2] <- 362.95; tau.c[3,4] <- 362.95; tau.c[3,5] <- 362.95
tau.c[4,1] <- 507.95; tau.c[4,2] <- 507.95; tau.c[4,3] <- 507.95; tau.c[4,5] <- 507.95
tau.c[5,1] <- 254.86; tau.c[5,2] <- 254.86; tau.c[5,3] <- 254.86; tau.c[5,4] <- 254.86
```

```
mu.phi[3,1] <- 0.6583; mu.phi[3,2] <- 0.6583; mu.phi[3,4] <- 0.6583; mu.phi[3,5] <- 0.6583
mu.phi[5,1] <- 0.6128; mu.phi[5,2] <- 0.6128; mu.phi[5,3] <- 0.6128; mu.phi[5,4] <- 0.6128
tau.phi[3,1] <- 5.8008; tau.phi[3,2] <- 5.8008; tau.phi[3,4] <- 5.8008; tau.phi[3,5] <- 5.8008
tau.phi[5,1] <- 4.4735; tau.phi[5,2] <- 4.4735; tau.phi[5,3] <- 4.4735; tau.phi[5,4] <- 4.4735

mu.th[4,1] <- -0.4058; mu.th[4,2] <- -0.4058; mu.th[4,3] <- -0.4058; mu.th[4,5] <- -0.4058
mu.th[5,1] <- -0.2710; mu.th[5,2] <- -0.2710; mu.th[5,3] <- -0.2710; mu.th[5,4] <- -0.2710
tau.th[4,1] <- 7.9082; tau.th[4,2] <- 7.9082; tau.th[4,3] <- 7.9082; tau.th[4,5] <- 7.9082
tau.th[5,1] <- 5.8514; tau.th[5,2] <- 5.8514; tau.th[5,3] <- 5.8514; tau.th[5,4] <- 5.8514

r.tau[1,2] <- 13.037; r.tau[1,3] <- 13.037; r.tau[1,4] <- 13.037; r.tau[1,5] <- 13.037
r.tau[2,1] <- 13.024; r.tau[2,3] <- 13.024; r.tau[2,4] <- 13.024; r.tau[2,5] <- 13.024
r.tau[3,1] <- 12.746; r.tau[3,2] <- 12.746; r.tau[3,4] <- 12.746; r.tau[3,5] <- 12.746
r.tau[4,1] <- 12.666; r.tau[4,2] <- 12.666; r.tau[4,3] <- 12.666; r.tau[4,5] <- 12.666
r.tau[5,1] <- 12.630; r.tau[5,2] <- 12.630; r.tau[5,3] <- 12.630; r.tau[5,4] <- 12.630

m.tau[1,2] <- 0.2358; m.tau[1,3] <- 0.2358; m.tau[1,4] <- 0.2358; m.tau[1,5] <- 0.2358
m.tau[2,1] <- 0.2189; m.tau[2,3] <- 0.2189; m.tau[2,4] <- 0.2189; m.tau[2,5] <- 0.2189
m.tau[3,1] <- 0.2170; m.tau[3,2] <- 0.2170; m.tau[3,4] <- 0.2170; m.tau[3,5] <- 0.2170
m.tau[4,1] <- 0.2196; m.tau[4,2] <- 0.2196; m.tau[4,3] <- 0.2196; m.tau[4,5] <- 0.2196
m.tau[5,1] <- 0.2137; m.tau[5,2] <- 0.2137; m.tau[5,3] <- 0.2137; m.tau[5,4] <- 0.2137

# Forecasts for t = n+1 … N
for (t in n+1:N) {  mu[t,1] <- c[1]
                    mu[t,2] <- c[2] + m.new[t-1,2]
                    mu[t,3] <- c[3] + phi[3] * m.new[t-1,3]
                    mu[t,4] <- c[4] - theta[4] * e[t-1,4]
                    mu[t,5] <- c[5] + phi[5] * m.new[t-1,5] - theta[5] * e[t-1,5]
                    e[t,4] <- m.new[t,4] - mu[t,4]; e[t,5] <- m.new[t,5] - mu[t,5]
                    for (k in 1:5) { m.new[t,k] ~ dnorm(mu[t,k], tau[k]) }  }

# for (k in 1:5) { prob[k] <- step(mod-k)-step(mod-k-1) }  # posterior p(M|x), phase 2
# for (t in n+1:N) { mav.new[t] <- m.new[t,mod] }          # averaged forecast, phase 3

}

# The following three simulation phases are sequentially performed:

# 1 - to obtain posterior estimates of parameters and use them in pseudo-priors (underlined)
# 2 - to estimate posterior model probabilities p(M|x) using the Carlin and Chib (1995) method
# 3 - to mix individual forecasts using p(M|x) in order to obtain averaged ones

# 250,000 iterations in the MCMC sample for each run, burn-in length - 100,000 iterations


# Sample data set

# Migration CH-DE
list(n = 20, N = 31, MR = c(0.9442, 1.0071, 1.0237, 1.0424, 1.0874, 1.1061, 1.1804, 1.2833,
1.2131, 1.0884, 1.1282, 1.1225, 1.0856, 1.0812, 1.0932, 1.1149, 1.1458, 1.1714, 1.1646,
1.2346))

# Inits
list(c = c(0,0,0,0,0), phi = c(NA,NA,0.5,NA,0.5), theta = c(NA,NA,NA,0.5,0.5), tau =
c(10,10,10,10,10), mod = 3)
```

B.2a. VAR model selection: HPD tests in the 'general to specific' approach

```
model {
                                                             # Priors for parameters:
c[1] ~ dnorm(0,0.0001); c[2] ~ dnorm(0,0.0001); c[3] ~ dnorm(0,0.0001) # constants
alpha[1,1] ~ dnorm(1,1); alpha[2,2] ~ dnorm(1,1); alpha[3,3] ~ dnorm(1,1) # own-variable lags
alpha[1,2] ~ dnorm(0.5,1); alpha[2,1] ~ dnorm(0,100); alpha[1,3] ~ dnorm(0.5,1) # cross-lags
alpha[3,1] ~ dnorm(0,9); alpha[2,3] ~ dnorm(0.1,0.25); alpha[3,2] ~ dnorm(0.1,0.25)
T[1:3,1:3] ~ dwish(P[1:3,1:3],3)                             # precision matrix (T = Sigma^-1)
for (k in 1:3) { P[k,k] <- 1.5 }                             # Wishart distribution parameters
P[1,2]<- -0.03; P[2,1]<- -0.03; P[1,3]<- -0.03; P[3,1]<- -0.03; P[2,3]<- -0.03; P[3,2]<- -0.03

# Data definitions
for (t in 1:n) { y[t,1] <- log(VAR1[t]); y[t,2] <- log(VAR2[t]); y[t,3] <- log(VAR3[t]) }
for (t in n:N) { x2[t] <- log(CVAR2[t-n+1]); x3[t] <- log(CVAR3[t-n+1]) }

# Model update
for (t in 2:n) { for (i in 1:3) {  mu[t,i] <- alpha[i,1] * y[t-1,1] + alpha[i,2] * y[t-1,2] +
                                         + alpha[i,3] * y[t-1,3] + c[i] }
                 y[t,1:3] ~ dmnorm(mu[t,1:3],T[1:3,1:3]) }
for (k in 1:3) { y.new[n,k] <- y[n,k]; cy.new[n,k] <- y[n,k] }

# Calculations of the test statistics, manual variance inversions due to numerical problems

b[1] <- alpha[1,2]; b[2] <- alpha[1,3] # Vector b for the impact of variables 2 and 3 jointly
Sigma[1:3,1:3] <- inverse(T[,]) # Regression parameters for the conditional model y[t]|x[t]
beta[1] <- (Sigma[1,2] * Sigma[3,3] - Sigma[1,3] * Sigma[3,2]) / (Sigma[2,2] * Sigma[3,3] +
                                                               - Sigma[2,3] * Sigma[3,2])
beta[2] <- (Sigma[1,3] * Sigma[2,2] - Sigma[1,2] * Sigma[2,3]) / (Sigma[2,2] * Sigma[3,3] +
                                                               - Sigma[2,3] * Sigma[3,2])
ctau <- 1/(Sigma[1,1] - beta[1] * Sigma[2,1] - beta[2] * Sigma[3,1])

# The following lines are to be run after obtaining expected values of b[] and beta[] from
MCMC summaries, and inserting them into the respective places of the code (underlined)

# Eb[1] <- 0.0000; Eb[2] <- 0.0000; Ebeta[1] <- 0.0000; Ebeta[2] <- 0.0000
# for (i in 1:2) { for (j in 1:2) { bvar[i,j] <- b[i] * b[j] - Eb[i] * Eb[j]   # Variances
                                    betavar[i,j] <- beta[i] * beta[j] - Ebeta[i] * Ebeta[j]}}

# Lines to be run after obtaining covariance matrices of b[] and beta[] from MCMC summaries

# Db[1,1] <- 1.0000; Db[2,2] <- 1.0000; Db[1,2] <- 0.0000; Db[2,1] <- 0.0000
# Dbeta[1,1] <- 1.0000; Dbeta[2,2] <- 1.0000; Dbeta[1,2] <- 0.0000; Dbeta[2,1] <- 0.0000

# Tb[1,1] <- Db[2,2] / (Db[1,1] * Db[2,2] - Db[2,1] * Db[1,2])
# Tb[1,2] <- - Db[2,1] / (Db[1,1] * Db[2,2] - Db[2,1] * Db[1,2])
# Tb[2,1] <- - Db[1,2] / (Db[1,1] * Db[2,2] - Db[2,1] * Db[1,2])
# Tb[2,2] <- Db[1,1] / (Db[1,1] * Db[2,2] - Db[2,1] * Db[1,2])
# Tbeta[1,1] <- Dbeta[2,2] / (Dbeta[1,1] * Dbeta[2,2] - Dbeta[2,1] * Dbeta[1,2])
# Tbeta[1,2] <- - Dbeta[2,1] / (Dbeta[1,1] * Dbeta[2,2] - Dbeta[2,1] * Dbeta[1,2])
# Tbeta[2,1] <- - Dbeta[1,2] / (Dbeta[1,1] * Dbeta[2,2] - Dbeta[2,1] * Dbeta[1,2])
# Tbeta[2,2] <- Dbeta[1,1] / (Dbeta[1,1] * Dbeta[2,2] - Dbeta[2,1] * Dbeta[1,2])

# for (i in 1:2) { ctrb[i] <- b[i] - Eb[i]; ctrbeta[i] <- beta[i] - Ebeta[i] }
# HPDtest[1] <- pow(ctrb[1],2) / Db[1,1]                     # Impact of lagged variable 2 only
# HPDtest[2] <- pow(ctrb[2],2) / Db[2,2]                     # Impact of lagged variable 3 only
# HPDtest[3] <- ctrb[1] * (ctrb[1] * Tb[1,1] + ctrb[2] * Tb[2,1])
     + ctrb[2] * (ctrb[1] * Tb[1,2] + ctrb[2] * Tb[2,2])  # Impact of lagged variables 2 & 3
# HPDtest[4] <- pow(ctrbeta[1],2) / Dbeta[1,1]               # Conditional impact of variable 2
# HPDtest[5] <- pow(ctrbeta[1],2) / Dbeta[1,1]               # Conditional impact of variable 3
# HPDtest[6] <- ctrbeta[1] * (ctrbeta[1] * Tbeta[1,1] + ctrbeta[2] * Tbeta[2,1])
     + ctrbeta[2] * (ctrbeta[1] * Tbeta[1,2] + ctrbeta[2] * Tbeta[2,2]) # Conditional, 2 & 3

# Forecasts for t = n+1 … N
for (t in n+1:N) { for (i in 1:3) { mu.new[t,i] <- alpha[i,1] * y.new[t-1,1] +
                        + alpha[i,2] * y.new[t-1,2] + alpha[i,3] * y.new[t-1,3] + c[i] }
                   y.new[t,1:3] ~ dmnorm(mu.new[t,1:3], T[1:3,1:3])
#                  cy.new[t,1:3] ~ dmnorm(cmu.new[t,1:3], T[1:3,1:3]) # Conditional forecasts
#                  cmu.new[t,1] <- alpha[1,1] * cy.new[t-1,1] + alpha[1,2] * x2[t-1] +
#                  + alpha[1,3] * x3[t-1] + c[1]; cmu.new[t,2] <- x2[t]; cmu.new[t,3] <- x3[t]
}

# The procedure consists in two preliminary phases, aimed at obtaining Eb, Ebeta, Db and Dbeta
# third phase with HPD tests and unconditional forecasts, and fourth - conditional forecasts

# 250,000 iterations in the MCMC sample for each run, burn-in length - 200,000 iterations
# Preliminary phases: 100,000 iterations in the sample, after 150,000 burn-in
```

B.2b. VAR model selection: estimation of restricted AR(1) models

```
model {
                                                                # Priors for parameters
c[1] ~ dnorm(0,0.0001); c[2] ~ dnorm(0,0.0001); c[3] ~ dnorm(0,0.0001) # constants
alpha[1] ~ dnorm(1,1); alpha[2] ~ dnorm(1,1); alpha[3] ~ dnorm(1,1) # own variable lags
for (k in 1:3) { P[k,k] <- 1.5 }                                # Wishart distribution parameters
P[1,2]<- -0.03; P[2,1]<- -0.03; P[1,3]<- -0.03; P[3,1]<- -0.03; P[2,3]<- -0.03; P[3,2]<- -0.03
V[1:3,1:3] <- inverse(P[,]) # precision matrix
for (k in 1:3) { margtau[k] ~ dchisqr(3); tau[k] <- margtau[k] * V[k,k] } # marginal precision

# Data definitions
for (t in 1:n) { y[t,1] <- log(VAR1[t]); y[t,2] <- log(VAR2[t]); y[t,3] <- log(VAR3[t]) }

# Model update
for (t in 2:n) { for (i in 1:3) { mu[t,i] <- alpha[i] * y[t-1,i] + c[i]
                                  y[t,i] ~ dnorm(mu[t,i],tau[i]) }}
for (i in 1:3) { y.new[n,i] <- y[n,i] }

# Forecasts for t = n+1 … N
for (t in n+1:N) { for (i in 1:3) { mu.new[t,i] <- alpha[i] * y.new[t-1,i] + c[i]
                                    y.new[t,i] ~ dnorm(mu.new[t,i], tau[i]) }}
}

# 250,000 iterations in the MCMC sample for each run, burn-in length - 100,000 iterations
```

 # Sample data set

```
# Migration CH-DE (common for HPD tests and restricted VAR estimation and for forecasts)
# CVAR2 and CVAR3 are filled in from the medians of VAR2 and VAR3 obtained from the MCMC
# summaries from the third run of the MCMC algorithm (underlined values)

list(n = 20, N = 31, VAR1 = c(0.9442, 1.0071, 1.0237, 1.0424, 1.0874, 1.1061, 1.1804, 1.2833,
1.2131, 1.0884, 1.1282, 1.1225, 1.0856, 1.0812, 1.0932, 1.1149, 1.1458, 1.1714, 1.1646,
1.2346), VAR2 = c(0.7284, 0.7318, 0.7402, 0.7489, 0.7450, 0.7545, 0.8081, 0.8280, 0.8295,
0.8473, 0.8698, 0.8769, 0.8935, 0.8984, 0.9105, 0.9267, 0.8996, 0.9032, 0.9146, 0.8567), VAR3
= c(0.8000, 0.7000, 0.7000, 0.6000, 0.5000, 0.5000, 1.9000, 3.0000, 3.9000, 3.9000, 3.5000,
3.9000, 4.2000, 3.6000, 3.0000, 2.7000, 2.6000, 3.2000, 4.2000, 4.4000), CVAR2 = c(0.8567,
0.8625, 0.8653, 0.8670, 0.8683, 0.8703, 0.8718, 0.8735, 0.8751, 0.8763, 0.8779, 0.8791),
CVAR3 = c(4.4000, 4.3362, 4.306, 4.2759, 4.2589, 4.2759, 4.2974, 4.3189, 4.3319, 4.3667,
4.3973, 4.4327))

# Inits (HPD tests)
list(alpha = structure(.Data = c(1,0,0,0,1,0,0,0,1), .Dim=c(3,3)), T = structure(.Data =
c(1,0,0,0,1,0,0,0,1), .Dim = c(3,3)), c = c(0,0,0))

# Inits (estimation of the restricted VARs)
list(alpha = c(1,1,1), margtau = c(1,1,1), c = c(0,0,0))
```

B.3a. Capturing discontinuity in trends: forecasting by analogy

```
model {
                                        # Model probabilities
mod ~ dcat(p[])                         # categorical distribution over the model space, phase 2
p[1] <- 0.4; p[2] <- 0.2; p[3] <- 0.2; p[4] <- 0.2 # Occam's razor, dummy parameter not counted

# Data definitions
for (t in 2:n) { for (k in 1:4) { m[k,t] <- log(MR[t-1]) }
#                mav[t] <- log(MR[t-1]) }  # stage 2 only - Carlin-Chib estimation
for (t in 2:N) { mref[2,t] <- log(MRPT[t-1]); mref[3,t] <- log(MRIB[t-1]) # for definitions,
                 mref[4,t] <- log(MRES[t-1]); z[t] <- Dummy[t-1] } # see 'Sample data' below

# Priors for parameters - phase 1
for (k in 1:4) { c[k] ~ dnorm(0,0.0001); phi[k] ~ dnorm(0.5,1); tau[k] ~ dgamma(4,0.4);
                                                                  e[k,1] ~ dnorm(0,tau[k])}
for (k in 2:3) { alpha[k] ~ dnorm(0.5,1); beta[k] ~ dnorm(0,1) }; alpha[4] ~ dnorm(0.5,1)

# Priors for parameters - phase 2
# for (k in 1:4) { c[k] ~ dnorm(mu.c[k,mod],tau.c[k,mod]); e[k,1] ~ dnorm(0,tau[k]);
#                  phi[k] ~ dnorm(mu.phi[k,mod],tau.phi[k,mod]);
#                  tau[k] ~ dgamma(r.tau[k,mod], m.tau[k,mod]) }
# for (k in 2:3) { alpha[k] ~ dnorm(mu.alpha[k,mod],tau.alpha[k,mod]);
#                  beta[k] ~ dnorm(mu.beta[k,mod],tau.beta[k,mod]) }
# alpha[4] ~ dnorm(mu.alpha[4,mod],tau.alpha[4,mod])

# Input for priors and pseudo-priors, phase 2
for (k in 1:4) { mu.c[k,k] <- 0; tau.c[k,k] <- 0.0001; mu.phi[k,k] <- 0.5; tau.phi[k,k] <- 1;
                                                      r.tau[k,k] <- 4; m.tau[k,k] <- 0.4 }
for (k in 2:3) { mu.alpha[k,k]<-0.5; tau.alpha[k,k]<-1; mu.beta[k,k] <- 0; tau.beta[k,k]<-1 }
mu.alpha[4,4] <- 0.5; tau.alpha[4,4] <- 1
... # For pseudo-priors, structure analogous to the one presented in Code B.1 (omitted)

# Model update
for (t in 2:n) { for (k in 2:3) { mu[k,t] <- c[k] + alpha[k] * mref[k,t] + beta[k] * z[t] +
                                                       + phi[k] * e[k,t-1]
                 mu[1,t] <- c[1] + phi[1] * e[1,t-1]
                 mu[4,t] <- c[4] + alpha[4] * mref[4,t] + phi[4] * e[4,t-1]
                 for (k in 1:4) { m[k,t] ~ dnorm(mu[k,t], tau[k]) } # phase 1 only
#                mav[t] ~ dnorm(mu[mod,t], tau[mod])                # phase 2 only, P(M|x)
                 for (k in 1:4) { e[k,t] <- m[k,t] - mu[k,t] }}
for (k in 1:4) { e.new[k,n] <- e[k,n]; m.new[k,n] <- m[k,n] }

# Forecasts for t = n+1 … N
for (t in n+1:N) { for (k in 2:3) { mu.new[k,t] <- c[k] + alpha[k] * mref[k,t] +
                                    beta[k] * z[t] + phi[k] * e.new[k,t-1] }
                   mu.new[1,t] <- c[1] + phi[1] * e.new[1,t-1]
                   mu.new[4,t] <- c[4] + alpha[4] * mref[4,t] + phi[4] * e.new[4,t-1]
                   for (k in 1:4) { e.new[k,t] <- m.new[k,t] - mu.new[k,t];
                                    m.new[k,t] ~ dnorm(mu.new[k,t], tau[k]) }}
# for (i in 1:4) { prob[i] <- step(mod-i)-step(mod-i-1) } # posterior p(M|x), phase 2
}

# Phase 1 - estimation and forecasting, phase 2 - calculation of P(M|x) only
# 250,000 iterations in the MCMC sample for each phase, burn-in length - 100,000 iterations


# Sample data set

# Migration PL-DE
list(n = 14, N = 25, MR = c(3.6538, 2.1421, 2.3050, 2.6049, 2.3845, 2.2346, 2.1410, 2.3537,
2.4580, 2.6273, 2.6409, 2.7464, 3.6478), MRPT = c(1.5798, 0.9838, 0.6771, 0.5451, 0.5068,
0.4952, 0.4375, 0.3538, 0.2407, 0.2038, 0.1993, 0.2124, 0.3797, 0.3689, 0.4244, 0.6157,
0.7885, 1.1580, 1.0997, 1.4975, 3.0760, 3.3894, 3.6934, 2.6959), MRES = c(0.4469, 0.2750,
0.2325, 0.2163, 0.2009, 0.1945, 0.2052, 0.2070, 0.1568, 0.1322, 0.1426, 0.1496, 0.1785,
0.1737, 0.1859, 0.2049, 0.2076, 0.2189, 0.2411, 0.2497, 0.2539, 0.2921, 0.3184, 0.3081), MRIB
= c(0.6728, 0.4195, 0.3243, 0.2841, 0.2640, 0.2566, 0.2532, 0.2374, 0.1742, 0.1470, 0.1544,
0.1626, 0.2200, 0.2139, 0.2348, 0.2886, 0.3255, 0.4096, 0.4157, 0.5031, 0.8270, 0.9213,
1.0044, 0.7938), Dummy = c(0, 0, 0, 0, 0, 0, 0, 0, 0, 0, 0, 0, 0, 0, 0, 0, 0, 0, 0, 0, 0,
0, 0)) # MRPT, MRES and MRIB are migration rates from Portugal, Spain, and both Iberian
countries together, lagged 18 years. Dummy removes atypical observation from 1984 for Portugal

# Inits
list(c = c(0,0,0,0), alpha = c(NA,0.25,0.25,0.25), beta = c(NA,0,0), phi = c(0,0,0,0), tau =
c(10,10,10,10), mod = 3)
```

B.3b. Capturing discontinuity in trends: model selection for extended AR(1)

```
model {
                                        # Model probabilities
mod ~ dcat(p[])                         # categorical distribution over the model space, phase 2
p[1] <- 0.5; p[2] <- 0.25; p[3] <- 0.125; p[4] <- 0.125 # Occam's razor prior

# Artificial initial values for t = 1, models M[2]: AR-ARCH, M[3]: AR-GARCH and M[4]: AR-SV
t12 ~ dgamma(1,0.1); tau[1,2] <- t12; sigma[1,2] <- 1/t12; e[1,2] ~ dnorm(0,tau[1,2]) # M[2]
t13 ~ dgamma(1,0.1); tau[1,3] <- t13; sigma[1,3] <- 1/t13; e[1,3] ~ dnorm(0,tau[1,3]) # M[3]
t14 ~ dnorm(0,1); tmp[1,4] <- t14 # M[4]

# Priors for parameters - phase 1
for (i in 1:4) { c[i] ~ dnorm(0,0.0001); phi[i] ~ dnorm(0.5,1) } # main-level constants and AR
for (i in 2:4) { k[i] ~ dgamma(1,0.1) }                          # 2nd-level constants
alpha ~ dgamma(10,20); beta ~ dgamma(10,20); gamma ~ dgamma(10,20)I(0,1) # 2nd-level params. &
delta ~ dgamma(10,20); theta ~ dgamma(0.25,0.25); psi ~ dgamma(10,1) # random-term precisions

# Priors for parameters - phase 2
# for (i in 1:4) { c[i] ~ dnorm(mu.c[i,mod],tau.c[i,mod])
#                  phi[i] ~ dnorm(mu.phi[i,mod],tau.phi[i,mod]) }
# for (i in 2:4) { k[i] ~ dgamma(r.k[i,mod],m.k[i,mod]) }
# alpha ~ dgamma(r.a[mod],m.a[mod])I(0,1); beta ~ dgamma(r.b[mod],m.b[mod])I(0,1)
# gamma ~ dgamma(r.g[mod],m.g[mod])I(0,1); delta ~ dgamma(r.d[mod],m.d[mod])I(0,1)
# theta ~ dgamma(r.theta[mod],m.theta[mod]); psi ~ dgamma(r.psi[mod],m.psi[mod])

# Input for priors and pseudo-priors, phase 2
for (i in 1:4) { mu.c[i,i] <- 0; tau.c[i,i] <- 0.0001; mu.phi[i,i] <- 0.5; tau.phi[i,i] <- 1 }
for (i in 2:4) { r.k[i,i] <- 1; m.k[i,i] <- 0.1 }
r.theta[1] <- 0.25; m.theta[1] <- 0.25; r.psi[4] <- 10; m.psi[4] <- 1
r.a[3] <- 10; r.b[3] <- 10; r.g[4] <- 10; r.d[2] <- 10
m.a[3] <- 20; m.b[3] <- 20; m.g[4] <- 20; m.d[2] <- 20
... # For pseudo-priors, structure analogous to the one presented in Code B.1 (omitted)

# Definitions of data
for (t in 1:n) { for (i in 1:4) { m[t,i] <- log(MR[t]) }
                 mav[t] <- log(MR[t]) }

# Model update
for (t in 2:n) { for (i in 1:4) { mu[t,i] <- c[i] + phi[i] * m[t-1,i]
                                  m[t,i] ~ dnorm(mu[t,i], tau[t,i]) } # phase 1: estimation
                 e[t,2] <- m[t,2] - mu[t,2]; e[t,3] <- m[t,3] - mu[t,3]; tau[t,1] <- theta
                 sigma[t,2] <- k[2] + delta * pow(e[t-1,2],2); tau[t,2] <- 1/sigma[t,2]
                 sigma[t,3] <- k[3] + alpha * pow(e[t-1,3],2) + beta * sigma[t-1,3]
                 tau[t,3] <- 1/sigma[t,3] ; z[t]~dnorm(0,psi); tau[t,4] <- k[4]*exp(-tmp[t,4])
                 logvar[t,4] <- gamma * tmp[t-1,4]; tmp[t,4] <- logvar[t,4] + z[t] }
#                mav[t] ~ dnorm(mu[t,mod], tau[t,mod]) } # phase 2: calculation of P(M|x) only
for (i in 1:4) { m.new[n,i] <- m[n,i]; tau.new[n,i] <- tau[n,i] }
e.new[n,2] <- e[n,2]; e.new[n,3] <- e[n,3]

# Forecasts for t = n+1 … N
for (t in n+1:N) { for (i in 1:4) { mu.new[t,i] <- c[i] + phi[i] * m.new[t-1,i]
                                    m.new[t,i] ~ dnorm(mu.new[t,i], tau.new[t,i]) }
                   e.new[t,2] <- m.new[t,2]-mu.new[t,2]; e.new[t,3] <- m.new[t,3]-mu.new[t,3]
                   tau.new[t,1] <- theta; tau.new[t,2] <- 1/sigma[t,2]
                   sigma[t,2] <- k[2]+delta*pow(e.new[t-1,2],2); tau.new[t,3] <- 1/sigma[t,3]
                   sigma[t,3] <- k[3]+alpha*pow(e.new[t-1,3],2) + beta * sigma[t-1,3]
                   logvar[t,4] <- gamma * tmp[t-1,4]; tmp[t,4] <- logvar[t,4] + z[t]
                   z[t] ~ dnorm(0,psi); tau.new[t,4] <- k[4] * exp(-tmp[t,4]) }
# for (i in 1:4) { prob[i] <- step(mod-i)-step(mod-i-1) } # posterior p(M|x), phase 2
}

# Phase 1 - estimation and forecasting, phase 2 - calculation of P(M|x) only
# 250,000 iterations in the MCMC sample for each phase, burn-in length - 100,000 iterations


# Sample data set

# Migration PL-DE
list(n = 20, N = 31, MR = c(2.4108, 2.8135, 4.2005, 8.1751, 11.8977, 7.8165, 3.7151, 3.6538,
2.1421, 2.3050, 2.6049, 2.3845, 2.2346, 2.1410, 2.3537, 2.4580, 2.6273, 2.6409, 2.7464,
3.6478))

# Inits
list(c = c(0,0,0,0), phi = c(0.5,0.5,0.5,0.5), k = c(NA,10,10,10), alpha2 = 0.5, alpha3 = 0.5,
beta = 0.5, gamma = 0.5, theta = 1, t12 = 10, t13 = 10, t14 = 0, psi = 10, mod = 1)
```

Annex C
Selected Results of Presented Migration Forecasts

Annex C supplements the empirical results of Bayesian forecasts from Chapters 5, 6, 7 and 8 (Tables C.1, C.2, C.3a, C.3b, C.4, C.5, C.6 and C.7).

Table C.1 Summaries of posterior distributions of the parameters of general VAR(1) models from Chapter 6: median, 2.5 and 97.5% quantiles (estimated by MCMC)

Migration flow	From Italy to Germany			From Poland to Germany			From Switzerland to Germany		
Parameters	2.5%	Median	97.5%	2.5%	Median	97.5%	2.5%	Median	97.5%
Elements of vector **c**: c_1	–2.94	–0.33	2.08	–4.32	–0.52	2.91	–0.12	0.15	0.43
c_2	–2.65	0.00	2.46	–3.89	–0.14	3.65	–0.27	–0.02	0.24
c_3	–2.00	0.37	2.75	–3.82	0.58	4.40	–0.41	0.32	1.04
Elements of matrix **A**: α_{11}	**0.40**	**1.05**	**1.68**	–0.48	0.60	1.70	–0.19	0.69	1.58
α_{12}	–1.37	0.49	2.36	–1.35	0.05	1.51	–0.62	0.41	1.45
α_{13}	–0.89	0.13	1.24	–0.67	0.34	1.40	–0.17	–0.05	0.08
α_{21}	–0.19	0.00	0.19	–0.19	0.00	0.19	–0.20	0.00	0.19
α_{22}	–0.85	0.98	2.82	–0.37	1.05	2.40	–0.13	0.87	1.86
α_{23}	–1.07	0.00	1.16	–1.06	0.02	1.11	–0.11	0.00	0.12
α_{31}	–0.39	0.07	0.52	–0.54	0.04	0.61	–0.45	0.20	0.84
α_{32}	–2.82	0.40	3.58	–1.85	–0.15	1.78	–2.11	0.73	3.48
α_{33}	–0.18	0.84	1.86	–0.15	0.84	1.97	**0.49**	**0.82**	**1.16**
Elements of matrix **T**: τ_{11}	7.56	15.33	27.25	2.96	7.08	14.12	51.09	105.20	189.60
τ_{22}	9.46	19.07	33.64	3.60	8.54	16.76	57.81	118.00	210.40
τ_{33}	8.80	17.74	31.42	3.19	7.61	15.11	4.61	9.43	17.07
$\tau_{12}=\tau_{21}$	–7.44	0.32	8.24	–4.28	0.07	4.48	–44.89	5.84	59.29
$\tau_{13}=\tau_{31}$	–8.37	–0.67	6.82	–4.12	0.04	4.25	–21.91	–3.83	12.35
$\tau_{23}=\tau_{32}$	–7.99	0.34	8.85	–4.44	0.08	4.65	–20.94	–4.21	10.71

Migration flow	From Germany to Italy			From Germany to Poland			From Germany to Switzerland		
Parameters	2.5%	Median	97.5%	2.5%	Median	97.5%	2.5%	Median	97.5%
Elements of vector **c**: c_1	–1.79	–0.49	0.80	–1.22	0.77	2.70	**1.69**	**2.32**	**3.22**
c_2	–0.99	0.00	0.97	–1.88	–0.19	1.54	–2.28	–1.12	0.87
c_3	**0.04**	**1.19**	**2.26**	–0.91	1.06	2.87	–0.46	0.17	1.60
Elements of matrix **A**: α_{11}	–0.48	0.78	1.97	–0.11	0.74	1.63	**0.69**	**1.14**	**1.56**
α_{12}	–1.45	0.48	2.41	–0.52	0.47	1.46	–1.55	0.14	1.85
α_{13}	–0.35	0.16	0.69	–0.85	–0.13	0.65	–0.36	0.07	0.52
α_{21}	–0.20	0.00	0.19	–0.19	0.00	0.19	–0.16	0.01	0.19
α_{22}	–0.91	1.01	2.95	**0.18**	**0.97**	**1.76**	–0.61	0.83	2.28
α_{23}	–0.47	0.00	0.48	–0.49	0.10	0.66	–0.39	–0.01	0.40
α_{31}	–0.65	–0.04	0.56	–0.37	0.16	0.67	–0.29	0.14	0.54
α_{32}	–2.92	0.79	4.45	–0.89	0.06	0.95	–1.51	0.90	3.18
α_{33}	–0.05	0.45	0.95	–0.09	0.55	1.18	**0.03**	**0.52**	**0.99**
Elements of matrix **T**: τ_{11}	16.54	39.56	78.20	11.00	27.04	54.50	28.49	70.90	142.70
τ_{22}	18.72	44.40	86.82	16.94	41.06	81.98	34.25	82.02	162.00
τ_{33}	16.26	38.32	75.38	14.60	35.93	72.43	26.71	65.35	131.80
$\tau_{12}=\tau_{21}$	–21.77	1.23	25.06	–19.80	–0.44	18.63	–41.75	1.27	45.29
$\tau_{13}=\tau_{31}$	–21.11	0.52	22.55	–19.77	–1.08	16.85	–33.04	5.29	46.64
$\tau_{23}=\tau_{32}$	–21.44	1.18	24.60	–21.74	0.44	22.99	–41.64	–0.07	41.99

Note: **Boldface** denotes 95 per cent credible intervals for the estimates of c and α, which do not cover zero.
Source: own elaboration in WinBUGS

Table C.2 Summaries of predictive distributions of emigration rates forecasted for 2005–2015 for models from Chapter 5: median, 10 and 90% quantiles (estimated by MCMC)

Model / Year	Migration Italy-Germany, $m_{IT\text{-}DE}$			Migration Poland-Germany, $m_{PL\text{-}DE}$			Migration Switzerland-Germany, $m_{CH\text{-}DE}$		
	10%	Median	90%	10%	Median	90%	10%	Median	90%
M_1: oscillations around the constant									
2005	* 0.41	* 0.65	* 1.04	**1.77**	**2.57**	**3.72**	0.94	1.12	1.35
2006	* 0.41	* 0.65	* 1.03	**1.78**	**2.57**	**3.72**	0.94	1.12	1.35
2007	* 0.41	* 0.65	* 1.04	**1.78**	**2.57**	**3.72**	0.94	1.12	1.35
2008	* 0.41	* 0.65	* 1.03	**1.78**	**2.57**	**3.72**	0.94	1.12	1.35
2009	* 0.41	* 0.65	* 1.04	**1.78**	**2.57**	**3.71**	0.94	1.12	1.35
2010	* 0.41	* 0.65	* 1.04	**1.78**	**2.57**	**3.71**	0.94	1.12	1.35
2011	* 0.41	* 0.65	* 1.04	**1.78**	**2.57**	**3.72**	0.94	1.12	1.35
2012	* 0.41	* 0.65	* 1.04	**1.78**	**2.57**	**3.72**	0.94	1.12	1.35
2013	* 0.41	* 0.65	* 1.04	**1.78**	**2.57**	**3.72**	0.94	1.12	1.35
2014	* 0.41	* 0.65	* 1.04	**1.78**	**2.57**	**3.72**	0.94	1.12	1.35
2015	* 0.41	* 0.65	* 1.04	**1.78**	**2.57**	**3.72**	0.94	1.12	1.35
M_2: random walk with drift									
2005	**0.24**	**0.35**	**0.52**	2.49	3.64	5.33	**1.05**	**1.25**	**1.49**
2006	**0.20**	**0.34**	**0.59**	2.09	3.64	6.35	**0.99**	**1.27**	**1.64**
2007	**0.17**	**0.33**	**0.65**	1.80	3.63	7.35	**0.94**	**1.29**	**1.77**
2008	**0.14**	**0.32**	**0.71**	1.58	3.64	8.39	**0.90**	**1.31**	**1.90**
2009	**0.12**	**0.31**	**0.77**	1.39	3.64	9.45	**0.86**	**1.32**	**2.03**
2010	**0.11**	**0.29**	**0.82**	1.23	3.62	10.71	**0.83**	**1.34**	**2.17**
2011	**0.09**	**0.28**	**0.88**	1.09	3.61	12.04	**0.80**	**1.36**	**2.30**
2012	**0.08**	**0.27**	**0.94**	0.97	3.62	13.50	**0.78**	**1.38**	**2.45**
2013	**0.07**	**0.26**	**1.00**	0.87	3.62	15.09	**0.75**	**1.40**	**2.60**
2014	**0.06**	**0.25**	**1.05**	0.77	3.61	16.83	**0.73**	**1.42**	**2.76**
2015	**0.05**	**0.24**	**1.12**	0.69	3.61	18.80	**0.71**	**1.44**	**2.93**
M_3: autoregression AR(1)									
2005	0.22	0.35	0.56	1.99	3.00	4.53	1.00	1.21	1.45
2006	0.14	0.34	0.65	1.86	2.85	4.72	0.96	1.20	1.54
2007	0.08	0.32	0.72	1.76	2.75	4.79	0.94	1.19	1.64
2008	0.03	0.31	0.77	1.71	2.71	4.88	0.92	1.19	1.73
2009	0.01	0.29	0.81	1.67	2.68	4.93	0.91	1.19	1.85
2010	0.00	0.28	0.84	1.64	2.67	5.02	0.90	1.19	1.99
2011	0.00	0.27	0.87	1.61	2.65	5.04	0.89	1.19	2.15
2012	0.00	0.26	0.90	1.60	2.64	5.09	0.89	1.19	2.35
2013	0.00	0.25	0.92	1.58	2.64	5.11	0.88	1.19	2.59
2014	0.00	0.24	0.94	1.57	2.64	5.17	0.88	1.19	2.90
2015	0.00	0.23	0.96	1.56	2.63	5.17	0.87	1.19	3.31
M_4: moving average MA(1)									
2005	* 0.34	* 0.51	* 0.79	1.80	2.80	4.21	0.97	1.16	1.39
2006	* 0.39	* 0.65	* 1.07	1.72	2.60	3.95	0.92	1.12	1.38
2007	* 0.39	* 0.65	* 1.07	1.73	2.60	3.95	0.92	1.12	1.38
2008	* 0.39	* 0.65	* 1.07	1.72	2.60	3.94	0.92	1.12	1.38
2009	* 0.39	* 0.65	* 1.07	1.72	2.60	3.95	0.92	1.12	1.38
2010	* 0.39	* 0.65	* 1.07	1.72	2.60	3.94	0.92	1.12	1.37
2011	* 0.39	* 0.65	* 1.07	1.72	2.60	3.94	0.92	1.12	1.38
2012	* 0.39	* 0.65	* 1.07	1.72	2.60	3.95	0.92	1.12	1.38
2013	* 0.39	* 0.65	* 1.07	1.72	2.61	3.96	0.92	1.12	1.38
2014	* 0.39	* 0.65	* 1.07	1.72	2.60	3.95	0.92	1.12	1.38
2015	* 0.39	* 0.65	* 1.07	1.72	2.60	3.96	0.92	1.12	1.38
M_5: ARMA(1,1) process									
2005	* 0.23	* 0.35	* 0.55	* 1.97	* 3.09	* 4.81	* 1.01	* 1.22	* 1.48
2006	* 0.14	* 0.35	* 0.73	* 1.80	* 2.84	* 4.99	* 0.94	* 1.20	* 1.61
2007	* 0.07	* 0.35	* 0.83	* 1.70	* 2.75	* 5.06	* 0.91	* 1.19	* 1.72
2008	* 0.03	* 0.34	* 0.91	* 1.67	* 2.72	* 5.17	* 0.89	* 1.19	* 1.84
2009	* 0.01	* 0.34	* 0.97	* 1.61	* 2.69	* 5.20	* 0.88	* 1.19	* 1.99
2010	* 0.00	* 0.33	* 1.02	* 1.60	* 2.68	* 5.29	* 0.87	* 1.19	* 2.16
2011	* 0.00	* 0.33	* 1.06	* 1.57	* 2.67	* 5.31	* 0.86	* 1.18	* 2.37
2012	* 0.00	* 0.32	* 1.10	* 1.56	* 2.67	* 5.34	* 0.85	* 1.18	* 2.63
2013	* 0.00	* 0.32	* 1.13	* 1.54	* 2.66	* 5.37	* 0.85	* 1.18	* 2.94
2014	* 0.00	* 0.32	* 1.16	* 1.54	* 2.65	* 5.46	* 0.84	* 1.18	* 3.34
2015	* 0.00	* 0.31	* 1.19	* 1.53	* 2.65	* 5.46	* 0.84	* 1.18	* 3.85

Table C.2 (continued)

Model / Year	Migration Italy-Germany, m_{IT-DE}			Migration Poland-Germany, m_{PL-DE}			Migration Switzerland-Germany, m_{CH-DE}		
	10%	Median	90%	10%	Median	90%	10%	Median	90%
M_{avg}(A): averaged forecast, uniform prior									
2005	0.24	0.36	0.54	1.92	2.95	4.55	1.00	1.21	1.46
2006	0.18	0.35	0.64	1.82	2.81	4.77	0.96	1.20	1.56
2007	0.14	0.34	0.71	1.76	2.75	4.95	0.93	1.20	1.65
2008	0.11	0.32	0.77	1.71	2.73	5.12	0.91	1.20	1.75
2009	0.09	0.31	0.82	1.67	2.71	5.28	0.90	1.20	1.85
2010	0.07	0.30	0.87	1.64	2.70	5.45	0.88	1.20	1.96
2011	0.05	0.29	0.91	1.61	2.68	5.60	0.87	1.20	2.07
2012	0.04	0.29	0.95	1.58	2.68	5.74	0.86	1.20	2.18
2013	0.03	0.28	1.00	1.55	2.67	5.86	0.85	1.20	2.30
2014	0.02	0.27	1.04	1.52	2.66	6.00	0.84	1.20	2.43
2015	0.02	0.26	1.09	1.50	2.66	6.13	0.83	1.20	2.56
M_{avg}(B): averaged forecast, 'Occam's razor' prior									
2005	0.24	0.36	0.53	1.92	2.94	4.57	1.00	1.21	1.46
2006	0.19	0.35	0.62	1.83	2.83	4.85	0.96	1.21	1.57
2007	0.16	0.33	0.68	1.77	2.77	5.09	0.94	1.21	1.67
2008	0.13	0.32	0.74	1.72	2.75	5.33	0.92	1.21	1.77
2009	0.11	0.31	0.79	1.67	2.72	5.57	0.90	1.21	1.87
2010	0.09	0.30	0.84	1.63	2.71	5.80	0.88	1.21	1.98
2011	0.07	0.29	0.89	1.59	2.70	6.02	0.87	1.21	2.09
2012	0.06	0.28	0.94	1.55	2.69	6.25	0.85	1.21	2.20
2013	0.05	0.27	1.00	1.51	2.68	6.49	0.84	1.22	2.32
2014	0.04	0.26	1.04	1.48	2.67	6.75	0.83	1.22	2.45
2015	0.04	0.25	1.09	1.45	2.67	7.00	0.81	1.22	2.58

Model / Year	Migration Germany-Italy, m_{DE-IT}			Migration Germany-Poland, m_{DE-PL}			Migration Germany-Switzerland, m_{DE-CH}		
	10%	Median	90%	10%	Median	90%	10%	Median	90%
M_1: oscillations around the constant									
2005	0.34	0.45	0.59	0.72	1.00	1.39	* 0.08	* 0.13	* 0.22
2006	0.34	0.45	0.59	0.72	1.00	1.38	* 0.08	* 0.13	* 0.22
2007	0.34	0.45	0.59	0.72	1.00	1.38	* 0.08	* 0.13	* 0.22
2008	0.34	0.45	0.60	0.72	1.00	1.38	* 0.08	* 0.13	* 0.22
2009	0.34	0.45	0.59	0.72	1.00	1.38	* 0.08	* 0.13	* 0.22
2010	0.34	0.45	0.59	0.72	1.00	1.39	* 0.08	* 0.13	* 0.22
2011	0.34	0.45	0.60	0.72	1.00	1.38	* 0.08	* 0.13	* 0.22
2012	0.34	0.45	0.59	0.72	1.00	1.38	* 0.08	* 0.13	* 0.22
2013	0.34	0.45	0.59	0.72	1.00	1.38	* 0.08	* 0.13	* 0.22
2014	0.34	0.45	0.60	0.72	1.00	1.38	* 0.08	* 0.13	* 0.22
2015	0.34	0.45	0.60	0.72	1.00	1.38	* 0.08	* 0.13	* 0.22
M_2: random walk with drift									
2005	**0.33**	**0.44**	**0.58**	**0.91**	**1.25**	**1.71**	**0.16**	**0.22**	**0.30**
2006	**0.29**	**0.43**	**0.65**	**0.78**	**1.24**	**1.96**	**0.14**	**0.23**	**0.36**
2007	**0.25**	**0.43**	**0.72**	**0.68**	**1.22**	**2.20**	**0.13**	**0.24**	**0.43**
2008	**0.23**	**0.42**	**0.79**	**0.60**	**1.21**	**2.42**	**0.13**	**0.25**	**0.51**
2009	**0.21**	**0.42**	**0.86**	**0.54**	**1.19**	**2.65**	**0.12**	**0.27**	**0.59**
2010	**0.19**	**0.42**	**0.94**	**0.48**	**1.18**	**2.90**	**0.11**	**0.28**	**0.69**
2011	**0.17**	**0.41**	**1.02**	**0.43**	**1.16**	**3.16**	**0.11**	**0.29**	**0.80**
2012	**0.15**	**0.41**	**1.10**	**0.39**	**1.15**	**3.42**	**0.10**	**0.31**	**0.93**
2013	**0.14**	**0.41**	**1.18**	**0.35**	**1.14**	**3.71**	**0.10**	**0.33**	**1.08**
2014	**0.13**	**0.40**	**1.27**	**0.31**	**1.13**	**4.03**	**0.10**	**0.34**	**1.25**
2015	**0.12**	**0.40**	**1.36**	**0.28**	**1.12**	**4.36**	**0.09**	**0.36**	**1.44**
M_3: autoregression AR(1)									
2005	0.33	0.44	0.59	0.82	1.14	1.58	0.15	0.23	0.33
2006	0.30	0.44	0.63	0.74	1.09	1.65	0.14	0.25	0.55
2007	0.28	0.44	0.67	0.70	1.05	1.70	0.13	0.29	1.17
2008	0.25	0.44	0.70	0.67	1.04	1.73	0.12	0.33	3.65
2009	0.23	0.44	0.74	0.65	1.03	1.76	0.12	0.38	20.17
2010	0.21	0.44	0.77	0.64	1.02	1.79	0.11	0.45	265.60
2011	0.18	0.44	0.81	0.62	1.01	1.82	0.11	0.53	∞
2012	0.15	0.44	0.84	0.61	1.01	1.84	0.11	0.65	∞
2013	0.13	0.44	0.88	0.60	1.01	1.85	0.10	0.80	∞
2014	0.10	0.44	0.92	0.60	1.00	1.87	0.10	1.01	∞
2015	0.08	0.44	0.97	0.59	1.00	1.89	0.10	1.30	∞

Table C.2 (continued)

Model / Year	Migration Germany-Italy, m_{DE-IT}			Migration Germany-Poland, m_{DE-PL}			Migration Germany-Switzerland, m_{DE-CH}		
	10%	Median	90%	10%	Median	90%	10%	Median	90%
M_4: moving average MA(1)									
2005	0.34	0.46	0.61	0.77	1.09	1.52	* 0.11	* 0.17	* 0.26
2006	0.33	0.45	0.61	0.70	1.01	1.46	* 0.08	* 0.13	* 0.23
2007	0.33	0.45	0.61	0.70	1.00	1.45	* 0.08	* 0.13	* 0.23
2008	0.33	0.45	0.61	0.70	1.00	1.45	* 0.08	* 0.13	* 0.23
2009	0.33	0.45	0.62	0.70	1.00	1.45	* 0.08	* 0.13	* 0.23
2010	0.33	0.45	0.61	0.70	1.01	1.45	* 0.08	* 0.13	* 0.23
2011	0.33	0.45	0.61	0.70	1.00	1.45	* 0.08	* 0.13	* 0.23
2012	0.33	0.45	0.61	0.70	1.00	1.45	* 0.08	* 0.13	* 0.23
2013	0.33	0.45	0.61	0.70	1.00	1.45	* 0.08	* 0.13	* 0.23
2014	0.33	0.45	0.62	0.70	1.01	1.45	* 0.08	* 0.13	* 0.23
2015	0.33	0.45	0.62	0.70	1.01	1.45	* 0.08	* 0.13	* 0.23
M_5: ARMA(1,1) process									
2005	* 0.34	* 0.45	* 0.61	* 0.80	* 1.14	* 1.61	* 0.16	* 0.23	* 0.35
2006	* 0.30	* 0.45	* 0.67	* 0.72	* 1.06	* 1.66	* 0.13	* 0.26	* 0.66
2007	* 0.28	* 0.45	* 0.72	* 0.68	* 1.04	* 1.68	* 0.12	* 0.29	* 1.69
2008	* 0.25	* 0.45	* 0.76	* 0.66	* 1.02	* 1.70	* 0.11	* 0.33	* 7.39
2009	* 0.23	* 0.44	* 0.81	* 0.64	* 1.01	* 1.72	* 0.10	* 0.39	* 76.17
2010	* 0.21	* 0.44	* 0.87	* 0.63	* 1.01	* 1.72	* 0.10	* 0.45	* ∞
2011	* 0.18	* 0.44	* 0.93	* 0.62	* 1.00	* 1.73	* 0.09	* 0.54	* ∞
2012	* 0.16	* 0.44	* 1.01	* 0.61	* 1.00	* 1.74	* 0.09	* 0.64	* ∞
2013	* 0.14	* 0.44	* 1.08	* 0.61	* 1.00	* 1.75	* 0.09	* 0.79	* ∞
2014	* 0.12	* 0.44	* 1.18	* 0.60	* 0.99	* 1.75	* 0.08	* 0.98	* ∞
2015	* 0.09	* 0.44	* 1.27	* 0.59	* 0.99	* 1.76	* 0.08	* 1.24	* ∞
M_{avg}(A): averaged forecast, uniform prior									
2005	0.33	0.44	0.59	0.81	1.15	1.61	0.18	0.22	0.27
2006	0.31	0.44	0.63	0.74	1.10	1.73	0.17	0.23	0.33
2007	0.29	0.44	0.65	0.70	1.08	1.82	0.16	0.25	0.40
2008	0.28	0.44	0.68	0.66	1.06	1.91	0.16	0.26	0.49
2009	0.26	0.44	0.70	0.63	1.05	2.00	0.16	0.28	0.61
2010	0.25	0.44	0.72	0.60	1.04	2.09	0.16	0.29	0.76
2011	0.23	0.44	0.75	0.58	1.03	2.18	0.15	0.31	0.97
2012	0.22	0.44	0.77	0.55	1.03	2.28	0.15	0.33	1.26
2013	0.21	0.44	0.80	0.53	1.02	2.38	0.15	0.35	1.69
2014	0.20	0.44	0.82	0.50	1.02	2.47	0.15	0.37	2.40
2015	0.18	0.44	0.85	0.48	1.02	2.58	0.15	0.39	3.62
M_{avg}(B): averaged forecast, 'Occam's razor' prior									
2005	0.33	0.44	0.59	0.82	1.16	1.62	0.18	0.22	0.27
2006	0.31	0.44	0.62	0.74	1.11	1.76	0.17	0.23	0.32
2007	0.29	0.44	0.65	0.70	1.09	1.89	0.16	0.24	0.37
2008	0.28	0.44	0.68	0.66	1.07	2.01	0.16	0.26	0.43
2009	0.26	0.44	0.70	0.62	1.06	2.13	0.16	0.27	0.49
2010	0.25	0.44	0.73	0.59	1.05	2.25	0.16	0.29	0.57
2011	0.23	0.44	0.75	0.55	1.04	2.37	0.16	0.30	0.66
2012	0.22	0.44	0.77	0.52	1.04	2.49	0.15	0.32	0.76
2013	0.21	0.44	0.80	0.49	1.03	2.63	0.15	0.34	0.88
2014	0.20	0.44	0.83	0.46	1.03	2.77	0.15	0.36	1.01
2015	0.19	0.44	0.86	0.43	1.02	2.92	0.15	0.38	1.16

Notes: **Boldface** denotes values yielded by models selected on the basis of the posterior odds criteria (Table 4.2), in this case, the same under both types of priors. (A) and (B) stand for uniform and 'Occam's razor' priors, respectively.

Asterisks (*) indicate forecasts yielded by models hardly probable a posteriori, with $p(M_i|x) < 0.05$ under the 'Occam's razor' prior. Grey shading denotes implausible values, exceeding over 5 times maximum levels observed in samples. The (1985–)1991–2004 maxima are: $\max(m_{IT-DE}) = 0.89$, $\max(m_{DE-IT}) = 0.50$, $\max(m_{PL-DE}) = 3.72$, $\max(m_{DE-PL}) = 1.48$, $\max(m_{CH-DE}) = 1.28$ and $\max(m_{DE-CH}) = 0.20$ '∞' refers to nearly impossible values (rates of over 1,000 per 1,000 inhabitants).

NB: Observations for 2005 are: $m_{IT-DE} = 0.35$, $m_{DE-IT} = 0.35$, $m_{PL-DE} = 4.17$, $m_{DE-pl} = 1.28$, $m_{CH-DE} = 1.26$ and $m_{DE-CH} = 0.22$.

Source: own elaboration in WinBUGS, data for 2005: Eurostat/NewCronos

Table C.3a Summaries of predictive distributions of emigration rates for 2005–2015, general VAR(1) and reduced AR(1) models: median, 10 and 90% quantiles (estimated by MCMC)

Model	Migration Italy-Germany, m_{IT-DE}			Migration Poland-Germany, m_{PL-DE}			Migration Switzerland-Germany, m_{CH-DE}		
Year	10%	Median	90%	10%	Median	90%	10%	Median	90%
General VAR(1) – unconditional forecasts									
2005	0.22	0.33	0.51	1.76	3.47	6.85	1.01	1.18	1.37
2006	0.12	0.30	0.64	1.36	3.60	12.09	0.94	1.16	1.45
2007	0.04	0.26	0.93	0.88	3.63	29.25	0.86	1.15	1.54
2008	0.01	0.23	1.95	0.44	3.74	132.82	0.78	1.14	1.67
2009	0.00	0.19	8.57	0.12	3.79	∞	0.68	1.14	1.85
2010	0.00	0.15	152.63	0.01	3.90	∞	0.58	1.14	2.12
2011	0.00	0.12	∞	0.00	3.97	∞	0.46	1.14	2.53
2012	0.00	0.09	∞	0.00	4.12	∞	0.35	1.14	3.20
2013	0.00	0.06	∞	0.00	4.17	∞	0.23	1.14	4.36
2014	0.00	0.04	∞	0.00	4.39	∞	0.14	1.14	6.58
2015	0.00	0.02	∞	0.00	4.47	∞	0.07	1.14	11.35
Reduced AR(1)									
2005	0.23	0.34	0.51	1.83	3.15	5.46	1.06	1.21	1.39
2006	0.14	0.31	0.58	1.65	3.02	6.49	1.02	1.20	1.46
2007	0.08	0.29	0.63	1.50	2.91	7.60	1.00	1.20	1.52
2008	0.03	0.26	0.67	1.43	2.87	9.10	0.99	1.19	1.58
2009	0.01	0.24	0.71	1.35	2.84	10.98	0.98	1.19	1.65
2010	0.00	0.22	0.74	1.30	2.82	14.01	0.97	1.19	1.73
2011	0.00	0.20	0.76	1.25	2.80	17.89	0.97	1.19	1.82
2012	0.00	0.18	0.78	1.21	2.79	24.66	0.96	1.19	1.91
2013	0.00	0.16	0.80	1.17	2.78	34.19	0.96	1.19	2.03
2014	0.00	0.14	0.81	1.14	2.78	53.25	0.95	1.19	2.16
2015	0.00	0.13	0.83	1.11	2.77	82.27	0.95	1.19	2.31

Model	Migration Germany-Italy, m_{DE-IT}			Migration Germany-Poland, m_{DE-PL}			Migration Germany-Switzerland, m_{DE-CH}		
Year	10%	Median	90%	10%	Median	90%	10%	Median	90%
General VAR(1) – unconditional forecasts									
2005	0.36	0.46	0.59	0.87	1.31	2.00	0.18	0.23	0.30
2006	0.31	0.47	0.75	0.79	1.38	3.28	0.17	0.27	0.47
2007	0.25	0.48	1.10	0.72	1.46	6.86	0.16	0.33	0.95
2008	0.16	0.49	2.28	0.67	1.55	21.56	0.14	0.40	2.86
2009	0.08	0.49	9.32	0.62	1.64	123.10	0.11	0.51	16.66
2010	0.02	0.50	142.31	0.56	1.74	∞	0.09	0.67	290.32
2011	0.00	0.50	∞	0.51	1.84	∞	0.05	0.91	∞
2012	0.00	0.51	∞	0.45	1.95	∞	0.03	1.29	∞
2013	0.00	0.51	∞	0.38	2.06	∞	0.01	1.93	∞
2014	0.00	0.51	∞	0.32	2.17	∞	0.00	3.06	∞
2015	0.00	0.51	∞	0.25	2.30	∞	0.00	5.19	∞
Reduced AR(1)									
2005	0.36	0.44	0.54	0.88	1.15	1.51	0.19	0.23	0.28
2006	0.33	0.44	0.58	0.79	1.09	1.55	0.19	0.26	0.37
2007	0.30	0.44	0.60	0.75	1.06	1.58	0.19	0.30	0.55
2008	0.28	0.44	0.63	0.72	1.04	1.60	0.19	0.35	0.93
2009	0.25	0.44	0.66	0.70	1.02	1.61	0.19	0.42	1.85
2010	0.22	0.44	0.69	0.68	1.02	1.63	0.19	0.52	4.61
2011	0.18	0.44	0.72	0.67	1.01	1.64	0.20	0.65	15.55
2012	0.15	0.44	0.75	0.66	1.00	1.65	0.20	0.85	77.94
2013	0.11	0.44	0.79	0.65	1.00	1.66	0.20	1.13	659.84
2014	0.08	0.44	0.83	0.64	1.00	1.67	0.20	1.57	∞
2015	0.05	0.44	0.88	0.63	1.00	1.68	0.20	2.28	∞

Note: grey shading denotes implausible values, exceeding over 5 times the maximum levels observed in the sample. The (1985–)1991–2004 maxima are: $\max(m_{IT-DE}) = 0.89$, $\max(m_{DE-IT}) = 0.50$, $\max(m_{PL-DE}) = 3.72$, $\max(m_{DE-PL}) = 1.48$, $\max(m_{CH-DE}) = 1.28$, and $\max(m_{DE-CH}) = 0.20$. '∞' refers to nearly impossible values (rates of over 1,000 per 1,000 inhabitants).

NB: Observations for 2005 are: $m_{IT-DE} = 0.35$, $m_{DE-IT} = 0.35$, $m_{PL-DE} = 4.17$, $m_{DE-PL} = 1.28$, $m_{CH-DE} = 1.26$ and $m_{DE-CH} = 0.22$.

Source: own elaboration in WinBUGS, data for 2005: Eurostat/NewCronos

Table C.3b Summaries of predictive distributions of emigration rates for 2005–2015, conditional VAR(1)-based forecasts: median, 10 and 90% quantiles (estimated by MCMC), with input

Model	Migration Italy-Germany, m_{IT-DE}			Migration Poland-Germany, m_{PL-DE}			Migration Switzerland-Germany, m_{CH-DE}		
Year	10%	Median	90%	10%	Median	90%	10%	Median	90%
General VAR(1) – conditional forecasts									
2005	0.37	0.37	0.37	1.75	3.47	6.83	1.01	1.18	1.38
2006	0.22	0.33	0.51	1.57	3.56	10.59	0.96	1.16	1.43
2007	0.13	0.30	0.57	1.37	3.59	18.01	0.92	1.14	1.47
2008	0.06	0.26	0.61	1.26	3.68	35.37	0.88	1.13	1.52
2009	0.02	0.23	0.65	1.12	3.76	83.76	0.85	1.13	1.57
2010	0.01	0.20	0.68	1.03	3.87	252.65	0.82	1.13	1.63
2011	0.00	0.17	0.71	0.91	3.98	∞	0.79	1.12	1.69
2012	0.00	0.13	0.75	0.81	4.17	∞	0.75	1.12	1.75
2013	0.00	0.10	0.81	0.68	4.34	∞	0.72	1.12	1.82
2014	0.00	0.07	0.89	0.58	4.54	∞	0.69	1.12	1.90
2015	0.00	0.05	1.02	0.46	4.74	∞	0.65	1.12	1.98
Input trajectories of exogenous variables									
	Y_{DE}/Y_{IT}	$U_{IT}(t)$	Comment	Y_{DE}/Y_{PL}	$U_{PL}(t)$	Comment	Y_{DE}/Y_{CH}	$U_{CH}(t)$	Comment
2005	1.00	7.81	Both trajectories plausible, though extremely optimistic with respect to Italian economic development	2.09	19.79	Both trajectories implausible and inconsistent (extreme growth *vis à vis* doubling of unemployment)	0.86	4.34	Both trajectories are plausible, indicate stability of the two migration determinants
2006	1.00	7.60		1.95	20.47		0.87	4.31	
2007	0.99	7.46		1.86	21.26		0.87	4.28	
2008	0.97	7.06		1.74	22.11		0.87	4.26	
2009	0.96	6.53		1.64	23.20		0.87	4.28	
2010	0.93	5.81		1.50	24.68		0.87	4.30	
2011	0.91	5.18		1.37	26.55		0.87	4.32	
2012	0.88	4.48		1.19	28.56		0.88	4.33	
2013	0.86	3.94		1.07	30.88		0.88	4.37	
2014	0.83	3.26		0.91	33.28		0.88	4.40	
2015	0.80	2.79		0.80	36.56		0.88	4.43	

Model	Migration Germany-Italy, m_{DE-IT}			Migration Germany-Poland, m_{DE-PL}			Migration Germany-Switzerland, m_{DE-CH}		
Year	10%	Median	90%	10%	Median	90%	10%	Median	90%
General VAR(1) – conditional forecasts									
2005	0.36	0.46	0.59	0.86	1.30	1.98	0.18	0.23	0.30
2006	0.33	0.47	0.68	0.80	1.37	3.11	0.18	0.27	0.46
2007	0.31	0.49	0.80	0.76	1.44	5.66	0.17	0.32	0.83
2008	0.29	0.50	0.99	0.74	1.54	12.55	0.17	0.39	1.95
2009	0.27	0.51	1.33	0.73	1.64	35.62	0.16	0.49	6.40
2010	0.25	0.52	2.00	0.71	1.75	140.33	0.16	0.62	34.33
2011	0.23	0.53	3.52	0.70	1.87	828.82	0.15	0.83	364.31
2012	0.21	0.53	7.49	0.69	2.01	∞	0.14	1.14	∞
2013	0.19	0.54	21.33	0.67	2.16	∞	0.13	1.63	∞
2014	0.17	0.55	88.77	0.66	2.34	∞	0.13	2.47	∞
2015	0.15	0.56	632.70	0.65	2.55	∞	0.12	3.94	∞
Input trajectories of exogenous variables									
	Y_{IT}/Y_{DE}	$U_{DE}(t)$	Comment	Y_{PL}/Y_{DE}	$U_{DE}(t)$	Comment	Y_{CH}/Y_{DE}	$U_{DE}(t)$	Comment
2005	1.00	9.23	Both trajectories are plausible, indicate stability of the two migration determinants	0.48	9.67	Both trajectories plausible, assume GDP convergence and difficulties on the German labour market	1.16	9.93	Both trajectories are plausible, although very pessimistic (depict an economic crisis in Germany)
2006	1.01	9.04		0.51	9.82		1.18	10.12	
2007	1.02	8.99		0.53	9.95		1.18	10.35	
2008	1.03	9.03		0.56	10.12		1.20	10.65	
2009	1.04	9.14		0.59	10.29		1.20	11.07	
2010	1.05	9.27		0.62	10.44		1.22	11.58	
2011	1.05	9.41		0.65	10.61		1.23	12.17	
2012	1.07	9.49		0.69	10.77		1.24	12.90	
2013	1.07	9.63		0.73	10.94		1.25	13.71	
2014	1.08	9.69		0.78	11.07		1.26	14.60	
2015	1.08	9.80		0.84	11.21		1.26	15.60	

Note: grey shading denotes implausible values, exceeding over 5 times the maximum sample levels (see Table C.3a). '∞' refers to nearly impossible rates (over 1,000 per 1,000 inhabitants). Plausibility of predictors is discussed in Chapter 6.

Source: own elaboration in WinBUGS, data for 2005, Eurostat/NewCronos

Table C.4 Summaries of predictive distributions of emigration rates forecasted for 2005–2015 for models from Chapter 7: median, 10 and 90% quantiles (estimated by MCMC)

Model	M_6: no analogy (reference) *			M_7: Analogy to PT * ¶			M_8: Analogy to ES *			M_9: Analogy to PT+ES * ¶		
Year	10%	Median	90%	10%	Median	90%	10%	Median	90%	10%	Median	90%
Poland-Germany, m_{PL-DE}												
2005	1.96	2.99	4.41	1.99	2.99	4.42	1.97	2.94	4.31	1.98	2.97	4.37
2006	1.71	2.62	4.05	1.66	2.62	4.15	1.64	2.57	4.01	1.65	2.59	4.08
2007	1.71	2.63	4.05	1.73	2.73	4.37	1.69	2.64	4.12	1.72	2.70	4.25
2008	1.71	2.63	4.06	1.77	2.82	4.57	1.70	2.65	4.13	1.75	2.76	4.39
2009	1.71	2.62	4.06	1.79	2.94	4.96	1.73	2.69	4.20	1.81	2.89	4.70
2010	1.71	2.62	4.04	1.79	2.93	4.88	1.77	2.76	4.34	1.81	2.90	4.72
2011	1.71	2.63	4.06	1.79	3.03	5.28	1.78	2.79	4.41	1.83	3.01	5.03
2012	1.71	2.63	4.06	1.76	3.29	6.47	1.79	2.80	4.44	1.85	3.31	6.11
2013	1.71	2.62	4.05	1.75	3.33	6.65	1.83	2.92	4.68	1.85	3.38	6.42
2014	1.71	2.62	4.05	1.74	3.36	6.83	1.86	2.99	4.87	1.85	3.44	6.68
2015	1.71	2.63	4.05	1.77	3.24	6.22	1.85	2.96	4.80	1.85	3.28	6.03
Germany-Poland, m_{DE-PL}												
2005	0.80	1.12	1.55	0.73	1.12	1.69	0.74	1.06	1.50	0.71	1.05	1.54
2006	0.69	1.01	1.48	0.63	1.02	1.65	0.62	0.93	1.41	0.60	0.93	1.45
2007	0.69	1.01	1.47	0.65	1.01	1.58	0.63	0.94	1.41	0.62	0.94	1.45
2008	0.69	1.01	1.48	0.65	1.01	1.58	0.62	0.93	1.41	0.61	0.94	1.45
2009	0.69	1.01	1.48	0.65	1.01	1.58	0.59	0.90	1.39	0.59	0.93	1.46
2010	0.69	1.01	1.48	0.66	1.01	1.56	0.60	0.91	1.40	0.61	0.94	1.45
2011	0.69	1.01	1.47	0.67	1.01	1.53	0.61	0.92	1.40	0.62	0.95	1.45
2012	0.69	1.01	1.47	0.66	1.01	1.52	0.63	0.94	1.41	0.67	0.99	1.48
2013	0.69	1.01	1.47	0.65	1.00	1.55	0.62	0.93	1.40	0.68	1.01	1.50
2014	0.69	1.01	1.47	0.63	1.00	1.59	0.64	0.95	1.41	0.69	1.03	1.53
2015	0.69	1.01	1.48	0.63	1.00	1.58	0.65	0.96	1.42	0.69	1.03	1.53

Model	M_{10}: reference AR(1) **			M_{11}: AR(1)–ARCH(1) **			M_{12}: AR(1)–GARCH(1,1) **			M_{13}: AR(1)–SV **		
Year	10%	Median	90%	10%	Median	90%	10%	Median	90%	10%	Median	90%
Poland-Germany, m_{PL-DE}												
2005	2.23	3.62	5.88	2.04	3.18	4.91	2.36	3.46	4.86	2.44	3.54	5.14
2006	1.93	3.61	6.81	1.82	2.92	4.92	2.07	3.31	5.24	2.15	3.46	5.74
2007	1.77	3.60	7.50	1.71	2.79	4.93	1.91	3.21	5.66	1.98	3.40	6.17
2008	1.66	3.60	8.06	1.65	2.71	4.93	1.80	3.14	6.06	1.87	3.36	6.56
2009	1.59	3.60	8.58	1.60	2.67	4.95	1.72	3.08	6.50	1.79	3.34	6.87
2010	1.53	3.60	9.01	1.57	2.64	4.96	1.67	3.05	6.94	1.74	3.31	7.18
2011	1.48	3.59	9.40	1.54	2.62	4.99	1.62	3.02	7.37	1.70	3.30	7.45
2012	1.45	3.59	9.77	1.52	2.61	4.99	1.57	3.00	7.85	1.67	3.29	7.71
2013	1.42	3.60	10.11	1.51	2.61	5.00	1.54	2.99	8.32	1.64	3.29	7.93
2014	1.39	3.60	10.46	1.50	2.60	5.03	1.51	2.97	8.81	1.62	3.29	8.18
2015	1.37	3.60	10.73	1.49	2.59	5.03	1.49	2.97	9.37	1.60	3.28	8.40
Germany-Poland, m_{DE-PL}												
2005	0.90	1.26	1.76	0.84	1.17	1.60	0.87	1.19	1.59	0.94	1.25	1.67
2006	0.81	1.25	1.94	0.77	1.11	1.61	0.78	1.13	1.65	0.85	1.23	1.83
2007	0.76	1.24	2.08	0.73	1.07	1.61	0.73	1.10	1.69	0.80	1.23	1.95
2008	0.73	1.24	2.18	0.71	1.05	1.61	0.69	1.07	1.73	0.77	1.22	2.05
2009	0.70	1.24	2.28	0.69	1.03	1.61	0.66	1.06	1.76	0.74	1.22	2.15
2010	0.69	1.24	2.36	0.68	1.02	1.61	0.64	1.04	1.79	0.72	1.21	2.24
2011	0.67	1.24	2.44	0.67	1.02	1.61	0.63	1.03	1.81	0.71	1.21	2.32
2012	0.66	1.24	2.52	0.66	1.01	1.61	0.61	1.03	1.83	0.70	1.21	2.40
2013	0.65	1.24	2.59	0.66	1.01	1.61	0.60	1.02	1.85	0.69	1.21	2.48
2014	0.64	1.24	2.65	0.66	1.01	1.61	0.59	1.02	1.88	0.68	1.21	2.55
2015	0.63	1.24	2.71	0.65	1.00	1.61	0.58	1.02	1.89	0.68	1.21	2.62

Notes: †PT-Portugal, ES-Spain; ¶Dummy variable for 1984; *Sample = 13 (1992–2004); **Sample = 20 (1985–2004).

No implausible values, exceeding over 5 times the maximum levels observed in the sample, have been forecasted.

NB: Observations for 2005 are: $m_{PL-DE} = 4.17$, $m_{DE-PL} = 1.28$.

Source: own elaboration in WinBUGS, data for 2005, Eurostat/NewCronos

Table C.5 Distribution summaries of τ_t estimated for 1987–2004 and forecasted for 2005–2015 for models with changes in conditional variance: median, 10 and 90% quantiles (MCMC)

Model	M_{11}: AR(1)–ARCH(1)			M_{12}: AR(1)–GARCH(1,1)			M_{13}: AR(1)–SV		
Year	2.5%	median	97.5%	2.5%	median	97.5%	2.5%	median	97.5%
Poland-Germany, m_{PL-DE}									
1987	5.84	16.46	38.85	1.46	10.80	33.24	2.67	8.46	22.52
1988	3.46	6.33	11.12	2.51	6.99	14.38	2.66	7.33	16.65
1989	1.14	2.18	4.58	1.33	2.82	5.84	2.84	7.69	17.88
1990	0.89	2.19	8.25	0.99	2.85	9.35	3.18	8.84	21.97
1991	2.31	9.55	26.72	1.69	5.05	10.58	3.34	9.05	21.94
1992	2.28	10.94	38.33	1.58	4.45	10.49	3.78	10.47	27.25
1993	5.75	15.43	32.85	3.21	8.61	20.30	3.95	10.70	26.77
1994	3.20	7.66	15.87	2.52	5.85	12.84	4.37	12.77	37.87
1995	6.02	18.98	63.04	4.45	11.33	27.54	4.64	14.08	46.93
1996	6.00	18.13	49.85	6.14	17.52	42.79	4.74	14.77	53.97
1997	5.82	17.99	60.16	6.92	21.55	56.88	4.83	15.32	61.69
1998	5.77	17.47	56.18	7.27	24.19	69.59	4.88	15.74	69.26
1999	5.73	17.19	54.12	7.41	26.02	80.24	4.95	16.25	77.60
2000	6.02	18.95	61.13	8.00	30.43	86.71	4.95	16.40	82.46
2001	6.08	19.21	62.38	8.29	33.66	106.70	4.96	16.39	82.29
2002	6.09	18.85	54.97	8.43	34.45	114.80	4.89	16.05	78.62
2003	6.11	19.28	60.52	8.50	35.72	117.10	4.77	15.40	70.54
2004	6.11	18.83	53.36	8.55	35.72	124.10	4.52	14.26	58.44
2005	4.68	9.45	16.94	6.11	14.87	37.07	4.19	13.39	53.32
2006	2.29	11.42	42.59	3.35	15.93	50.20	3.99	13.00	53.37
2007	1.88	12.25	47.05	2.50	15.89	62.02	3.90	12.75	52.77
2008	1.81	12.54	48.98	2.04	15.77	70.63	3.82	12.56	52.36
2009	1.80	12.64	49.66	1.73	15.65	76.77	3.77	12.44	51.79
2010	1.79	12.68	50.15	1.51	15.55	81.20	3.72	12.35	51.26
2011	1.78	12.67	50.19	1.33	15.39	83.87	3.67	12.29	51.09
2012	1.78	12.73	50.36	1.20	15.32	86.06	3.63	12.24	50.61
2013	1.78	12.73	50.02	1.09	15.18	87.48	3.60	12.20	50.36
2014	1.78	12.72	50.57	0.99	15.12	88.79	3.58	12.17	50.35
2015	1.77	12.72	50.28	0.92	15.06	89.84	3.56	12.13	50.08
Germany-Poland, m_{DE-PL}									
1987	11.94	32.60	74.97	3.04	17.46	48.45	6.49	20.45	57.84
1988	10.96	25.59	49.57	5.75	20.24	46.10	6.42	17.69	42.23
1989	4.83	9.06	16.56	4.63	9.36	18.11	6.21	16.30	37.22
1990	2.67	5.91	12.13	2.63	5.90	12.55	6.44	17.16	40.48
1991	2.98	9.40	24.14	2.51	7.41	19.37	6.60	17.28	40.16
1992	5.37	18.04	63.68	3.89	8.53	16.97	7.30	19.98	51.12
1993	10.90	29.48	65.94	6.22	15.04	31.79	7.55	20.77	53.78
1994	11.57	32.12	75.71	8.45	22.84	52.96	7.46	19.64	47.37
1995	6.10	13.00	26.80	5.60	12.33	25.17	8.09	22.57	62.62
1996	11.90	33.58	85.50	8.41	20.42	45.26	8.41	24.14	72.72
1997	12.03	34.60	94.78	10.39	29.10	73.01	8.55	24.81	77.33
1998	11.97	34.52	97.46	11.47	35.71	101.00	8.59	24.65	77.79
1999	10.37	26.98	72.31	10.54	31.11	91.65	8.64	25.31	85.13
2000	11.21	31.42	91.56	11.00	35.07	120.80	8.77	25.74	90.37
2001	11.64	33.40	97.15	11.58	39.10	139.00	8.75	25.91	92.20
2002	11.95	34.21	92.60	12.09	41.44	132.50	8.69	25.69	90.24
2003	12.02	34.53	93.83	12.38	43.18	142.70	8.46	24.84	83.54
2004	12.05	33.94	83.77	12.52	42.80	136.20	8.00	23.21	72.12
2005	8.73	17.40	32.01	9.21	20.34	45.50	7.54	22.22	68.30
2006	4.40	21.46	69.61	4.88	21.43	58.30	7.24	21.78	67.80
2007	3.79	22.88	75.89	3.62	21.08	71.36	7.03	21.51	67.53
2008	3.65	23.40	78.47	2.90	20.82	80.93	6.92	21.31	66.98
2009	3.64	23.54	79.14	2.49	20.49	87.91	6.89	21.23	66.58
2010	3.67	23.63	79.59	2.15	20.24	93.19	6.80	21.14	66.20
2011	3.63	23.65	79.48	1.90	19.99	96.58	6.73	21.04	65.76
2012	3.62	23.59	79.85	1.71	19.82	99.63	6.72	20.99	65.82
2013	3.66	23.57	79.53	1.55	19.64	101.70	6.73	21.00	65.72
2014	3.65	23.61	79.67	1.43	19.44	103.40	6.66	20.98	65.28
2015	3.63	23.61	79.40	1.30	19.35	105.00	6.64	21.00	65.21

Notes: Sample = 20 (1985–2004). In the reference constant-variance model (M_{10}) precision summaries are as follows: for m_{PL-DE}, $q_{0.025} = 3.61$, median = 7.70 and $q_{0.975} = 14.08$; for m_{DE-PL}, $q_{0.025} = 8.59$, median = 15.86 and $q_{0.975} = 26.46$. Grey shading indicates the system transformation period (1989–1991).
Source: own elaboration in WinBUGS

Table C.6 Summaries of predictive distributions of emigration rates forecasted for 2005–2015 by models with diffuse priors (Chapter 8): median, 10 and 90% quantiles (MCMC estimates)

Model / Year	Migration Italy-Germany, m_{IT-DE}			Migration Poland-Germany, m_{PL-DE}			Migration Switzerland-Germany, m_{CH-DE}		
	10%	Median	90%	10%	Median	90%	10%	Median	90%
Random walk with drift, *a priori* $\tau \sim \Gamma(0.001, 0.001)$									
2005	0.30	0.35	0.41	2.80	3.64	4.74	1.18	1.25	1.33
2006	0.27	0.34	0.43	2.47	3.64	5.34	1.16	1.27	1.39
2007	0.25	0.33	0.44	2.23	3.63	5.91	1.15	1.29	1.45
2008	0.23	0.32	0.44	2.03	3.62	6.46	1.14	1.31	1.50
2009	0.21	0.31	0.45	1.86	3.62	7.03	1.13	1.32	1.55
2010	0.19	0.29	0.45	1.70	3.61	7.65	1.13	1.34	1.60
2011	0.18	0.28	0.45	1.57	3.61	8.27	1.13	1.36	1.65
2012	0.17	0.27	0.46	1.45	3.60	8.94	1.12	1.38	1.70
2013	0.15	0.26	0.46	1.34	3.59	9.65	1.12	1.40	1.76
2014	0.14	0.25	0.46	1.24	3.59	10.40	1.12	1.42	1.81
2015	0.13	0.25	0.46	1.15	3.58	11.21	1.11	1.44	1.87
AR(1), *a priori* $\phi \sim N(0, 100^2)$, $\tau \sim \Gamma(0.001, 0.001)$									
2005	0.28	0.34	0.41	2.29	2.99	3.90	1.14	1.21	1.29
2006	0.22	0.31	0.43	2.13	2.79	3.80	1.11	1.20	1.30
2007	0.16	0.28	0.44	2.05	2.70	3.72	1.09	1.19	1.30
2008	0.11	0.26	0.45	2.01	2.66	3.68	1.08	1.18	1.31
2009	0.07	0.23	0.46	1.98	2.63	3.65	1.08	1.17	1.31
2010	0.04	0.21	0.47	1.96	2.62	3.63	1.07	1.17	1.31
2011	0.02	0.18	0.47	1.94	2.61	3.62	1.07	1.17	1.31
2012	0.01	0.16	0.48	1.94	2.60	3.61	1.07	1.17	1.31
2013	0.00	0.14	0.49	1.93	2.60	3.61	1.07	1.17	1.32
2014	0.00	0.12	0.49	1.92	2.59	3.60	1.07	1.17	1.32
2015	0.00	0.10	0.49	1.92	2.59	3.60	1.07	1.17	1.32

Model / Year	Migration Germany-Italy, m_{DE-IT}			Migration Germany-Poland, m_{DE-PL}			Migration Germany-Switzerland, m_{DE-CH}		
	10%	Median	90%	10%	Median	90%	10%	Median	90%
Random walk with drift, $\tau \sim \Gamma(0.001, 0.001)$									
2005	0.40	0.44	0.47	1.02	1.25	1.54	0.20	0.22	0.23
2006	0.38	0.43	0.49	0.92	1.24	1.67	0.20	0.23	0.25
2007	0.37	0.43	0.50	0.84	1.22	1.79	0.21	0.24	0.27
2008	0.35	0.42	0.51	0.77	1.21	1.90	0.21	0.25	0.30
2009	0.34	0.42	0.52	0.71	1.19	2.00	0.22	0.27	0.32
2010	0.33	0.42	0.53	0.66	1.18	2.12	0.23	0.28	0.35
2011	0.32	0.41	0.54	0.61	1.17	2.23	0.23	0.29	0.37
2012	0.31	0.41	0.55	0.57	1.15	2.34	0.24	0.31	0.40
2013	0.30	0.41	0.56	0.53	1.14	2.46	0.25	0.33	0.43
2014	0.29	0.40	0.57	0.49	1.12	2.58	0.26	0.34	0.46
2015	0.28	0.40	0.57	0.46	1.11	2.70	0.26	0.36	0.50
AR(1), *a priori* $\phi \sim N(0, 100^2)$, $\tau \sim \Gamma(0.001, 0.001)$									
2005	0.41	0.44	0.48	0.94	1.14	1.39	0.21	0.23	0.25
2006	0.40	0.44	0.48	0.86	1.08	1.37	0.23	0.26	0.30
2007	0.40	0.44	0.49	0.82	1.04	1.36	0.25	0.30	0.39
2008	0.40	0.44	0.49	0.80	1.02	1.35	0.27	0.36	0.52
2009	0.40	0.44	0.49	0.78	1.01	1.34	0.29	0.44	0.75
2010	0.40	0.44	0.49	0.77	1.00	1.34	0.32	0.54	1.17
2011	0.40	0.44	0.49	0.76	1.00	1.33	0.35	0.69	2.00
2012	0.39	0.44	0.49	0.76	0.99	1.33	0.38	0.92	3.85
2013	0.39	0.44	0.49	0.75	0.99	1.33	0.42	1.26	8.56
2014	0.39	0.44	0.49	0.75	0.99	1.32	0.47	1.79	22.76
2015	0.39	0.44	0.49	0.74	0.99	1.33	0.53	2.68	75.04

Note: grey shading denotes implausible values, exceeding over 5 times the maximum levels observed in the sample. The (1985–)1991–2004 maxima are: $\max(m_{IT-DE}) = 0.89$, $\max(m_{DE-IT}) = 0.50$, $\max(m_{PL-DE}) = 3.72$, $\max(m_{DE-PL}) = 1.48$, $\max(m_{CH-DE}) = 1.28$, and $\max(m_{DE-CH}) = 0.20$.

NB: Observations for 2005 are: $m_{IT-DE} = 0.35$, $m_{DE-IT} = 0.35$, $m_{PL-DE} = 4.17$, $m_{DE-PL} = 1.28$, $m_{CH-DE} = 1.26$ and $m_{DE-CH} = 0.22$.

Source: own elaboration in WinBUGS, data for 2005: Eurostat/NewCronos

Table C.7 Bayesian predictive medians and frequentist point forecasts of emigration rates for 2000–2007 yielded by models estimated on truncated time series (Chapter 8)

Year \ Approach and models	Bayesian median forecasts					Observed values	Frequentist point forecasts				
	M_1: const.	M_2: RWD	M_3: AR(1)	M_4: MA(1)	M_5: ARMA		M_1: const.	M_2: RWD	M_3: AR(1)	M_4: MA(1)	M_5: ARMA
Migration from Italy to Germany, m_{IT-DE}											
2000	0.73	**0.65**	0.68	0.72	0.69	0.62	0.73	0.65	0.68	0.71	**0.68**
2001	0.73	**0.64**	0.67	0.73	0.68	0.55	0.73	0.64	0.70	0.73	**0.71**
2002	0.73	**0.64**	0.67	0.73	0.69	0.47	0.73	0.64	0.71	0.73	**0.72**
2003	0.73	**0.63**	0.66	0.73	0.68	0.41	0.73	0.63	0.72	0.73	**0.72**
2004	0.73	**0.63**	0.67	0.73	0.69	0.37	0.73	0.63	0.72	0.73	**0.72**
2005	0.73	**0.62**	0.66	0.73	0.68	0.35	0.73	0.62	0.72	0.73	**0.72**
2006	0.73	**0.62**	0.66	0.73	0.69	0.34	0.73	0.61	0.72	0.73	**0.72**
2007	0.73	**0.61**	0.66	0.73	0.68	0.35	0.73	0.61	0.72	0.73	**0.72**
Migration from Germany to Italy, m_{DE-IT}											
2000	**0.46**	0.46	0.46	0.47	0.46	0.45	0.46	**0.45**	0.47	0.46	0.47
2001	**0.46**	0.44	0.45	0.47	0.45	0.44	0.46	**0.43**	0.48	0.46	0.48
2002	**0.46**	0.43	0.45	0.47	0.45	0.44	0.46	**0.42**	0.49	0.46	0.48
2003	**0.46**	0.42	0.44	0.47	0.45	0.41	0.46	**0.40**	0.49	0.46	0.49
2004	**0.46**	0.41	0.44	0.47	0.45	0.44	0.46	**0.39**	0.50	0.46	0.50
2005	**0.46**	0.40	0.44	0.47	0.45	0.35	0.46	**0.38**	0.50	0.46	0.50
2006	**0.46**	0.39	0.44	0.47	0.45	0.33	0.46	**0.36**	0.50	0.46	0.51
2007	**0.46**	0.38	0.44	0.47	0.44	0.31	0.46	**0.35**	0.51	0.46	0.51
Migration from Switzerland to Germany, m_{CH-DE}											
2000	1.11	**1.10**	1.10	1.11	1.11	1.11	1.10	1.10	1.09	1.11	**1.11**
2001	1.11	**1.12**	1.12	1.11	1.12	1.15	1.10	1.12	1.08	1.09	**1.09**
2002	1.11	**1.13**	1.12	1.11	1.13	1.17	1.10	1.13	1.08	1.09	**1.09**
2003	1.11	**1.14**	1.13	1.11	1.13	1.16	1.10	1.14	1.08	1.09	**1.08**
2004	1.11	**1.15**	1.13	1.11	1.14	1.23	1.10	1.15	1.08	1.09	**1.08**
2005	1.11	**1.16**	1.14	1.11	1.14	1.26	1.10	1.16	1.07	1.09	**1.08**
2006	1.11	**1.18**	1.14	1.11	1.14	1.39	1.10	1.18	1.07	1.09	**1.08**
2007	1.11	**1.19**	1.14	1.11	1.14	1.49	1.10	1.19	1.07	1.09	**1.08**
Migration from Germany to Switzerland, m_{DE-CH}											
2000	**0.11**	0.13	0.12	0.12	0.12	0.14	0.11	**0.13**	0.13	0.12	0.13
2001	**0.11**	0.13	0.12	0.11	0.11	0.16	0.11	**0.14**	0.13	0.11	0.13
2002	**0.11**	0.13	0.11	0.11	0.11	0.18	0.11	**0.14**	0.13	0.11	0.13
2003	**0.11**	0.13	0.11	0.11	0.11	0.18	0.11	**0.15**	0.13	0.11	0.12
2004	**0.11**	0.13	0.11	0.11	0.11	0.20	0.11	**0.15**	0.12	0.11	0.12
2005	**0.11**	0.13	0.11	0.11	0.11	0.22	0.11	**0.15**	0.12	0.11	0.12
2006	**0.11**	0.14	0.11	0.11	0.11	0.27	0.11	**0.16**	0.12	0.11	0.12
2007	**0.11**	0.14	0.11	0.11	0.11	0.34	0.11	**0.16**	0.12	0.11	0.12

Note: no implausible values, exceeding over 5 times the maximum levels observed in the sample, have been detected. **Boldface** indicates forecasting models yielded by formal selection criteria: the highest-posterior probability models under uniform model priors (Bayesian approach), or highest-likelihood models (frequentist approach).

Source: own elaboration in WinBUGS and R, data: Eurostat/NewCronos (see Annex A)

References

Abel, G. (2009). *International migration flow table estimation*. PhD Dissertation. University of Southampton, Southampton.

Abler, R., Adams, J. S., & Gould, P. (1972). *Spatial organization: The geographer's view of the world*. Prentice-Hall, London.

Abraham, Ch. (2009). A computation method in robust Bayesian decision theory. *International Journal of Approximate Reasoning, 50*(2), 289–302.

Adam, K. (2004). On the relation between robust and Bayesian decision making. *Journal of Economic Dynamics & Control, 28*(10), 2105–2117.

Ahlburg, D. A. (1995). Simple versus complex models: Evaluation, accuracy and combining. *Mathematical Population Studies, 5*(3), 281–290.

Ahlburg, D. A., & Land, K. C. (1992). Population forecasting: Guest editors' introduction. *International Journal of Forecasting, 8*(3), 289–299.

Ahlburg, D. A., Lutz, W., & Vaupel, J. W. (1998). Ways to improve population forecasting: What should be done differently in the future? *Population and Development Review, 24*(Suppl.), 191–198.

Akaike, H. (1973). Information theory and an extension of the maximum likelihood principle. In B. N. Petrov & F. Csáki (Eds.), *Second international symposium in information theory* (pp. 267–281). Budapest: Akadémiai Kiadó.

Albert, J. (2007). *Bayesian computation with R*. 'Use R!' series, Berlin-Heidelberg: Springer.

Alecke, B., Huber, P., & Untiedt, G. (2001). What difference a constant makes? How predictable are international migration flows? In *Migration policies and EU enlargement. The case of central and eastern Europe* (pp. 63–78) Paris: OECD.

Alho, J. (1990). Stochastic methods in population forecasting. *International Journal of Forecasting, 6*(4), 521–530.

Alho, J. (1998). *A stochastic forecast of the population of Finland*. Reviews 1998/4. Helsinki: Statistics Finland.

Alho, J. (1999). *On probabilistic forecasts of population and their uses*. Paper for the 52nd session of the international statistical institute, Helsinki, 10–18 August.

Alho, J. (2008). Aggregation across countries in stochastic population forecasts. *International Journal of Forecasting, 24*(3), 343–353.

Alho, J., Alders, M., Cruijsen, H., Keilman, N., & Nikander, T. (2005). *New forecast: Population decline postponed in Europe*. Paper for the joint UNECE-eurostat work session on demographic projections, Vienna, 21–23 September.

Alho, J., & Spencer, B. (1985). Uncertain population forecasting. *Journal of the American Statistical Association, 80*(390), 306–314.

Alho, J., & Spencer, B. (2005). *Statistical demography and forecasting*. Berlin-Heidelberg: Springer.

Alkema, L., Raftery, A. E., & Brown, T. (2009). Bayesian melding for estimating uncertainties in national HIV prevalence estimates. *Sexually Transmitted Infections, 84*(Suppl. I), i11–i16.

Alvarez-Plata, P., Brücker, H., & Siliverstovs, B. (2003). *Potential migration from central and eastern Europe into the EU-15 – An update*. Report for the EC DG of employment and social affairs. Deutsches Institut für Wirtschaftsforschung, Berlin.

Anderson, J. L. (1998). Embracing uncertainty: The interface of Bayesian statistics and cognitive psychology. *Conservation Ecology, 2*(1), 2. Available online from www.consecol.org/vol2/iss1/art2/. Accessed 10 Dec 2008.

Anderson, T. W. (2003). *An introduction to multivariate statistical analysis* (3rd ed.). New York: Wiley.

Arango, J. (2000). Explaining migration: A critical view. *International Social Science Journal, 52*(165), 283–296.

Armstrong, J. S. (1985). *Long-range forecasting*. New York: Wiley.

Armstrong, J. S., & Green, K. C. (2007). Methodology tree. In J. S. Armstrong & K. C. Green (Eds.), *Forecasting principles* (online resource), Philadelphia: Marketing Department of The Wharton School, University of Pennsylvania; www.forecastingprinciples.com/methodologytree.html (version from Nov 2007).

Arrow, K. J. (1950). A difficulty in the concept of social welfare. *Journal of Political Economy, 58*(4), 328–346.

Barnard, G. A. (1947). The meaning of a significance level. *Biometrika, 34*(2), 179–182.

Barnard, G. A. (1949). Statistical inference. *Journal of the Royal Statistical Society B, 11*(1), 115–149.

Barnard, G. A. (1951). A theory of mathematical statistics independent of the calculus of probability. In R. Bayer (Ed.), *XV Congrés international de Philosophie des Sciences. Paris 1949* (pp. 115–124). Paris: Hermann.

Barrell, R., Gottschalk, S., Kirby, S., & Orazgani, A. (2009). *Projections of migration inflows under alternative scenarios for the UK and world economies* (Economics Paper No. 3). Department for Communities and Local Government, London.

Barro, R. J., & Sala-i-Martin, X. (1995). *Economic growth*. New York: McGraw-Hill.

Bartholomew, D. J. (1995). What is statistics? *Journal of the Royal Statistical Society A, 158*(1), 1–20.

Bates, J. M., & Granger, C. W. J. (1969). The combination of forecasts. *Operations Research Quarterly, 20*(4), 451–468.

Bauman, Z. (2007). *Liquid times: Living in an age of uncertainty*. Cambridge, MA: Polity Press.

Bauer, D., Feichtinger, G., Lutz, W., & Sanderson, W. (1999). *Variances of population projections: Comparison of two approaches*. IIASA Interim Report IR-99-63. International Institute for Applied Systems Analysis, Laxenburg.

Bauer, T., & Zimmermann, K. F. (1999). *Assessment of possible migration pressure and its labour market impact following EU following EU enlargement to central and eastern Europe*. IZA Research Report 3. Institut zur Zukunft der Arbeit, Bonn.

Baum, L. E., & Petrie, T. (1966). Statistical inference for probabilistic functions of finite state Markov chains. *The Annals of Mathematical Statistics, 37*(6), 1554–1563.

Bauwens, L., Lubrano, M., & Richard, J.-F. (1999). *Bayesian inference in dynamic econometric models*. Oxford, MA: Oxford University Press.

Bayarri, M. J., & Berger, J. O. (2004). The interplay of bayesian and frequentist analysis. *Statistical Science, 19*(1), 58–80.

Baydar, N. (1984). *Issues in multiregional demographic forecasting*. PhD Thesis, Vrije Universiteit, Brussels.

Bayes, T. (1763). An essay towards solving a problem in the doctrine of chances. *Philosophical Transactions of the Royal Society, 53*, 370–418 (www.stat.ucla.edu/history/essay.pdf. Accessed on 5 Nov 2005).

Becker, R. A., Chambers, J. M., & Wilks, A. R. (1988). *The new S language: A programming environment for data analysis and graphics*. Pacific Grove, CA: Wadsworth & Brooks/Cole.

Ben-Haim, Y. (2001). *Information-gap decision theory. Decisions under severe uncertainty*. San Diego, CA: Academic Press.

Berger, J., & Berliner, L. M. (1986). Robust Bayes and empirical Bayes analysis with ε-contaminated priors. *The Annals of Statistics, 14*(2), 461–486.

Bernardelli, H. (1941). Population waves. *Journal of the Burma Research Society, 31*, 1–18.

Bernardo, J. M. (2003). Bayesian statistics. In R. Viertl (Ed.), *Probability and statistics*. Encyclopaedia of life support systems (EOLSS): An integrated virtual library. Oxford, MA: UNESCO; www.eolss.net (www.uv.es/~bernardo/BayesStat2.pdf. Accessed 24 Aug 2006).

Bernardo, J. M., & Muñoz, J. (1993). Bayesian analysis of population evolution. *The Statistician, 42*(4), 541–550.

Bernardo, J. M., & Smith, A. F. M. (2000). *Bayesian theory*. Chichester: Wiley.

Bernstein, P. L. (1996). *Against the gods: Remarkable story of risk*. New York: Wiley [(1997). *Przeciw bogom: niezwykłe dzieje ryzyka*. Warszawa: WIG Press].

Berzuini, C., Clayton, D., & Bernardinelli, L. (1993). Bayesian inference on the lexis diagram. *Bulletin of the International Statistical Institute, 55*(1), 149–165.

Bijak, J. (2006). *Forecasting international migration: Selected theories, models, and methods* (CEFMR Working Paper 4/2006). Central European Forum for Migration Research, Warsaw.

Bijak, J. (2008a). Bayesian methods in international migration forecasting. In J. Raymer & F. Willekens (Eds.), *Estimation of international migration in Europe: Issues, models and assessment* (pp. 255–281). Chichester: Wiley.

Bijak, J. (2008b). Prognozowanie migracji zagranicznych w Europie: podejście bayesowskie [Forecasting international migration in Europe: The Bayesian approach]. In A. Welfe (Ed.), *Metody ilościowe w naukach ekonomicznych* [*Quantitative methods in economics*] (Vol. 8, pp. 77–98). Warsaw: Warsaw School of Economics.

Bijak, J., & Koryś, I. (2009). Poland. In H. Fassmann, U. Reeger, & W. Sievers (Eds.), *Statistics and reality: Concepts and measurements of migration in Europe* (pp. 195–215). IMISCOE Reports, Amsterdam: AUP.

Bijak, J., Kupiszewska, D., Kupiszewski, M., Saczuk, K., & Kicinger, A. (2007). Population and labour force projections for 27 European countries, 2002–2052: Impact of international migration on population ageing. *European Journal of Population, 23*(1), 1–31.

Bijak, J., & Wiśniowski, A. (2010). Bayesian forecasting of immigration to selected European countries by using expert knowledge. *Journal of the Royal Statistical Society A, 173*(4), 775–796.

Bijak, W. (1990). Estimation of the linear regression model in conditions of incomplete prior information. In S. Geisser, J. S. Hodges, S. J. Press, & A. Zellner (Eds.), *Bayesian and likelihood methods in statistics and econometrics: Essays in honor of George A. Barnard* (pp. 407–422) Amsterdam: North-Holland.

Bilsborrow, R., Hugo, G., Oberai, A. S., & Zlotnik, H. (1997). *International migration statistics. Guidelines for improving data collection systems*. Geneva: International Labour Organisation.

Blumen, I., Kogan, M., & McCarthy, P. J. (1955). *The industrial mobility of labour as a probability process*. Ithaca, NY: Cornell University Press.

Boeri, T., & Brücker, H. (2001). *Eastern enlargement and EU labour markets: Perceptions, challenges and opportunities* (IZA Discussion Paper 256). Institut zur Zukunft der Arbeit, Bonn.

Bollerslev, T. (1986). Generalized autoregressive conditional heteroscedasticity. *Journal of Econometrics, 31*(3), 307–327.

Booth, H. (2004). On the importance of being uncertain: Forecasting population futures for Australia. *People and Place, 12*(2), 1–12.

Booth, H. (2006). Demographic forecasting: 1980 to 2005 in review. *International Journal of Forecasting, 22*(3), 547–581.

Borjas, G. (1990). *Friends or strangers: The impact of immigrants on the US economy*. New York: Basic Books.

Boulding, K. E. (1969). Economics as a moral science. *The American Economic Review, 59*(1), 1–12.

Box, G. E. P., & Jenkins, G. M. (1976). *Time series analysis: Forecasting and control*. San Francisco: Holden-Day.

Box, G. E. P., & Tiao, G. C. (1973). *Bayesian inference in statistical analysis*. New York: Addison-Wesley.

Brierley, M., Forster, J., MacDonald, J. W., & Smith, P. W. F. (2008). Bayesian estimation of migration flows. In J. Raymer & F. Willekens (Eds.), *Estimation of international migration in Europe: Issues, models and assessment* (pp. 149–174) Chichester: Wiley.

Brown, L. A. (1970). On the use of Markov chains in movement research. *Economic Geography, 46*(Suppl.), 393–403.

Bryman, A. (1988). *Quantity and quality in social research*. Abingdon: Routledge.

Brücker, H., & Siliverstovs, B. (2006). On the estimation and forecasting of international migration: How relevant is heterogeneity across countries. *Empirical Economics, 31*(3), 735–754.

Carlin, B. P., & Chib, S. (1995). Bayesian model choice via Markov chain Monte Carlo methods. *Journal of the Royal Statistical Society B, 57*(3), 473–484.

Casella, G., & George, E. (1992). Explaining the Gibbs sampler. *The American Statistician, 46*(3), 167–174.

Castro, L. J., & Rogers, A. (1983). Patterns of family migration: Two melthodological approaches. *Environment and Planning A, 15*(2), 237–254.

Charemza, W. W., & Deadman, D. F. (1992). *New directions in econometric practice*. Aldershot: Edward Elgar [(1997). *Nowa ekonometria*. Warszawa: PWE].

Chatfield, C. (2002). Confessions of a pragmatic statistician. *The Statistician, 51*(1), 1–20.

Chojnicki, Z. (1977). Podstawy metodologiczne prognozowania w geografii ekonomicznej [Methodological foundations of forecasting in economic geography]. *Przegląd Geograficzny, 49*(2), 247–261. [A similar, earlier paper in English is available as: Chojnicki, Z. (1970). Prediction in economic geography. *Economic Geography, 46*(2 Suppl.), 213–222.]

Christoffersen, P. F., & Diebold, F. X. (1997). Optimal prediction under asymmetric loss. *Econometric Theory, 13*(6), 808–817.

Clark, J. S. (2003). Uncertainty and variability in demography and population growth: A hierarchical approach. *Ecology, 84*(6), 1370–1381.

Clark, P. K. (1973). A subordinated stochastic process model with finite variance for speculative prices. *Econometrica, 41*(1), 135–155.

Coale, A., & Demeny, P. (1967). *Manual IV: Methods of estimating basic demographic measures from incomplete data*. Manuals on Methods of Estimating Population. New York: United Nations.

Cohen, J. E. (1986). Population forecasts and confidence intervals for Sweden: A comparison of model-based and empirical approaches. *Demography, 23*(1), 105–126.

Cohen, J. E. (1998). Should population projections consider "limiting factors" – And if so, how? *Population and Development Review, 24*(Suppl.), 118–138.

Cohen, J. E., Roig, M., Reuman D. C., & GoGwilt, C. (2008). International migration beyond gravity: A statistical model for use in population projections. *Proceedings of the National Academy of Sciences USA, 105*(40), 15268–15274.

Collopy, F., & Armstrong, J. S. (1992). Rule-based forecasting: Development and validation of an expert systems approach to combining time series extrapolations. *Management Science, 38*(14), 1394–1414.

Congdon, P. (2000). A Bayesian approach to prediction using the gravity model, with an application to patient flow modelling. *Geographical Analysis, 32*(3), 205–224.

Congdon, P. (2001a). *Bayesian statistical modelling*. Chichester: Wiley.

Congdon, P. (2001b). The development of gravity models for hospital patient flows under system change: A Bayesian modelling approach. *Health Care Management Science, 4*(4), 289–304.

Congdon, P. (2003). *Applied Bayesian modelling*. Chichester: Wiley.

Congdon, P. (2005). *Bayesian models for categorical data*. Chichester: Wiley.

Congdon, P. (2008). Models for migration age schedules: A Bayesian perspective with an application to flows between Scotland and England. In J. Raymer & F. Willekens (Eds.), *Estimation of international migration in Europe: Issues, models and assessment* (pp. 193–206). Chichester: Wiley.

Constant, A., & Zimmermann, K. F. (2003). *The dynamics of repeat migration: A Markov chain analysis* (IZA Discussion Paper 885). Institut zur Zukunft der Arbeit, Bonn.

Cooley, T. F., & LeRoy, S. F. (1985). Atheoretical macroeconomics: A critique. *Journal of Monetary Economics, 16*(3), 283–308.

Council of Europe. (2004). *Demographic yearbook 2004: Recent demographic developments in Europe*. Strasbourg: Council of Europe.

Courbis, R. (1979). The REGINA model, a regional-national model for French planning. *Regional Science and Urban Economics, 9*(2–3), 117–139.

Courgeau, D. (1985). Interaction between spatial mobility, family and career life cycle: A French survey. *European Sociological Review, 1*(2), 139–162.

Courgeau, D. (1995). Migration theories and behavioural models. *International Journal of Population Geography, 1*(1), 19–27.

Courgeau, D. (2003). From the macro-micro opposition to multilevel analysis. In D. Courgeau (Ed.), *Methodology and epistemology of multilevel analysis* (pp. 43–92). Dordrecht : Kluwer.

Courgeau, D. (2004). Probabilités, demographie et sciences sociales. *Mathématiques et Sciences humaines, 42*(3), 27–50.

Courgeau, D., & Baccaïni, B. (1998). Multilevel analysis in the social sciences. *Population–E, 10*(1), 39–71.

Courgeau, D., & Lelièvre, E. (1992). *Event history analysis in demography*. Oxford: Oxford University Press.

CPB. [Centraal Planbureau] (2004). *Arbeidsmigratie uit de Midden- en Oost-Europese toetredingslanden* [*Labour migration from the Central and Eastern European accession countries*]. Notice to the Dutch Ministry of Social Affairs and Employment. The Hague: CPB.

Cruijsen, H., & Keilman, N. (1992). A comparative analysis of the forecasting process. In N. Keilman & H. Cruijsen (Eds.), *National population forecasting in industrialised countries*. NIDI & CBGS publication 24 (pp. 3–25). Amsterdam: Swets & Zeitlinger.

Dahlhaus, R., & Eichler, M. (2003). Causality and graphical models in time series analysis. In P. J. Green, N. L. Hjort, & S. Richardson (Eds.), *Highly structured stochastic systems*. Oxford Statistical Science Series 27 (pp. 115–145). Oxford: Oxford University Press (with discussion).

Dalkey, N. C. (1969). *The Delphi method: An experimental study of group opinion*. Research Memorandum RM-5888-PR. Santa Monica, CA: RAND Corporation.

Daponte, B. O., Kadane, J. B., & Wolfson, L. J. (1997). Bayesian demography: Projecting the Iraqi Kurdish population, 1977–1990. *Journal of the American Statistical Association, 92*(440), 1256–1267.

Dawid, A. P. (1984). Statistical theory: The prequential approach. *Journal of the Royal Statistical Society A, 147*(2), 278–292.

de Beer, J. (1990a). *Uncertainty of international-migration projections for the 12 EC-countries*. Department of Population Statistics, Central Bureau of Statistics, Voorburg.

de Beer, J. (1990b). Voorspelbaarheid van de buitenlandse migratie [Predictability of international migration]. *Maandstatistiek van de bevolking, 38*(5), 14–25.

de Beer, J. (1997). The effect of uncertainty of migration on national population forecasts: The case of the Netherlands. *Journal of Official Statistics, 13*(3), 227–243.

de Beer, J. (2000). *Dealing with uncertainty in population forecasting*. Department of Population Statistics, Central Bureau of Statistics, Voorburg.

de Beer, J. (2008). Forecasting international migration: Time series projections vs argument-based forecasts. In J. Raymer & F. Willekens (Eds.), *Estimation of international migration in Europe: Issues, models and assessment* (pp. 283–306). Chichester: Wiley.

de Finetti, B. (1937). La prévision: ses lois logiques, ses sources subjectives. *Annales del'Institute Henri Poincaré, 7*, 1–68.

DeGroot, M. H. (1970). *Optimal statistical decisions*. New York: McGraw-Hill [(1981). *Optymalne decyzje statystyczne*. Warszawa: PWN].

DeJong, G. F., & Fawcett, J. T. (1981). Motivations for migration: An assessment and a value-expectancy research model. In G. F. DeJong & R. W. Gardener (Eds.), *Migration decision making: Multidisciplinary approaches to microlevel studies in developed and developing countries* (pp. 13–57). New York: Pergamon Press.

Dellaportas, P., Forster, J. J., & Ntzoufras, I. (2002). On Bayesian model and variable selection using MCMC. *Statistics and Computing, 12*(1), 27–36.

Diebold, F. X., & Kilian, L. (2001). Measuring predictability: Theory and macroeconomic applications. *Journal of Applied Econometrics, 16*(6), 657–669.

Draper, D. (1995). Inference and hierarchical modelling in the social sciences. *Journal of Educational and Behavioral Statistics, 20*(2), 115–147, 233–239 (with discussion).

Drbohlav, D. (1996). The probable future of European east-west international migration – Selected aspects. In F. W. Carter, P. Jordan, & V. Rey (Eds.), *Central Europe after the fall of the iron curtain; geopolitical perspectives, spatial patterns and trends* (pp. 269–296). Frankfurt: Wiener Osteuropa Studien 4. Peter Lang & Co.

Duchêne, J., & Wanner, P. (1999). *Uncertainty in demographic projections and its consequences for the user.* Paper for the Joint UNECE-Eurostat Work Session on Demographic Projections, Perugia, Italy, 3–7 May.

Dustmann, C. (1997). Return migration, uncertainty and precautionary savings. *Journal of Development Economics, 52*(2), 295–316.

Dustmann, C., Casanova, M., Fertig, M., Preston, I., & Schmidt, C. M. (2003). *The impact of EU enlargement on migration flows.* Home Office Report 25/03. London: Home Office.

Eade, J., Drinkwater, S., & Garapich, M. P. (2006). *Class and ethnicity: Polish migrant workers in London.* ESRC Project Report. CRONEM, Guildford: University of Surrey.

Engle, R. (1982). Autoregressive conditional heteroscedasticity with estimates of the variance of United Kingdom inflation. *Econometrica, 50*(4), 987–1008.

Eurostat (1997). *Comparing data sources for measuring international migration in central and eastern Europe* (Eurostat Working Paper 1997/1). Eurostat, Luxembourg.

Eurostat (2005). *Population projections for the European Union 2004–2050.* Luxembourg: Eurostat.

Fachin, S., & Venanzoni, G. (2002). *IDEM: An integrated demographic and economic model of Italy.* Paper for the 14th International Conference on Input-Output Techniques, Montreal, Canada, 10–15 Oct.

Faist, T. (2000). *The volume and dynamics of international migration and transnational social spaces.* Oxford, MA: Oxford University Press.

Fama, E. F. (1965). The behavior of stock market prices. *Journal of Business, 38*(1), 34–105.

Fassmann, H., & Hintermann, C. (1997). *Migrationspotential ostmitteleuropa.* ISR-Forschungsberichte 15. Institut für Stadt- und Regionalforschung, Wien.

Fertig, M., & Schmidt, C. M. (2000). *Aggregate-level migration studies as a tool for forecasting future migration streams* (IZA Discussion paper 183). Institut zur Zukunft der Arbeit, Bonn.

Fisher, R. A. (1922). On the mathematical foundations of theoretical statistics. *Philosophical Transactions of the Royal Society A, 222*, 309–368 (Accessed via the University of Adelaide R. A. Fisher Digital Archive, digital.library.adelaide.edu.au/coll/special/fisher, as of 5 Feb 2008).

Franzmeyer, F., & Brücker, H. (1997). *Europäische Union: Osterweiterung und Arbeitskräftemigration.* DIW Wochenbericht 5/97 (pp. 89–96). Deutsches Institut für Wirtschaftsforschung, Berlin.

Frydman, R., & Goldberg, M. D. (2007). *Imperfect knowledge economics. Exchange rates and risk.* Princeton, NJ/Oxford, MA: Princeton University Press.

Gawryszewski, A. (1989). *Przestrzenna ruchliwość ludności Polski 1952–1985* [*Spatial mobility of the population of Poland, 1952–1985*]. Wrocław: Ossolineum.

Gelfand, A. E., & Smith, A. F. M. (1990). Sampling-based approaches to calculating marginal densities. *Journal of the American Statistical Association, 85*(410), 398–409.

Geman, S., & Geman, D. (1984). Stochastic relaxation, Gibbs distribution and the Bayesian restoration of images. *IEEE Transactions on Pattern Analysis and Machine Intelligence, 6*, 721–741.

Geweke, J. (1989). Bayesian inference in econometric models using Monte Carlo integration. *Econometrica, 57*(6), 1317–1339.

Gilboa, I. (2009). *Theory of decision under uncertainty*. Cambridge, MA: Cambridge University Press.

Gilks, W. R., & Wild, P. (1992). Adaptive rejection sampling for Gibbs sampling. *Applied Statistics, 41*(2), 337–348.

Ginsberg, R. B. (1971). Semi-Markov processes and mobility. *Journal of Mathematical Sociology, 1*(1), 233–262.

Ginsberg, R. B. (1978). Probability models of residence histories: Analysis of times between moves. In W. A. V. Clark, & E. G. Moore (Eds.), *Population mobility and residential change*. Studies in Geography 25. Evanston, IL: Northwestern University.

Girosi, F., & King, G. (2008). *Demographic forecasting*. Princeton, NJ: Princeton University Press.

Gjaltema, T. A. (2001). *Judgement in population forecasting*. Paper for the European Population Conference, Helsinki, 7–9 June.

Goldstein, H. (2003). *Multilevel statistical models*. London: Arnold.

Goodman, L. A. (1961). Statistical methods for the mover-stayer model. *Journal of the American Statistical Association, 56*(296), 841–868.

Gorbey, S., James, D., & Poot, J. (1999). Population forecasting with endogenous migration: An application to trans-tasman migration. *International Regional Science Review, 22*(1), 69–101.

Granger, C. W. J., & Morris, M. J. (1976). Time series modelling and interpretation. *Journal of the Royal Statistical Society A, 139*(2), 246–257.

Green, P. J. (1995). Reversible jump Markov chain Monte Carlo computation and Bayesian model determination. *Biometrika, 82*(4), 711–732.

Greene, W. H. (2000). *Econometric analysis*. Upper Saddle River, NJ: Prentice Hall.

Greenwood, M. J. (1973). Urban economic growth and migration: Their interaction. *Environment and Planning A, 5*(1), 91–112.

GUS. [Główny Urząd Statystyczny/Central Statistical Office] (2003). *Raport z wyników Narodowego Spisu Powszechnego Ludności i Mieszkań 2002* [*Report from the results of the National census of population and housing*]. Warsaw: GUS.

Gustafson, P., Aeschliman, D., & Levy, A. R. (2003). A simple approach to fitting Bayesian survival models. *Lifetime Data Analysis, 9*(1), 5–19.

Hägerstrand, T. (1957). Migration and area. Survey of a sample of Swedish migration fields and hypothetical considerations on their genesis. In D. Hannerberg, T. Hägerstrand, & B. Odeving (Eds.), *Migration in Sweden. A Symposium*. "Lund Studies in Geography", Series B, no. 13 (pp. 27–158). Lund: Gleerup.

Hägerstrand, T. (1970). What about people in regional science? *Papers of the Regional Science Association, 24*(1), 7–21.

Hajnal, J. (1955). The prospects for population forecasts. *Journal of the American Statistical Association, 50*(270), 309–322.

Harker, P. T. (1986). The use of expert judgments in predicting interregional migration patterns: An analytic hierarchy approach. *Geographical Analysis, 18*(1), 62–80.

Harris, J. R., & Todaro, M. P. (1970). Migration, unemployment and development: A two-sector analysis. *American Economic Review, 60*(1), 126–142.

Hart, R. A. (1975). Interregional economic migration: Some theoretical considerations (Part I). *Journal of Regional Science, 15*(2), 127–138.

Hastings, W. K. (1970). Monte Carlo sampling methods using Markov chains and their applications. *Biometrika, 57*(1), 97–109.

Hatton, T. J., & Williamson, J. G. (1998). *The age of mass migration: Causes and economic impact*. Oxford, MA: Oxford University Press.

Hawking, S. W. (1988). *A brief history of time: From the big bang to black holes*. New York: Bantam Books [(1990). *Krótka historia czasu: od Wielkiego Wybuchu do czarnych dziur*. Warszawa: Alfa].

Hendry, D. F. (1995). *Dynamic econometrics*. Oxford, MA: Oxford University Press.

Hoeting, J. A., Madigan, D., Raftery, A. E., & Volinsky, C. T. (1999). Bayesian model averaging: A tutorial. *Statistical Science, 14*(4), 382–417.

Holzer, J. Z. (1959). *Prognoza demograficzna Polski na lata 1960–1975 wg województw* [*Demographic forecast of Poland for 1960–1975, by voivodships*]. Warsaw: PWG State Economic Publishers.

Huff, J. O., & Clark, W. A. V. (1978). The role of stationarity in Markov and opportunity models of intraurban migration. In W. A. V. Clark & E. G. Moore (Eds.), *Population mobility and residential change*. Studies in Geography 25, Evanston, IL: Northwestern University.

Hyndman, R. J., & Booth, H. (2008). Stochastic population forecasts using functional data models for mortality, fertility and migration. *International Journal of Forecasting, 24*(3), 323–342.

Ibrahim, J. G., Chen, M.-H., & Sinha, D. (2001). *Bayesian survival analysis*. Berlin-Heidelberg: Springer.

IOM. [International Organization for Migration] (1998). *Migration potential in central and eastern Europe*. Geneva: IOM.

IOM. [International Organization for Migration] (2008). *World migration report 2008*. Geneva: IOM.

IPPR. (2008). *Floodgates or turnstiles? Post-EU enlargement migration flows to (and from) the UK*. London: Institute for Public Policy Research.

Isard, W. (1960). *Methods of regional analysis: An introduction to regional science*. New York: Wiley [(1965). *Metody analizy regionalnej. Wprowadzenie do nauki o regionach*. Warszawa: PWN].

Jacquier, E., Polson, N. G., & Rossi, P. E. (1994). Bayesian analysis of stochastic volatility models. *Journal of Business and Economic Statistics, 12*(4), 371–417 (with discussion).

Janicki, W. (2005). *Migracje wewnętrzne w Unii Europejskiej i ich uwarunkowania* [*Internal migration in the European Union and its determinants*]. PhD Dissertation. Maria Curie-Skłodowska University, Lublin.

Jaynes, E. T. (1976). Confidence intervals vs. Bayesian intervals. In W. L. Harper & C. A. Hooker (Eds.), *Foundations of probability theory, statistical inference and statistical theories of science 2* (pp. 175–257). Dordrecht: Reidel (with discussion).

Jeffreys, H. (1939). *Theory of probability*. Oxford, MA: Oxford University Press [(1961). 3rd ed.].

Jeffreys, H. (1946). An invariant form for the prior probability in estimation problems. *Proceedings of the Royal Society of London, 186*, 453–461.

Jeffreys, W. H., & Berger, J. O. (1992). Ockham's razor and Bayesian analysis. *American Scientist, 80*(1), 64–72.

Jennissen, R. (2004). *Macro-economic determinants of international migration in Europe*. Amsterdam: Dutch University Press.

Joseph, G. (1975). A Markov analysis of age/sex differences in inter-regional migration in Britain. *Regional Studies, 9*(2), 69–78.

Jóźwiak, J. (1987). On the problem of the accuracy of population projections. Manuscript. Published in Polish as: (1988). Problem oceny dokładności projekcji demograficznych. *Studia Demograficzne, 92*(2), 59–74.

Jóźwiak, J. (1992). *Mathematical models of population*. NIDI Report 26. Netherlands Interdisciplinary Demographic Institute, The Hague.

Kaczmarczyk, P., & Okólski, M. (2008). Demographic and labour-market impacts of migration on Poland. *Oxford Review of Economic Policy, 24*(3), 599–624.

Kadane, J. B., & Wolfson, L. J. (1998). Experiences in elicitation. *The Statistician, 47*(1), 3–19.

Kalman, R. E. (1960). A new approach to linear filtering and prediction problems. *Transaction of the ASME – Journal of Basic Engineering, 82*(D), 35–45.

Kędelski, M. (1990). Fikcja demograficzna w Polsce i RFN (Ze studiów nad migracjami zagranicznymi) [Demographic fiction in Poland and the FRG (From the studies on international migration)]. *Studia Demograficzne, 99*(1), 21–55.

Keilman, N. (1990). *Uncertainty in national population forecasting: Issues, backgrounds, analyses, recommendations.* Amsterdam: Swets & Zeitlinger.

Keilman, N. (2001). Demography: Uncertain population forecasts. *Nature, 412*(6846), 490–491.

Keilman, N., & Cruijsen, H. (Eds.). (1992). *National population forecasting in industrialised countries.* NIDI & CBGS publication 24. Amsterdam: Swets & Zeitlinger.

Keilman, N., & Pham, D. Q. (2004a). Time series based errors and empirical errors in fertility forecasts in the Nordic Countries. *International Statistical Review, 72*(1), 5–18.

Keilman, N., & Pham, D. Q. (2004b). *Empirical errors and predicted errors in fertility, mortality and migration forecasts in the European Economic Area* (Discussion Paper No. 386). Social and Demographic Research, Statistics Norway, Oslo.

Keilman, N., & Pham, D. Q. (2006). *Prediction intervals for Lee-Carter-based mortality forecasts.* Paper for the European Population Conference, Liverpool, 21–24 June.

Keilman, N., Pham, D. Q., & Hetland, A. (2001). *Norway's uncertain demographic future.* Norway, Oslo: Statistics.

Keilman, N., Pham, D. Q., & Hetland, A. (2002). Why population forecasts should be probabilistic – illustrated by the case of Norway. *Demographic Research, 6*, 409–454; www.demographic-research.org. Accessed 5 Nov 2005.

Keyfitz, N. (1968). *Introduction to the mathematics of population.* Berkeley, CA: University of California Press.

Keyfitz, N. (1972). On future population. *Journal of the American Statistical Association, 67*(338), 347–363.

Keyfitz, N. (1981). The limits of population forecasting. *Population and Development Review, 7*(4), 579–593.

Keyfitz, N. (1982). Can knowledge improve forecasts? *Population and Development Review, 8*(4), 729–751.

Keynes, J. M. (1921/1973). *A treatise on probability.* London: Macmillan and The Royal Economic Society (*The Collected Writings of John Maynard Keynes*, vol. VIII).

Kim, H.-M., & Mallick, B. K. (2003). A note on Bayesian spatial prediction using the elliptical distribution. *Statistics and Probability Letters, 64*(1), 271–276.

King, R. (2002). Towards a new map of European migration. *International Journal of Population Geography, 8*(2), 89–106.

Kloek, T., & van Dijk, H. K. (1978). Bayesian estimates of equation system parameters. An application of integration by Monte Carlo. *Econometrica, 46*(1), 1–19.

Knight, F. H. (1921). *Risk, uncertainty and profit.* Hart, Schaffner and Marx Prize Essays No. 31. Boston: Houghton Mifflin. Available from the online Library of Economics and Liberty, www.econlib.org/library/ Knight/knRUP1.html. Accessed 15 Oct 2005.

Koop, G. (2003). *Bayesian econometrics.* Chichester: Wiley.

Korčak-Čepurkivs'kij, Ju. A. (1970). Rozrachunok tempu zrastanja kilkosti i vikovoho skladu tych, ščo živut' u ustabilizovanomu naselenni z postijnym rivnem mihracij. *Demohrafična Doslidžennija, 1*, 24–41.

Korcelli, P. (1994). On interrelations between internal and international migration. *Innovation, 2*, 151–163.

Kritz, M., Lim, L. L., & Zlotnik, H. (Eds.). (1992). *International migration systems: A global approach.* Oxford: Clarendon Press.

Kuhn, T. S. (1962). *The structure of scientific revolutions.* Chicago: University of Chicago Press.

Kulu, H. (2005). Migration and fertility: Competing hypotheses re-examined. *European Journal of Population, 21*(1), 51–87.

Kupiszewska, D., & Kupiszewski, M. (2005). *A revision of a traditional multiregional model to better capture international migration: The MULTIPOLES model and its applications* (CEFMR Working Paper 10/2005). Central European Forum for Migration Research, Warsaw.

Kupiszewska, D., & Nowok, B. (2005). *Comparability of statistics on international migration flows in the European Union* (CEFMR Working Paper 7/2005). Central European Forum for Migration Research, Warsaw.

Kupiszewski, M. (1998). Czy na podstawie doświadczeń innych krajów można przewidywać zmiany migracji międzynarodowych po przyjęciu Polski do Unii Europejskiej? [Can the experience of other countries help forecast international migration changes after Poland's accession to the EU?]. In P. Korcelli (Ed.), *Przemiany w zakresie migracji ludności jako konsekwencja przystąpienia Polski do Unii Europejskiej* [*Changes in migration as a consequence of Poland's accession to the European Union*] (pp. 67–78). Warsaw: Bulletin of the Committee of Spatial Organization of the Polish Academy of Sciences, no. 184.

Kupiszewski, M. (2001). Demograficzne aspekty wybranych prognoz migracji zagranicznych [Demographic aspects of selected forecasts of international migration]. In A. Stępniak (Ed.), *Swobodny przepływ pracowników w kontekście wejścia Polski do Unii Europejskiej* [*Free flow of workers in the context of Poland's accession to the European Union*] (pp. 73–98). Warsaw: Office of the Committee of European Integration.

Kupiszewski, M. (2002a). How trustworthy are forecasts of international migration between Poland and the European Union? *Journal of Ethnic and Migration Studies, 28*(4), 627–645.

Kupiszewski, M. (2002b). *Modelowanie dynamiki przemian ludności w warunkach wzrostu znaczenia migracji międzynarodowych* [*The role of international migration in the modelling of population dynamics*]. Warsaw: Institute of Geography and Spatial Organisation, Polish Academy of Sciences.

Kupiszewski, M. (2006). Migration in Poland in the period of transition – The adjustment to the labour market change. In M. Kuboniwa & Y. Nishimura (Eds.), *Economics of intergenerational equity in transnational economies* (pp. 255–276). Tokyo: Maruzen.

Lanzieri, G. (2004). *EUROPOP 2003: Methodology for drafting international migration assumptions in the EU-15 member states*. Paper for the meeting of the Working Group on Population Projections, Luxembourg, 15–16 July.

Laplace, P.-S. (1812). *Théorie Analytique des Probabilités*. Paris: Veuve Courcier.

Lawrence, M., Goodwin, P., O'Connor, M., & Önkal, D. (2006). Judgemental forecasting: A review of progress over the last 25 years. *International Journal of Forecasting, 22*(3), 493–518.

Layard, R., Blanchard, O., Dornbusch, R., & Krugman, P. (1992). *East-West migration: The alternatives*. Cambridge, MA: MIT Press.

Ledbury, M., Miller, N., Lee, A., Fairman, T., & Clifton, C. (2006). *Understanding policy options*. Home Office Report 06/06. Home Office, London.

Ledent, J. (1981). Constructing multiregional life tables using place-of-birth-specific migration data. In A. Rogers (Ed.), *Advances in multiregional demography*. IIASA Research Report RR-81-6. Laxenburg: International Institute for Applied Systems Analysis.

Lee, E. S. (1966). A theory of migration. *Demography, 3*(1), 47–57.

Lee, R. D. (1998). Probabilistic approaches to population forecasting. *Population and Development Review, 24*(Suppl.), 156–190.

Lee, R. D., & Tuljapurkar, S. (1994). Stochastic population projections for the United States: Beyond high, medium and low. *Journal of the American Statistical Association, 89*(419), 1175–1189.

LeSage, J. P. (1990). Forecasting turning points in metropolitan employment growth rates using Bayesian techniques. *Journal of Regional Science, 30*(4), 533–548.

LeSage, J. P., & Pan, Z. (1995). Using spatial contiguity as Bayesian prior information in regional forecasting models. *International Regional Science Review, 18*(1), 33–53.

Leslie, P. H. (1945). On the use of matrices in certain population mathematics. *Biometrika, 33*(3), 183–212.

Lewis, A. W. (1954). *Economic development with unlimited supplies of labour*. Manchester: School of Economic and Social Studies.

Lindley, D. V. (1991). *Is our view of Bayesian statistics too narrow?* Paper for the 4th Valencia International Meeting on Bayesian Statistics, Peñiscola, Spain, 15–20 April.

Lindley, D. V. (2000). The philosophy of statistics. *The Statistician, 49*(3), 293–337.
Lindsay, J., & Barr, B. M. (1972). Two stochastic approaches to migration: Comparison of Monte Carlo simulation and Markov chain models. *Geografiska Annaler B, 54*(1), 56–67.
Litterman, R. B. (1979). *Techniques of forecasting using vector autoregressions* (Working Paper No. 115). Federal Reserve Bank of Minneapolis, Minneapolis, MN.
Lowry, I. S. (1966). *Migration and metropolitan growth: Two analytical models*. San Francisco: Chandler.
Lucas, R. E. Jr. (1976). Econometric policy evaluation: A critique. In K. Brunner & A. Meltzer (Eds.), The Phillips curve and labor markets. *Carnegie-Rochester Conference Series on Public Policy, 1*, 19–46.
Lunn, D. J., Thomas, A., Best, N., & Spiegelhalter, D. (2000). WinBUGS – a Bayesian modelling framework: Concepts, structure, and extensibility. *Statistics and Computing, 10*(4), 325–337.
Lutz, W., & Goldstein, J. R. (2004). Introduction: How to deal with uncertainty in population forecasting? *International Statistical Review, 72*(1), 1–4.
Lutz, W., Saariluoma, P., Sanderson, W., & Scherbov, S. (2000). *New developments in the methodology of expert- and argument-based probabilistic population forecasting*. IIASA Interim Report IR-00-20. International Institute for Applied Systems Analysis, Laxenburg.
Lutz, W., Sanderson, W., & Scherbov, S. (1996). Probabilistic population projections based on expert opinion. In W. Lutz (Ed.), *The future population of the World. What can we assume today?* (pp. 397–428). London: Earthscan.
Lutz, W., Sanderson, W. C., & Scherbov, S. (1997). Doubling of world population unlikely. *Nature, 387*(6635), 803–805.
Lutz, W., Sanderson, W. C., & Scherbov, S. (1998). Expert based probabilistic population projections. *Population and Development Review, 24*(Suppl.), 139–155.
Lutz, W., Sanderson, W. C., & Scherbov, S. (Eds.). (2004). *The end of world population growth in the 21st century: New challenges for human capital formation and sustainable development*. London: Earthscan.
Lynch, S. M. (2007). *Introduction to applied Bayesian statistics and estimation for social scientists*. New York: Springer.
Mabogunje, A. L. (1970). Systems approach to a theory of rural-urban migration. *Geographical Analysis, 2*(1), 1–18.
MacInnes, J., & Pérez Díaz, J. (2009). The reproductive revolution. *The Sociological Review, 57*(2), 262–284.
Makridakis, S, & Hibon, M. (2000). The M3-competition: Results, conclusions and implications. *International Journal of Forecasting, 16*(4), 451–476.
Makridakis, S., Hogarth, R. M., & Gaba, A. (2009). Forecasting and uncertainty in the economic and business world. *International Journal of Forecasting, 25*(4), 794–812.
Makridakis, S., & Taleb, N. (2009). Living in a world of low levels of predictability. *International Journal of Forecasting, 25*(4), 840–844.
Mandelbrot, B. B. (1963). The variation of certain speculative prices. *Journal of Business, 36*(4), 394–419.
Mandelbrot, B. B., & Hudson, R. L. (2004). *The (mis)behaviour of markets: A fractal view of risk, ruin and reward*. London: Profile Books.
Mankiw, N. G., & Shapiro, M. D. (1985). Trends, random walks and tests of the permanent income hypothesis. *Journal of Monetary Economics, 16*(2), 165–174.
Maritz, J. S. (1970). *Empirical Bayes methods*. London: Methuen and Co.
Markov, A. A. (1906). Rasprostranenie zakona bol'ših čisel na veličiny, zavisjaščie drug ot druga [Extension of the law of large numbers to dependent variables]. *Izvestija Fiziko-matematičeskogo obščestva pri Kazanskom universitete*, II Series, 15, 135–156.
Massey, D. S. (1987). The ethnosurvey in theory and practice. *International Migration Review, 21*(4), 1498–1522.
Massey, D. S. (1990). Social structure, household strategies, and the cumulative causation of migration. *Population Index, 56*, 3–26.

Massey, D. S. (2002). A synthetic theory of international migration. In V. Iontsev (Ed.), *World in the mirror of international migration* (pp. 142–152). Moscow: MAX Press.

Massey, D. S., Arango, J., Hugo, G., Kouaouci, A., Pellegrino, A., & Taylor, J. E. (1993). Theories of international migration: Review and appraisal. *Population and Development Review, 19*(3), 431–466.

Massey, D. S., Arango, J., Hugo, G., Kouaouci, A., Pellegrino, A., & Taylor, J. E. (1994). International migration theory: The North American case. *Population and Development Review, 20*(4), 699–751.

Massey, D. S., & Zenteno, R. M. (1999). The dynamics of mass migration. *Proceedings of the National Academy of Sciences USA, 96*(9), 5328–5335.

Mazurkiewicz, L. (1986). *Teoretyczne podstawy modeli przestrzennego oddziaływania* [*Theoretical foundations of the models of spatial interactions*]. Ossolineum, Wrocław: Institute of Geography and Spatial Organisation, Polish Academy of Sciences.

McDonald, P., & Kippen, R. (2002). Projecting future migration levels: Should rates or numbers be used? *People and Place, 10*(1), 82–83.

McGinnis, R., Myers, G. C., & Pilger, J. (1963). *Internal migration as a stochastic process*. Paper for the 34th session of the International Statistical Institute, Ottawa, 21–29 August.

Męczarski, M. (1998). *Problemy odporności w bayesowskiej analizie statystycznej* [*Robustness issues in Bayesian statistical analysis*]. "Monografie i opracowania" no. 446, Warsaw School of Economics, Warsaw.

Metropolis, N., Rosenbluth, A. W., Rosenbluth, M. N., Teller, A. H., & Teller, E. (1953). Equation of state calculations by fast computing machines. *Journal of Chemical Physics, 21*(6), 1087–1092.

Micheas, A. C., & Dey, D. K. (2004). On measuring loss robustness using maximum a posteriori estimate. *Communications in Statistics – Theory and Methods, 33*(5), 1069–1085.

Milne, W. J. (1993). Macroeconomic influences on migration. *Regional Studies, 27*(4), 365–373.

Min, C., & Zellner, A. (1993). Bayesian and non-Bayesian methods for combining models and forecasts with applications to forecasting international growth rates. *Journal of Econometrics, 56*(1–2), 89–118.

Mincer, J. (1978). Family migration decisions. *Journal of Political Economy, 86*(5), 749–773.

Morrison, P. A. (1973). Theoretical issues in the design of population mobility models. *Environment and Planning A, 5*(1), 125–134.

Mulder, C. H. (1993). *Migration dynamics: A life course approach*. Amsterdam: Thesis Publishers.

Myrdal, G. (1957). *Rich lands and poor*. New York: Harper and Row.

Nakamura, T. (1986). Bayesian cohort models for general cohort table analyses. *Annals of the Institute of Statistical Mathematics, 38*, 353–370.

Neal, R. M. (1993). *Probabilistic inference using Markov chain Monte Carlo methods*. Technical Report CRG-TR-93-1, University of Toronto, Toronto.

Neal, R. M. (1998). Suppressing random walks in Markov chain Monte Carlo using ordered overrelaxation. In M. I. Jordan (Ed.), *Learning in graphical models* (pp. 205–230). Dordrecht: Kluwer Academic Publishers.

Neal, R. M. (2003). Slice sampling. *Annals of Statistics, 31*(3), 705–767.

Nelson, C., & Plosser, C. I. (1982). Trends and random walks in macroeconomic time series: Some evidence and implications. *Journal of Monetary Economics, 10*(2), 139–162.

Ní Bhrolcháin, M., & Dyson, T. (2007). On causation in demography: Issues and illustrations. *Population and Development Review, 33*(1), 1–36.

Nowok, B. (2005). *Evolution of international migration statistics in selected central European countries* (CEFMR Working Paper 8/2005). Central European Forum for Migration Research, Warsaw.

Nowok, B., Kupiszewska, D., & Poulain, M. (2006). Statistics on international migration flows. In M. Poulain, N. Perrin, & A. Singleton, (Eds.), *Towards harmonised European statistics on international migration* (pp. 203–231). Louvain-la-Neuve: Presses Universitaires de Louvain.

NRC. [National Research Council] (2000). *Beyond six billion. Forecasting the world's population*. Washington, DC: National Academies Press.

Öberg, S. (1996). Spatial and economic factors in future south-north migration. In W. Lutz (Ed.), *The future population of the world: What can we assume today?* (pp. 336–357). London: Earthscan.

Öberg, S., & Wils, A. B. (1992). East-west migration in Europe. Can migration theories help estimate the numbers. *Popnet, 22*, 1–7.

O'Hagan, A., Buck, C. E., Daneshkhah, A., Eiser, J. R., Garthwaite, P. H., Jenkinson, D. J., Oakley, J. E., & Rakow, T. (2006). *Uncertain judgements – Eliciting experts' probabilities*. New York: Wiley.

Okun, B. (1968). Interstate population migration and state income inequality: A simultaneous equation approach. *Economic Development and Cultural Change, 16*(2), 297–313.

Orłowski, W. (2000). Migration from central and eastern European countries after the accession: Effects for regions, labour markets, and social security systems. *Research Bulletin, 2*(9), 7–21. Central Statistical Office, Warsaw.

Orrell, D. (2007). *The future of everything. The science of prediction*. New York: Thunder's Mouth Press.

Orrell, D., & McSharry, P. (2009). System economics: Overcoming the pitfalls of forecasting models via a multidisciplinary approach. *International Journal of Forecasting, 25*(4), 734–743.

OpenBUGS Manual. (2009). On-line resource, Available from: mathstat.helsinki.fi/openbugs/ Manuals/ Manual.html. Accessed 28 Nov 2009.

Osiewalski, J. (2000). Marek Męczarski – Problemy odporności w bayesowskiej analizie statystycznej, "Monografie i opracowania" 446, wyd. SGH, Warszawa, 1998, ss. 191 (A review). *Przegląd Statystyczny, 47*(1–2), 249–254.

Osiewalski, J. (2001). *Ekonometria bayesowska w zastosowaniach* [*Bayesian econometrics in applications*]. Cracow: Cracow University of Economics.

Osiewalski, J., & Steel, M. F. J. (1993a). A Bayesian perspective on model selection. Manuscript. published in Spanish as: Una perspectiva bayesiana en selección de modelos. *Cuadernos Economicos, 55*(3), 327–351.

Osiewalski, J., & Steel, M. F. J. (1993b). Robust Bayesian inference in elliptical regression models. *Journal of Econometrics, 57*(1–3), 345–363.

Osiewalski, J., & Steel, M. F. J. (1993c). Bayesian marginal equivalence of elliptical regression models. *Journal of Econometrics, 59*(3), 391–403.

Osiewalski, J., & Steel, M. F. J. (1996). A Bayesian analysis of exogeneity in models pooling time-series and cross-sectional data. *Journal of Statistical Planning and Inference, 50*(2), 187–206.

Palm, F. C., & Zellner, A. (1992). To combine or not to combine? Issues of combining forecasts. *Journal of Forecasting, 11*(8), 687–701.

Paradysz, J. (2004). Marek Kupiszewski – Modelowanie dynamiki przemian ludności w warunkach wzrostu znaczenia migracji międzynarodowych, Wydawnictwo PAN, IGiPZ, Prace Geograficzne nr 181, Warszawa, 2002, ss. 174 (A review). *Studia Demograficzne, 145*(1), 129–133.

Paradysz, J. (2006). Migracje ludności [Population migration]. In M. Kędelski & J. Paradysz (Eds.), *Demografia* [*Demography*] (pp. 231–254). Poznań: Poznań University of Economics.

Park, J. H, Martin, A. D., & Quinn, M. K. (2009). The Bayesian inference task view. On-line resource, Available from cran.r-project.org/web/views/Bayesian.html, version of 31 Aug 2009.

Petty, W. (1682). *An essay concerning the multiplication of mankind, together with another essay in political arithmetick concerning the growth of the City of London with the causes and consequences thereof*. London. Reprinted within the framework of the Gutenberg Project, www.gutenberg.org/etext/5619. Accessed on 15 March 2006.

Pijpers, R. (2008). Problematising the 'orderly' aesthetic assumptions of forecasts of East–West migration in the European Union. *Environment and Planning A, 40*(1), 174–188.

Piore, M. J. (1979). *Birds of passage: Migrant labour in industrial societies*. Cambridge, MA: Cambridge University Press.

Pittenger, D. (1980). Some problems in forecasting population for government planning purposes. *The American Statistician, 34*(3), 135–138.

Plane, D. A. (1993). Demographic influences on migration. *Regional Studies, 27*(4), 375–383.

Plane, D. A., & Rogerson, P. A. (1985). Economic-demographic models for forecasting interregional migration. *Environment and Planning A, 17*(2), 185–198.

Poirier, D. J., & Tobias, J. L. (2004). Bayesian econometrics. Manuscript. University of California, Irvine, CA, and Iowa State University, Ames, IA; www.econ.iastate.edu/faculty/tobias/documents/palgrave_pt.pdf. Accessed on 14 Dec 2007 [later published in, Mills, T. C., & Patterson, K. (Eds.). (2006), *Palgrave Handbook of Econometrics, Volume 1: Econometric Theory* (pp. 841–870). Basingstoke, UK: Palgrave Macmillan].

Poole, D., & A. E. Raftery (2000). Inference for deterministic simulation models: The Bayesian melding approach. *Journal of the American Statistical Association, 95*(452), 1244–1255.

Popper, K. R. (1935). *Logik der Forschung*. Julius Springer Verlag, Wien [(1959). *The logic of scientific discovery*. London: Hutchinson. (2003). *Logika odkrycia naukowego*. Antyk, Warszawa].

Popper, K. R. (1982). *The open universe. An argument for indeterminism*. London: Hutchinson [(1996). *Wszechświat otwarty. Argument na rzecz indeterminizmu*. Kraków: Znak].

Popper, K. R. (1990). *A world of propensities*. Bristol: Thoemmes [(1996). *Świat skłonności*. Kraków: Znak].

Poulain, M. (1993). Confrontation des statistiques de migration intra-Européennes: Vers plus d'harmonisation. *European Journal of Population, 9*(4), 353–381.

Poulain, M. (1994). La mobilité interne en Europe. Quelles données statistiques? *Espace, Populations, Sociétés, 1*, 13–30.

Poulain, M., Perrin, N., & Singleton, A. (Eds.). (2006). *Towards harmonised European statistics on international migration*. Louvain-la-Neuve: Presses Universitaires de Louvain.

Prais, S. J. (1955). Measuring social mobility. *Journal of the Royal Statistical Society A, 118*, 56–66.

Pries, L. (1999). *Migration and transnational social spaces*. Aldershot: Ashgate.

Quinn, M. A. (2006). Relative deprivation, wage differentials and Mexican migration. *Review of Development Economics, 10*(1), 135–153.

Raftery, A. E. (1995). Bayesian model selection in social research. *Sociological Methodology, 25*, 111–163.

Raftery, A. E., & Zheng, Y. (2003). Discussion: Performance of Bayesian model averaging. *Journal of the American Statistical Association, 98*(464), 931–938.

Ramos, F. F. R. (1996). *VAR priors: Success or lack of a decent macroeconomic theory?* Econometrics, Economics Working Paper Archive EconWPA, Internet Resource; econwpa.wustl.edu/eps/em/papers/9601/ 9601002.pdf. Accessed 31 July 2006.

Ramsey, F. P. (1926). Truth and probability. In R. B. Braithwaite (Ed.), (1931). *F. P. Ramsey: The foundations of mathematics and other logical essays* (pp. 156–196). London: Kegan, Paul, Trench, Trubner & Co.; New York: Harcourt, Brace and Co.

Rao, C. R. (1989). *Statistics and truth: Putting chance to work*. Fairland, MD: International Cooperative Publishing House [(1994). *Statystyka i prawda*. Warszawa: PWN].

Ravenstein, E. G. (1885). The laws of migration. *Journal of the Royal Statistical Society, 48*(2), 167–227.

Ravenstein, E. G. (1889). The laws of migration. *Journal of the Royal Statistical Society, 52*(2), 214–301.

Raymer, J. (2008). Obtaining an overall picture of population movement in the European Union. In J. Raymer & F. Willekens (Eds.), *Estimation of international migration in Europe: Issues, models and assessment* (pp. 209–234). Chichester: Wiley.

Rees, P. H. (1977). The measurement of migration from census data and other sources. *Environment and Planning A, 9*(3), 247–272.

Rees, P. H. (1996). Projecting national and regional populations of the European Union using the migration information. In P. H. Rees, J. S. C. Stillwell, A. Convey, & M. Kupiszewski (Eds.), *Population migration in the European Union* (pp. 330–364). Chichester: Wiley.

Rees, P. H., Kupiszewski, M., Eyre, H., Wilson, T., & Durham, H. (1999). *The evaluation of regional population projections for the European Union*. Research report for the ERDF Study 97/00/74/018. University of Leeds, Leeds.

Rees, P. H., Stillwell, J. S. C., & Convey, A. (1992). *Intra-community migration and its impact on the development of the demographic structure at regional level* (Working Paper 92/1). School of Geography, University of Leeds, Leeds.

Rees, P. H., & Turton, I. (1998). *Investigation of the effects of input uncertainty on population forecasting*. Paper for the 3rd International GeoComputation Conference, Bristol, UK, 17–19 Sep.

Rees, P. H., & Willekens, F. (1986). Data and accounts. In A. Rogers & F. Willekens (Eds.), *Migration and settlement. A multiregional comparative study* (pp. 15–59). Dordrecht: Reidel.

Rees, P. H., and Wilson, A. G. (1973). Accounts and models for spatial demographic analysis 1: Aggregate population. *Environment and Planning A*, 5(1): 61–90.

Regan, H. M., Ben-Haim, Y., Langford, B., Wilson, W. G., Lundberg, P., Andelman, S. J., & Burgman, M. A. (2005). Robust decision-making under severe uncertainty for conservation management. *Ecological Applications, 15*(4), 1471–1477.

Robert, C. P. (2001). Bayesian choice. From decision-theoretic foundations to computational implementation. New York: Springer.

Robert, C. P., & Casella, G. (2005). Monte Carlo statistical methods (2nd ed.). Springer, New York.

Roberts, G. O. (1995). Markov chain concepts related to sampling algorithms. In W. R. Gilks, D. J. Spiegelhalter, & S. Richardson (Eds.), *MCMC in practice* (pp. 45–58). London: Chapman and Hall.

Rogers, A. (1966). A Markovian policy model of interregional migration. *Papers of the Regional Science Association, 17*, 205–224.

Rogers, A. (1973). Estimating internal migration from incomplete data using model multiregional life tables. *Demography, 10*(2), 277–287.

Rogers, A. (1975). *Introduction to multiregional mathematical demography*. New York: Wiley.

Rogers, A. (1980). Introduction to multistate mathematical demography. *Environment and Planning A, 12*(5), 489–498.

Rogers, A. (1990). Requiem for the net migrant. *Geographical Analysis, 22*(4), 283–300.

Rogers, A. (Ed.) (1999). The indirect estimation of migration. *Mathematical Population Studies, 7*(3), 181–309.

Rogers, A., & Castro, L. J. (1981). *Model migration schedules*. IIASA Report RR-81-30. International Institute for Applied Systems Analysis, Laxenburg.

Rogers, A., Raquillet, R., & Castro, L. J. (1978). Model migration schedules and their applications. In A. Rogers (Ed.), *Migration and settlement: Selected essays*. IIASA Report RR-78-6. Laxenburg: International Institute for Applied Systems Analysis.

Rogers, A., & Raymer, J. (2005). Origin dependence, secondary migration, and the indirect estimation of migration flows from population stocks. *Journal of Population Research, 22*(1), 1–19.

Rogers, A., & Willekens, F. (1978). *Spatial population analysis: Methods and computer programs*. IIASA Report RR-78-18. Laxenburg: International Institute for Applied Systems Analysis.

Rogerson, P. A. (1984). New directions in the modelling of interregional migration. *Economic Geography, 60*(2), 111–121.

Roseveare, D., Leibniz, W., Fore, D., & Wurzel, A. (1996). *Ageing populations, pensions systems and government budgets: Simulations for 20 OECD countries*. Economics Department Working Papers 168. OECD, Paris.

Russell, B. (1912). *The problems of philosophy*, London: Oxford University Press [(2004). *Problemy filozofii*. Warszawa: PWN].

Rykiel, Z. (1986). Ograniczenia meldunkowe jako bariery przestrzenne [Administrative control of migration as spatial barriers]. *Przegląd Geograficzny, 58*(3), 395–409.

Sakson, B. (2002). *Wpływ „niewidzialnych" migracji zagranicznych lat osiemdziesiątych na struktury demograficzne Polski* [*Impact of the "invisible" international migration of the 1980s on the demographic structures of Poland*]. "Monografie i opracowania" no. 481, Warsaw School of Economics, Warsaw.

Salt, J., & Singleton, A. (1995). *Analysis and forecasting of international migration by major groups*. Report to the Working Party on Demographic Projections. Eurostat, Luxembourg.

Sanderson, W. (1998). Knowledge can improve forecasts: A review of selected socioeconomic population projection models. *Population and Development Review, 24*(Suppl.), 88–117.

Savage, L. J. (1954). *Foundations of statistics*. New York: Wiley.

Schmid, V., & Knorr-Held, L. (2001). *BAMP: Bayesian age-period-cohort modelling and prediction*. software documentation. Department of Statistics, Ludwig-Maximilian-University, Munich.

Schwarz, G. (1978). Estimating the dimension of a model. *The Annals of Statistics, 6*(2), 461–464.

Silvey, S. D. (1975). *Statistical inference*. London: Chapman & Hall [(1978). *Wnioskowanie statystyczne*. Warszawa: PWN].

Sims, C. A. (1980). Macroeconomics and Reality. *Econometrica, 48*(1), 1–48.

Sinn, H.-W., Flaig, G., Werding, M., Munz, S., Duell, N., & Hofmann, H. (2001). *EU-Erweiterung und Arbeitskräftemigration, Wege zu einer schrittweisen Annäherung der Arbeitsmärkte*. IFO-Institut für Wirtschaftsforschung, München.

Sjaastad, L. A. (1962). The costs and returns of human migration. *Journal of Political Economy, 70*(5), 80–93.

Skidelsky, R. (2009). *Keynes: The return of the master*. London: Allen Lane.

Smith, S. K. (1997). Further thoughts on simplicity and complexity in population projection models. *International Journal of Forecasting, 13*(4), 557–565.

Smith, S. K., & Sincich, T. (1992). Evaluating the forecast accuracy and bias of alternative population projections for states. *International Journal of Forecasting, 8*(3), 495–508.

Snowdon, B., & Vane, H. R. (2005). *Modern macroeconomics: Its origins, development and current state*. Cheltenham: Edward Elgar.

Spiegelhalter, D. J., Best, N. G., Carlin, B. P., & van der Linde, A. (2002). Bayesian measures of model complexity and fit. *Journal of the Royal Statistical Society B, 64*(4), 583–639 (with discussion).

Spiegelhalter, D. J., Thomas, A., Best, N. G., & Lunn, D. (2003). *WinBUGS Version 1.4 users manual*. Medical Research Centre, Cambridge; www.mrc-bsu.cam.ac.uk/bugs. Accessed 5 May 2006.

Spilerman, S. (1972). The analysis of mobility process by the introduction of independent variables into a Markov chain. *American Sociological Review, 37*(3), 277–294.

Stark, O. (1991). *The migration of labor.* Cambridge, MA: Basil Blackwell.

Stark, O. (2003). *Tales of migration without wage differentials: Individual, family and community contexts* (ZEF Discussion Paper on Development Policy No. 73). Zentrum für Entwicklungsforschung, Bonn.

Stark, O., & Bloom, D. E. (1985). The new economics of labor migration. *American Economic Review, 75*(2), 173–178.

Stark, O., & Taylor, J. E. (1989). Relative deprivation and international migration, *Demography, 26*(1), 1–14.

Stark, O., & Wang, Y. (2001). *Inducing human capital formation: Migration as a substitute for subsidies* (Economics Series No. 100). Institute for Advanced Studies, Vienna.

Stewart, J. Q. (1941). An inverse distance variation for certain social influences. *Science, 93*(2404), 89–90.

Stillwell, J. C. H., & Congdon, P. (Eds.) (1991). *Migration models: Macro and micro approaches*. London: Belhaven Press.

Stillwell, J. C. H., Eyre, H., and Rees, P. H. (1999). *Regional international migration and interregional migration within the European Union: A feasibility study*. Research report for the ERDF Study 98/00/27/174. University of Leeds, Leeds.

Stoto, M. A. (1983). The accuracy of population projections. *Journal of the American Statistical Association, 78*(381), 13–20.

Stouffer, S. A. (1940). Intervening opportunities: A theory relating mobility and distance. *American Sociological Review, 5*(6), 845–867.

Stouffer, S. A. (1960). Intervening opportunities and competing migrants. *Journal of Regional Studies, 2*(1), 187–208.

Taleb, N. N. (2009). Errors, robustness, and the fourth quadrant. *International Journal of Forecasting, 25*(4), 744–759.

Taylor, J. E. (1986). Differential migration, networks, information and risk. In O. Stark (Ed.), *Research in human capital and development*, Vol. 4: *Migration, human capital, and development* (pp. 141–171). Greenwich, CT JAI Press.

Thomas, A., O'Hara, B., Ligges, U., & Sturtz, S. (2006). Making BUGS open. *R News, 6*, 12–17.

Tierney, L. (1994). Markov chains for exploring posterior distributions. *The Annals of Statistics, 22*(4): 1701–1728.

Törnqvist, L. (1949). Näkökohdat, jotka ovat määränneet primääristen prognoosiolettamusten valinnan [Points of view that have determined the primary forecast assumptions]. In J. Hyppölä, A. Tunkelo, & L. Törnqvist (Eds.), *Suomen väestöä, sen uusiutumista ja tulevaa kehitystä koskevia laskelmia [Calculations concerning the population of Finland, its renewal and future development]*. Tilastollisia tiedonantoja 38. Statistics Finland, Helsinki.

Tuljapurkar, S. (1997). Demography. Taking the measure of uncertainty. *Nature, 387*(6635), 760–761.

Tuljapurkar, S., & Boe, C. (1999). Validation, probability-weighted priors and information in stochastic forecasts. *International Journal of Forecasting, 15*(3), 259–271.

Tuljapurkar, S., Lee, R. D., & Li, Q. (2004). Random scenario forecasts versus stochastic forecasts. *International Statistical Review, 72*(2), 185–199.

UNHCR. [United Nations High Commissioner for Refugees] (2000). *The state of the world's refugees 2000. Fifty years of humanitarian action*. Oxford, MA: Oxford University Press.

United Nations. (1970). *Methods of measuring internal migration*. New York: United Nations.

United Nations. (1978). *Population yearbook 1977*. New York: United Nations.

United Nations. (1998). *Recommendation on statistics of international migration*. New York: United Nations.

United Nations. (2000a). *Replacement migration: Is it a solution to declining and ageing populations?* New York: United Nations.

United Nations. (2000b). *United Nations economic survey of Europe 1/2000*. Geneva: United Nations ECE.

United Nations. (2005a). *World population prospects: 2004 revision*. New York: United Nations.

United Nations. (2005b). *United Nations economic survey of Europe 2/2005*. Geneva: United Nations ECE.

van der Gaag, N., de Beer, J., & Willekens, F. (2005). *MicMac. Combining micro and macro approaches in demographic forecasting*. Paper for the Joint UNECE-Eurostat Work Session on Demographic Projections, Vienna, 21–23 Sep.

van der Gaag, N., & van Wissen, L. (1999). *Analysis and forecasting of international migration by major groups (Part II)* (Eurostat Working Paper 3/1999/E/nº9). Eurostat, Luxembourg.

van Imhoff, E. (1990). The exponential multidimensional demographic projection model. *Mathematical Population Studies, 2*(3), 171–182.

van Imhoff, E., & Keilman, N. (1991). *LIPRO 2.0: An application of a dynamic demographic projection model to household structure in The Netherlands*. NIDI & CBGS Publication 23. Swets & Zeitlinger, Amsterdam.

van Imhoff, E., & Post, W. (1998). Microsimulation methods for population projection. *Population–E, 10*(1), 97–138.

van Imhoff, E., van der Gaag, N., van Wissen, L., & Rees, P. H. (1997). The selection of internal migration models for European regions. *International Journal of Population Geography, 3*(2), 137–159.

Varian, H. R. (1975). A Bayesian approach to real estate assessment. In S. E. Fienberg & A. Zellner (Eds.), *Studies in Bayesian econometrics and statistics. Essays in honor of Leonard J. Savage* (pp. 195–208). Amsterdam: North-Holland.

Veblen, T. (1898). Why is economics not an evolutionary science? *Quarterly Journal of Economics, 12*(4), 373–397.

Vishnevsky, A., & Zayonchkovskaya, Zh. [Višnevskij, A. G., & Zajončkovskaja, Ž. A.] (1994). Emigration from the former Soviet Union: The fourth wave. In H. Fassmann & R. Münz (Eds.), *European migration in the late twentieth century: Historical patterns, actual trends and social implications* (pp. 239–259). Aldershot: Edward Elgar.

von Mises, R. (1943). On the correct use of Bayes formula. *Annals of Mathematical Statistics, 13*, 156–165.

von Neumann, J., & Morgenstern, O. (1944). *Theory of games and economic behavior*. Princeton, NJ: Princeton University Press.

Wald, A. (1950). *Statistical decision functions*. New York: Wiley.

Wallerstein, I. (1974). *The modern world-system*, Vol. I: *Capitalist agriculture and the origins of European world-economy in the sixteenth century*. New York: Academic Press.

Wasserman, L. (1997). *Bayesian model selection and model averaging*. Department of Statistics Technical Report No. 666. Carnegie Mellon University, Pittsburgh, PA.

Wegren, S., & Cooper Drury, A. (2001). Patterns of internal migration during the Russian transition. *Journal of Communist Studies and Transition Politics, 17*(4), 15–42.

Weidlich, W., & Haag, G. (Eds.). (1988). *Interregional migration: Dynamic theory and comparative analysis*. Berlin-Heidelberg: Springer.

Welch, G., & Bishop, G. (2006). *An introduction to the Kalman filter*. Technical Report TR 95-041, Department of Computer Science, University of North Carolina, Chapel Hill, NC; www.cs.unc.edu/~welch/ media/pdf/kalman_intro.pdf. Accessed 18 Aug 2007.

Willekens, F. (1976). *Optimal migration policies*. IIASA Research Memorandum RM-76-50. International Institute for Applied Systems Analysis, Laxenburg.

Willekens, F. (1977). The recovery of the detailed migration patterns from aggregate data: An entropy maximizing approach. In A. Rogers (Ed.), *Advances in multiregional demography*. IIASA Report RR816. International Institute for Applied Systems Analysis, Laxenburg.

Willekens, F. (1982). Multidimensional population analysis with incomplete data. In K. C. Land & A. Rogers (Eds.), *Multidimensional mathematical demography* (pp. 43–111). New York: Academic Press.

Willekens, F. (1990). Demographic forecasting; state-of-the-art and research needs. In C. A. Hazeu & G. A. B. Frinking (Eds.), *Emerging issues in demographic research* (pp. 9–66). Amsterdam: Elsevier.

Willekens, F. (1994). Monitoring international migration flows in Europe. Towards a statistical data base combining data from different sources. *European Journal of Population, 10*(1), 1–42.

Willekens, F. (1995). *Multiregional demographic forecasting* (Population Research Centre Working Paper 95-1), University of Groningen, Groningen.

Willekens, F. (1999). Modeling approaches to the indirect estimation of migration flows: From entropy to EM. *Mathematical Population Studies, 7*(3), 239–278.

Willekens, F. (2008). Models of migration: Observations and judgments. In J. Raymer & F. Willekens (Eds.), *Estimation of international migration in Europe: Issues, models and assessment* (pp. 117–148). Chichester: Wiley.

Willekens, F., & Baydar, N. (1986). Forecasting place-to-place migration with generalized linear models. In R. Woods & P. H. Rees (Eds.), *Population structures and models* (pp. 203–244). London: Allen and Unwin.

Willekens, F., Pór, A., & Raquillet, R. (1981). Entropy, multiproportional and quadratic techniques for inferring patterns of migration from aggregate data. In A. Rogers (Ed.), *Advances in multiregional demography*. IIASA Report RR816. International Institute for Applied Systems Analysis, Laxenburg.

Wilson, A. G. (1967). A statistical theory of spatial distribution models. *Transportation Research, 1*, 253–269.

Wilson, A. G. (1970). *Entropy in urban and regional modelling*. London: Pion.

Wilson, A. G. (1981). *Catastrophe theory and bifurcation. Applications to urban and regional systems*. London: Croom Helm.

Wilson, G. A., & Rannala, B. (2003). Bayesian inference of recent migration rates using multilocus genotypes. *Genetics, 163*(3), 1177–1191.

Wilson, T., & Bell, M. (2004). *Australia's uncertain demographic future* (Discussion Paper 2003/04). Queensland Centre for Population Research, The University of Queensland, Brisbane.

Withers, S. D. (2002). Quantitative methods: Bayesian inference, Bayesian thinking. *Progress in Human Geography, 26*(4), 553–566.

World Bank. (2006). *2006 World development indicators CD-ROM*. World Bank.

Wright, G., & Goodwin, P. (2009). Decision making and planning under low levels of predictability: Enhancing the scenario method. *International Journal of Forecasting, 25*(4), 813–825.

Wróblewska, J. (2009). Bayesian model selection in the analysis of cointegration. *Central European Journal of Economic Modelling and Econometrics, 1*(1), 57–69; www.cejeme.eu.

Zelinsky, W. (1971). The hypothesis of the mobility transition. *Geographical Review, 61*(2), 219–249.

Zellner, A. (1971). *An introduction to Bayesian inference in econometrics*. New York: Wiley.

Zellner, A. (1986). Bayesian estimation and prediction using asymmetric loss functions. *Journal of the American Statistical Association, 81*(394), 446–451.

Zipf, G. K. (1946). The P_1P_2/D hypothesis: On the intercity movement of persons. *American Sociological Review, 11*(6), 677–686.

Zlotnik, H. (1998). *The theories of international migration*. Paper for the Conference on International Migration: Challenges for European Populations, Bari, Italy, 25–27 June.

Zolberg, A. R. (1989). The next waves: Migration theory for a changing world. *International Migration Review, 23*(3), 403–430.

Author Index

Subject Index

Kolibri 360 GmbH
Ferdinand-Jühlke-Straße 7
99095 Erfurt, Deutschland
produktsicherheit@kolibri360.de